ALLE ZEIT WACH 1842

Y. Dattée C. Dumas A. Gallais
(Eds.)

Reproductive Biology and Plant Breeding

Biologie de la Reproduction et Amélioration des Plantes

With 50 Figures

Springer-Verlag
Berlin Heidelberg New York
London Paris Tokyo
Hong Kong Barcelona
Budapest

Dr. YVETTE DATTÉE
G.E.V.E.S., La Minière, 78285 Guyancourt Cédex, France

Professor Dr. CHRISTIAN DUMAS
RCAP INRA 23879, Université Lyon I – Bât 741, 43 boulevard du 11 Novembre, 69622 Villeurbanne Cédex, France

Professor Dr. ANDRÉ GALLAIS
I.N.A.-P.G., Station de Génétique végétale, Ferme du Moulon, 91190 Gif-sur-Yvette, France

The XIIIth EUCARPIA Congress, from the 6th to the 11th July 1992 in Angers, France, was organized with the support of:

- The "Ministère de la Recherche et de la Technologie" (French Ministry of Research and Technology),
- The "Ministère de l'Agriculture et de la Forêt" (French Ministry of Agriculture and Forestry)

and with the sponsorship of:

- The "Institut National de la Recherche Agronomique" (I.N.R.A.),
- The "Groupement National Interprofessionnel des Semences"
- The "Regional Collective of Pays de la Loire."

ISBN-13: 978-3-642-77000-5 e-ISBN-13: 978-3-642-76998-6
DOI: 10.1007/978-3-642-76998-6

Softcover reprint of the hardcover 1st edition 1992

Typesetting: Camera-ready by author
31/3145-5 4 3 2 1 0 – Printed on acid-free paper

Preface

This volume has been produced for the XIIIth EUCARPIA Congress.

EUCARPIA (the European Association for Plant breeding) currently has 1.200 members, including scientists and staff of both public and private organizations. Its aim is to promote scientific and technical research and cooperation in the field of plant breeding, and thereby to contribute to the development of agriculture.

Every three years, EUCARPIA organizes a scientific congress. In 1992, the XIIIth EUCARPIA Congress will be held in ANGERS (France) and the theme is **"Reproductive biology and plant breeding"**.

Reproduction of plant material is central to selection. The geneticist, the plant breeder and the seed grower all use sexual and vegetative reproduction during the various stages of plant breeding and creation of variety.

The possibility of unlimited interspecific reproduction, the use of gametogenesis dysfunction, the creation of auto and allogamy, and the cloning of the best genotypes are the challenges before the plant breeder. To understand how the reproductive system conditions the genetic structure of a population, and to investigate the relationships between the reproductive mode and the organization of variability is a central key to genetic progress.

The articles presented in this book review the current state of knowledge of reproductive biology, and its impact on variety creation.

Molecular analyses and mutant studies have led to dramatic progress in several major areas of genetics and biology. For example :

. Some important genes (homeotic genes and the MADS box) have been identified and isolated from both *Arabidopsis thaliana* and *Antirrhinum majus* (in particular by the groups of Coen and Meyerowitz, 1991 ; Schwarz-Sommer, 1990). It is now possible to explain both actinomorphous and major genes (MADS box). Mutant studies have similarly shown that another small group of genes are responsible for the different inflorescences observed in Angiosperms.

. A set of at least nine genes have been identified that control pattern formation during embryogenesis in Arabidopsis thaliana (Mayer et al., 1991).

. Pollen genes have been shown to be specifically expressed during the male programme, and this has led to a better understanding of pollen wall formation and biosynthesis (Mascarenhas, 1990). A new technique based on inactivation of the tapetum using a specific gene expressing RNAse activity opens new possibilities for creating male sterility in plants (Goldberg's group and P.G.S. : Mariani et al., 1990).

. The effective transformation of a self-compatible line with a gene encoding a molecule governing self-incompatibility has been realized in a sporophytic system (Nasrallah et al., 1991). Futhermore, the possible involvement of RNAse in the gametophytic self-incompatibility system has recently emerged (McClure et al., 1989 and EC BRIDGE participants in this volume).

. In vitro fertilization and embryo rescue techniques for Angiosperms can be used to overcome incompatibility and are a good model for the analysis of the effects of stress on seed setting. Similarly novel techniques of in vitro plant gamete isolation and their electrofusion are now available (Zenkteler, this volume ; Kranz et al., 1991).

. Significant progresses have been made in several aspects related to the molecular control of somatic embryogenesis (for example : de Vries, this volume) and androgenesis. In addition, the better understanding of the cryopreservation allows new windows in cell, organs and plant storage.

These major advances in plant reproductive biology are the results of the emergence and use of new molecular genetic techniques (including cloning and gene sequencing, PCR, gene transfer techniques) and tools for cell biology (including in situ hibrydization, immunochemistry).

Major advances have been made in population studies and the creation of varieties :

. The growing interest in genetic resources, not only in cataloguing information, but also in their dynamic management (Allard, this volume), ensures effective use of biological diversity in plant breeding.

. The progress in cell biology has brought practical results in variety creation. The chapter dedicated to double haploids bears witness. Other applications will arise from cryoconservation (Kartha, this volume).

. The development of molecular methods (RFLP, RAPD) will lead to the establishment of genetic maps for cultivated plant species, and the identification of genes of interest.

. There promise to be many applications for molecular labelling, including marker-assisted selection (Helentjaris, this volume) and an understanding of, and consequently, analysis and prediction of heterosis (de Vienne, this volume).

Thus a variety of approaches have combined to give a better understanding of genetic diversity and above all, show that the plant breeding and the creation of varieties are activities that involve many elements, and take advantage of the most up to date progress in genetics, cell biology and molecular biology.

Yvette Dattée
INRA - GEVES

78285 Guyancourt

Christian Dumas
INRA - Université
Lyon I
69622 Villeurbanne

André Gallais
INRA - INAPG
Ferme du Moulon
91190 Gif-sur-Yvette

The editors thank Corinne Génermont for reading the manuscript.

References

Coen E.S. and Meyerowitz E.M. (1991).
The war of the whorls : genetic interactions controlling flower development. Nature 353, 31-36.

Kranz E., Bautor J. and Lörz H. (1991).
In vitro fertilization of single, isolated gametes, transmission of cytoplasmic organelles and cell reconstitution of maize (Zea mays L.). Sex. Plant. Reprod. 4, 12-16.

Mariani C., De Beuckeleer M., Tuettnar J., Leemans J. and Goldberg R.G. (1990).
Induction of male sterility in plants by a chimeric ribonuclease gene. Nature 347, 737-741.

Mascarenhas J.P. (1990).
Gene activity during pollen development. Annu. Rev. Plant Physiol. and Plant Mol. Biol. 41, 317-338.

Mayer U., Torres Ruiz R.A., Berleth T., Misera S. and Jürgens G. (1991).
Mutations affecting body organization in the Arabidopsis embryo. Nature 353, 402-407.

McClure B.A., Haring V., Ebert P.R., Anderson M.A., Simpson R.J., Sakiyama F. and Clarke A.E. (1989).
Style self-incompatibility gene products of Nicotiana alata are ribonucleases. Nature 342, 955-957.

Nasrallah J.B., Yu S.E. and Nasrallah M.E. (1991).
The self-incompatibility genes of Brassica oleracea : expression, isolation and structure. Proc. Nat. Acad. Sci. USA 35, 5551-5555.

Schwarz-Sommer Z., Huijser P., Nacken W., Saedler H. and Sommer H. (1990).
Genetic control of flower development by homeotic genes in Antirrhinum majus. Science 250, 931-936.

Préface

Le présent document est édité à l'occasion du XIIIème Congrès EUCARPIA.

EUCARPIA (Association Européenne pour l'Amélioration des Plantes) compte aujourd'hui 1.200 membres, scientifiques et ingénieurs d'établissements publics et privés. Elle a pour objectif de promouvoir la recherche et la coopération scientifique et technique en amélioration des plantes et de contribuer ainsi au développement de l'agriculture.

Tous les trois ans, EUCARPIA organise un congrès scientifique. En 1992, le XIIIème congrès EUCARPIA se tient à Angers sur le thème **"Biologie de la reproduction et amélioration des plantes"**.

La reproduction du matériel végétal est un processus de base de la sélection. Le généticien, le sélectionneur, le multiplicateur de semences utilisent la reproduction sexuée et végétative à toutes les étapes du schéma d'amélioration et de création variétale.

Pouvoir à volonté franchir les barrières de reproduction interspécifiques, utiliser les déviations de la gamétogenèse, provoquer l'auto et l'allogamie, cloner les génotypes les plus performants sont des défis permanents du sélectionneur. Comprendre comment le système de reproduction conditionne la structure génétique des populations et rechercher les relations entre le mode de reproduction et l'organisation de la variabilité constituent une clé fondamentale du progrès génétique.

Les articles présentés dans ce volume synthétisent les connaissances actuelles en biologie de la reproduction et leurs impacts sur la création de variétés.

Plusieurs domaines de la génétique et de la biologie ont considérablement progressé grâce à l'utilisation de l'analyse moléculaire et des mutants :

. Certains gènes majeurs (gènes homéotiques de la "MADS box") ont été caractérisés et isolés chez Arabidopsis thaliana et Antirrhinum majus (en particulier par les groupes de Coen et Meyerowitz, 1991 et Schwarz-Sommer et al., 1990). Il est maintenant possible d'expliquer tous les processus de la morphogenèse florale aussi bien pour les fleurs zygomorphes que pour les fleurs actinomorphes par le contrôle de quelques gènes majeurs. De même, des études de mutants ont montré que les différents types d'inflorescences observés chez les Angiospermes sont contrôlés par un petit nombre de gènes d'un autre groupe que les précédents.

. Une série d'au moins neuf gènes responsables du contrôle de la forme au cours de l'embryogenèse a été identifiée chez Arabidopsis thaliana (Mayer et al., 1991).

. Pour le pollen, il a été montré que certains gènes sont exprimés sélectivement pendant le programme mâle (Mascarenhas, 1990). Cette constatation a conduit à une meilleure compréhension de la formation et de la biosynthèse de la paroi pollinique. Une nouvelle technique, s'appuyant sur l'inactivation des cellules tapetales à l'aide d'un gène tissu spécifique, exprimant une activité RNAse, ouvre de nouvelles possibilités pour induire la stérilité mâle chez les plantes (groupe de Goldberg et P.G.S. : Mariani et al., 1990).

. La transformation d'une lignée avec un gène codant pour une molécule responsable de l'auto-incompatibilité a été réalisée avec succès dans un système sporophytique (Nasrallah et al., 1991). Par ailleurs, l'implication probable de la RNAse dans le système d'auto-incompatibilité gamétophytique a été récemment mise en évidence (McClure et al., 1989, et participants du programme BRIDGE dans ce volume).

. Les techniques de fécondation in vitro et de sauvetage d'embryon peuvent être utilisées pour surmonter l'incompatibilité ; elles constituent un bon modèle pour l'analyse des effets des stress sur la formation des graines. De même, des techniques nouvelles d'isolement *in vitro* et d'électrofusion des gamètes sont maintenant mises au point (Zenkteler dans ce volume; Kranz et al., 1991).

. Des progrès significatifs ont été faits sur plusieurs aspects de la connaissance du contrôle moléculaire de l'embryogenèse somatique (de Vries dans ce volume) et de l'androgenèse. De plus, une meilleure compréhension de la cryoconservation apporte des ouvertures dans la conservation des cellules, des organes végétaux et des plantes.

Ces importants progrès sur la connaissance de la biologie de la reproduction des plantes résultent de la mise au point et de l'utilisation de nouvelles techniques de biologie moléculaire (incluant le clonage et le séquençage de gènes, la PCR, la transformation génétique) et de la mise au point d'outils de biologie cellulaire (hybridation *in situ*, immunochimie).

Pour l'étude des populations et la création des variétés, des progrès importants sont réalisés aujourd'hui :

. L'intérêt accru porté aux ressources génétiques, non seulement pour leur recensement mais surtout pour leur gestion dynamique (Allard dans ce volume), garantit une utilisation effective de la diversité biologique en amélioration des plantes.

. Les progrès de la biologie cellulaire apportent des résultats pratiques en création variétale. Le chapitre consacré aux haploïdes doublés en est un témoignage. D'autres applications sont apportées par la cryoconservation (Kartha dans ce volume).

. Le développement des méthodes moléculaires (RFLP, RAPD) permet l'établissement de cartes génétiques chez les végétaux cultivés et la détection de gènes d'intérêt.

Le marquage moléculaire trouve de nombreuses autres applications : sélection assistée par marqueurs (Helentjaris dans ce volume) ; compréhension et, par conséquent à terme, analyse et prédiction de l'hétérosis (de Vienne dans ce volume).

Tous ces facteurs concourent à une meilleure connaissance de la diversité génétique mais, surtout, montrent que l'amélioration des plantes et la création variétale, disciplines de synthèse, intègrent maintenant à tous les niveaux les progrès les plus récents de la génétique, de la biologie cellulaire et de la biologie moléculaire.

Yvette Dattée
INRA - GEVES
78285 Guyancourt

Christian Dumas
INRA - Université
Lyon I
69622 Villeurbanne

André Gallais
INRA - INAPG
Ferme du Moulon
91190 Gif-sur-Yvette

Les éditeurs remercient Corinne Génermont pour la lecture du manuscrit.

Contents

Natural and artificial male-sterilities

Doubled haploids in plant breeding

Incompatibility and hybrid production

Fertilization and its potential

Zygotic embryogenesis

EC Research Programmes in the Reproductive Biology of Higher Plants

D. de Nettancourt and E. Magnien
Commission of the European Communities
Directorate-General for Science, Research and Development
Directorate Biology
Division Biotechnology
200, rue de la Loi
B-1049 Brussels, Belgium

Biotechnology R&D at Community level : the first programmes

BEP (Biomolecular Engineering Programme) was the first initiative of the Community in the area of biotechnology; it covered the period 1982–1985 and was followed by BAP (Biotechnology Action Programme) from 1985 to 1989. Examples of outstanding results in plant biology which resulted from BEP and BAP include the characterisation and isolation of numerous microbial and plant genes important for industry and agriculture, the development of new transfer techniques, the first discovery of a plant hormone receptor ... It was in the framework of BAP that a regulatory gene of cultivated higher plants (opaque 2 which control zein synthesis in maize) and a receptor protein gene governing auxin perception were cloned for the first time in the world and that successful gene transfer and expression was achieved in several key instances (soybean leghaemoglobin genes in *Lotus corniculatus*, a forage legume, under the control of nitrogen-fixing rhizobia which normally do not recognise soybean as a host; resistance genes to leaf roll virus from wild self-incompatible *S. brevidens* to the cultivated potato through somatic hybridisation ...).

BRIDGE, the current programme (1990-1993)

This programme, with a budget of 100 Mio ECU, took the succession of BAP and is subdivided, as was BAP, in Action I for Research and Training and Action II for Concertation. Four main sectors (Information infrastructure, Enabling technologies, Cellular biology, Pre-normative research) are covered by Action I. For each of these sectors, money was allocated, after the publication of detailed calls for proposals and evaluation of proposals by peer reviews, through cost-shared research contracts with public and private laboratories in the Community. Two different types of projects are presently being implemented in BRIDGE.

- Network N-projects (*), which aim at stimulating basic research in the Community and give rise to the organisation of highly integrated European Laboratories Without Walls (ELWWs) (*). ELWWs which are associations of research laboratories having decided to cooperate on the basis of flexible, informal rules and, when necessary, to establish joint collaboration links with industries interested by their research. ELWWs are devoted to the removal of specific bottlenecks in knowledge and know-how. More than 100 ELWWs have been created to date by the Commission services. The contribution of the Commission to each ELWW usually does not exceed 400.000 ECU/year.
- Targeted T-projects(*), for the elimination of

(*) These standard definitions of the purposes and structures of ELWWs, N-projects and T-projects are taken from articles, listed in "Suggested reading" and published in Biofutur (1989), Chimica Oggi (April 1991) and Agro-Industry Hi-Tech (March-April 1991), which should be consulted for further information.

barriers, in scale or in structure, which prevent the exploitation by agriculture or industry of the data, materials or methods originating from modern biology. Both N and T-projects are pre-competitive in nature, but T-projects base their formulations on the needs and requirements of agriculture or industry while N-projects aim at the elucidation or understanding, at molecular and cellular levels, of structures and processes essential for the expression and propagation of living organisms. A T-project starts where a N-project normally ends and is typical of a "Top-down" approach. It leads to the creation of bi-partite structures bringing together, for each T-project, a transnational team of contractors working together for the solution of one or several scientific problems and a monitoring unit composed of representatives of the Commission services and of the Member States who advise the group of contractors on all matters concerning the implementation of the project. In several instances, groupings or "platforms" have been created by industrial enterprises of the EC interested in the area covered by a given T-project and wishing to establish a dialogue, through the monitoring unit, with the contractors. Such platforms are open structures, depending on the initiatives of industries, and there is no involvement of the Commission services in their organisation or function. The data which the contractors decide to pass on to the outside world will be ultimately published by the laboratories or by the Commission services. Thus, the platforms do not have the exclusivity of the information and, in no case, would the information released by the contractors be withheld from other EC industries which either individually, or as a group, would request it.

The selection of research proposals in BRIDGE

For lack of funds, many good proposals for N and T-projects had to be rejected and less than 20% of the applications received could ultimately be selected. Statistics for funding are provided in Tables 1 and 2.

The procedure used for evaluating the proposals was identical for N and T-projects. For each sub-sector of the programme, experts nominated by the Member States and by the Commission services met during two or three days in Brussels for reading, assessing and rating each of the proposals submitted to their evaluation. The number of proposals (always transnational) assigned to each evaluation group varied between 10 and 25. The classification of proposals by the experts was based upon the criteria in Table 3.

Research in BRIDGE on the reproductive biology of crop plants

How the objectives and priorities were defined

This part of the programme, like all other sectors of BRIDGE, was prepared in 1988 and 1989 with the guidance of expert delegations in the Management and Coordination Committee responsible for biotechnology R&D at Community level and with the help of recommendations formulated by the Biotechnology Working Party of the Industrial Research and Development Advisory Committee (IRDAC).

The Management and Coordination Committee for the Biotechnology programme established 10 "brainstorming groups", one of which was responsible for defining new orientations of the plant biotechnology section in the

	Number of laboratories		Funding in Mio ECU	
	having applied	selected	requested	attributed
N-projects	1.900	338	350.00	42.54

Table 1: number of laboratories participating in the N-projects of BRIDGE and funding levels.

Title of T-project	Contribution EC (Mio ECU)	Number of laboratories participating
Sequencing of yeast genome *	5.0	30
Molecular identification of new plant genes	3.0	35
Biotechnology of Lactic Acid Bacteria	4.8	34
Industrial lipases	4.3	26
Regulation of plant cell regeneration	3.9	26
High resolution automated microbial identification	2.4	10
Animal cell technology	4.5	28

* Chromosome III entirely sequenced in the framework of BAP (1985-89)

Table 2: the seven "T"-projects of BRIDGE (1990-93).

- expected contribution of the proposed research to the removal of bottlenecks to applications which result from gaps in basic knowledge ("N"-projects) or from structural or scale constraints ("T"-projects)
- scientific interest of the proposal, its originality, feasibility and relevance to the scope of the programme
- technical competence of the proposers
- intention of the proposers to exploit converging contributions from different fields, disciplines and techniques in an integrated manner
- participation, through significant and balanced contributions, of at least two independent laboratories from two different Member States
- commitment of industry, in cash or in kind (particular attention was attributed, in relation to industrial involvement, to the necessity to encourage the participation of small and medium enterprises)
- when applicable, accuracy of the description of the risks possibly associated to the research proposed

Table 3 : criteria followed for the selection of research proposals in the BRIDGE programme.

new programme. The sub-group "Agriculture" of the IRDAC Working Party for Biotechnology recommended research priorities for consideration under BRIDGE.

In addition, the services of the Commission organised expert discussions on specific topics, which they considered important for the Communities, including:

- . RFLP technology for basic research and plant breeding
- . Control of differentiation and morphogenesis
- . Molecular and cellular biology of the reproduction of crop species.

All these consultations led to a few clear-cut conclusions, which pointed to opportunities associated with the applications of biotechnology to plant improvement. In all cases, observations were made indicating that:

- the seed industry, a pillar of European agriculture, was going to increase regularly its demands for new genetic techniques and for the skilled personnel which those techniques would require;
- the plant breeding areas likely to be affected by the progress of molecular biology were, primarily, the production of hybrid seeds, the control of flowering, the assistance to an early selection of predetermined traits;
- an important novel area, such as the transformation and regeneration of a wider range of crop species endowed with unique characteristics, was to be monitored in spite of the fact that breakthroughs, partly dependent on the availability of useful genes in a transferable form, were unpredictable.

What is being done

A response to these messages has been, under BRIDGE, the selection of strongly integrated research proposals which were shown to be central to plant breeding objectives. Hence, the title given to this new area entering a Community R&D programme for the first time: "The core biology of plant breeding". A total sum of 2.6 million ECU was earmarked for the period January 1991 - December 1993. Three shared cost research projects and one concerted action now form part of the programme outlined below.

Molecular and Genetic analysis of genes controlling flower development.

The aim is to isolate and study key genes controlling floral morphogenesis. These will be searched in an experimentally-suitable system via transposon-tagging or differential cDNA cloning, and their counterparts in other species such as <u>Pisum sativum</u> will be investigated.

Molecular genetics and physiology of self-incompatibility in Brassica.

Self-incompatibility represents one of the most efficient ways of producing F1 hybrid seeds in Brassica, but the degree of self-incompatibility to ensure exclusive cross-pollination is still not reliable under field conditions. The dominance relationship between S alleles will be investigated at molecular level to devise ways of circumventing the usual leakiness of the incompatibility reaction. In addition, attempts will be made to develop methods for transferring S alleles in important crops (rape), and to turn the incompatibility phenotype on and off.

The molecular basis of cell-cell interactions in self-incompatibility.

The gametophytic system of self-incompatibility will be investigated at molecular level in the Solanaceae. Pistil and pollen S-gene products will be identified, and a study made of the functions of proteins involved in the incompatibility reaction. S-specific substances have been isolated from *Solanum* styles.

Gamete differentiation and fertilisation.

This is a concerted action to establish a network of technology competences (cytology, gene expression, cell surface analysis, pollen tube kinetics, screening of mutants) with the aim of controlling fertilisation with isolated gametes *in vitro*.

All activities mentioned here are represented at this Eucarpia Conference by at least one of their coordinators.

These projects have thus been aligned strictly with the objectives of the area on "core processes for plant breeding". However, connections are established with the rest of the programme, for example, between the overlapping areas of the project on flower development, above, and work on floral mutations (Arabidopsis T-project); and between the projects on self-incompatibility and on cell signalling, as covered by the T-project on plant cell regeneration.

BIOTECH, a new Community programme for the period 1992-1994

Biotech, with a budget of 164 Mio ECU aims at the complementation and amplification of the research effort implemented in BRIDGE. Three different areas of research are foreseen:

Molecular biology

Protein structure and function
Gene structure
Expression of genes

Cellular biology

Cellular regeneration and differentiation in plant and animal cells
In vitro developmental toxicology
Analysis of essential physiological pathways in eukaryotes
Research on communication systems (including the development of human and veterinary vaccines, in vitro immune responses, in vitro immunotoxicology and the study of nerve and glial cell interactions)

Ecology and population biology

Ecological implications of biotechnology
Conservation of genetic resources

Horizontal activities, common to all of these three areas, cover:

- the development and research activities of information infrastructures important for biotechnology R&D in the Community

- the training, through training grants and through summer schools, of young European scientists specialising in one or several sectors of modern biotechnology

- the assessment of the ethical and socio-economic effects and technological risks from biotechnology.

The research efforts are implemented by means of cost-shared contracts and concerted actions. The breakdown of Community funding is foreseen as follows: 57.6 MECU for Area 1; 72.9 MECU for Area 2; 15.62 MECU for Area 3; 16.24 MECU devoted, in all areas, to the ethical and socio-economic aspects of biotechnology and to training.

Research foreseen by BIOTECH on the reproductive biology of crop plants

Important developments will result from the use of self-incompatibility or male sterility for breeding purposes. Industrial research has taken up very rapidly over the last few years key results and processes arising from the advances of molecular knowledge. This situation is highlighted by the annual budgetary figures for plant biotechnology R&D in the seed industry ($ million):

Du Pont 20; ICI 18; Monsanto 18; Sandoz 17; Ciba-Geigy 15; Rhône-Poulenc 13; Maribo 12; Sanofi 11; Bayer 8; Pioneer 7; Limagrain 5; KWS 5; etc. ...

Industry, and particularly in Europe, is now heavily engaged in designing reproductive systems for breeding purposes with all available molecular and cellular means.

Correspondingly, BIOTECH, during the period 1992-94 will shift emphasis from the mechanics of plant reproduction to the basic needs of the breeders, namely, the plant genes and their regulatory blue-prints. These pressing requirements of industry and practitioners explain why the four priority areas for the plant biotechnology part of the programme include:

- the systematic sequencing of contiguous genes in the *Arabidopsis* genome,
- the investigation of the genes, enzymes and transporters involved in key physiological traits,
- the study of cell signalling in relation to cell regeneration, and
- the development of rapid molecular genetic screening methods which link modern molecular biology and genetic resource assessments.

SUGGESTED READING

Biomolecular Engineering in the Community. Achievements of the Research Programme (1982-1986), the final report. Edited by E. Magnien; Martinus Nijhoff publishers, Dordrecht-Boston-Lancaster, 1986.

Core biology of plant breeding (a review prepared for the Commission of the European Communities in the framework of the preparation of BRIDGE). H. Dickinson with the contributions of R.J. Bino, C. Dumas, R.B. Flavell, A.J. Kool, E. Magnien, P. Olesen, W. Schuch. Designed by F. Holdsworth; printed by McLays, Cardiff, U.K., 1989.

Les laboratoires européens sans murs. A new tool for biotechnology R&D in the Community. E. Magnien, A. Aguilar, P. Wragg and D. de Nettancourt, Biofutur, Nov., 17 (1989).

Biotechnology R&D in the Community, Vol. I, Catalogue of BAP achievements. A. Vassarotti and E. Magnien eds. Elsevier (1990).

BRIDGE: launching of the research activities. D. de Nettancourt. Chimica Oggi, 9-11, April 1991.

The T-projects of BRIDGE, a new tool for technology transfer in the Community. Agro-Industry Hi-Tech, 2: 3-9, March-April 1991.

Plant Reproduction: Past, Present and Future

H.G. Dickinson
Department of Plant Sciences
Rodney Porter Building
South Parks Road
Oxford OX1 3RB
UK

INTRODUCTION

The study of plant reproductive processes presumably commenced as soon as farmers appreciated that events in the flower were in some way related to the properties of the plants that developed from the seed produced. The first reports in the literature of sex cell function in plants are, however, relatively recent (for review, see Sachs, 1890). Amici (1846, quoted in Sachs, 1890) observed germinating pollen tubes and guessed correctly at their function, and Hoffmeister (1851, quoted in Sachs, 1890) recognised that the female cells of lower plants in some way 'answered to the spermatozooids'. Nevertheless, the matter was not fully resolved until very late in the century - for even Sachs in 1975 (1890 translation) referred to fertilisation as the "blending" of cells, with the male contributing only fluid.

Investigation of the plant reproductive cycle has been hampered by the small number and location of the cells involved - and by the rapidity of their development. Thus, while recombination, male sterility, self-incompatibility and apomixis have long been recognised as being of interest to commerce and science, the technology to study them has not been available.

Recent years have seen the study of reproductive cells pursued with increased vigour for two new reasons. Firstly, they provide excellent paradigms for many aspects of cell development; thus, the generation of patterning may be studied on the surface of pollen grains (Sheldon and Dickinson 1983), and intercellular and intracellular signalling in self-incompatibility (SI) (Dickinson 1990). Secondly, the reproductive phase of the plant life cycle represents a natural system of "genetic engineering" in which new combinations of genes are generated in the gametes, tested in systems of gametophytic selection, and then assembled into new individuals - which are themselves subject to selective processes. The current technological, legal and ethical problems inherent in transferring new gene combinations into plants makes an ability to manipulate the plant's native system very attractive.

The advance of molecular methodology has enabled progress to be made in our understanding of a number of important reproductive processes in flowering plants. These are reviewed below, together with possibilities for the future considered.

FLORAL DEVELOPMENT

Some of the most spectacular advances in recent years have been made with the homeotic genes which direct floral development. Using both natural mutants (some known for over 100 years) and mutants induced by T-DNA and transposon insertion, the major genes controlling the development of sepals, petals stamens and carpels of *Arabidopsis* (Yanofsky *et al* 1990) and *Antirrhinum* (Carpenter and Coen 1990) have been identified. Surprisingly in view of their taxonomic distance, the genes of these two plants have proved to be highly homologous, and to act in an identical combinatorial fashion. Of the genes that have been cloned and sequenced, the DEF gene of *Antirrhinum* and the AG gene of *Arabidopsis* appear to be transcription factors containing sequences in common with mammalian serum response factors and yeast mating type activators (Coen and Meyerowitz 1991). Others may act in a different fashion (Coen, *pers.comm*).

These genes exert a gross effect on floral development and they thus could be effective tools in manipulating flower form and number. Once the molecular basis of floral induction itself has been established the plant breeder will have the power to determine when plants shall flower, the number of inflorescences produced and what sex organs they shall carry.

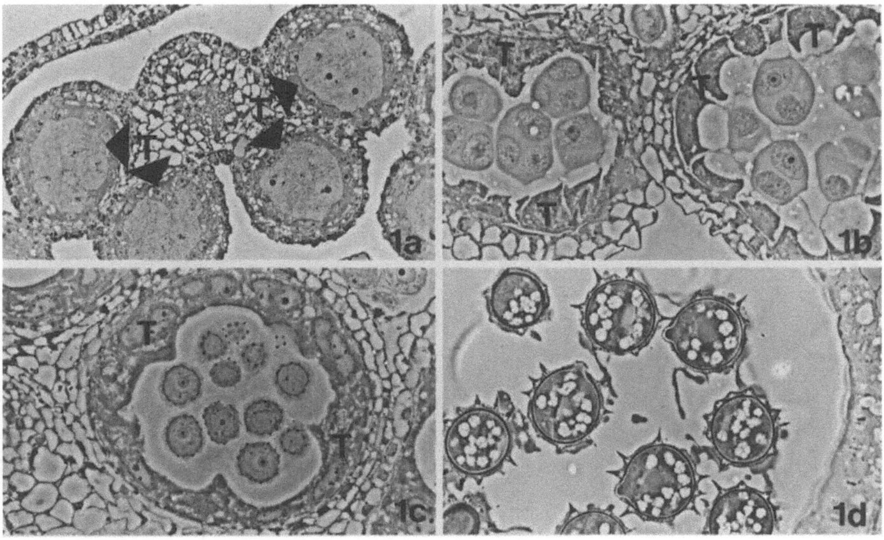

Figure 1a-d. Anther development in *Cosmos bipinnatus*. 1a. meiotic prophase; 1b. tetrad; 1c. young patterned microspores; 1d. maturing pollen. The tapetum (T) is visible at all stages except in 1d., where it has contributed to the coating of the pollen grains. Magn. 1a: 100X, 1b: 330X, 1c: 330X, 1d: 520X.

MEIOSIS, MICROSPOROGENESIS AND POLLEN DEVELOPMENT

The importance to the breeder of fertility, recombination, and male sterility makes it essential that anther development (See figure 1a-d) is understood. From the earliest studies, it has been clear that dramatic changes take place in the cytoplasm during early pollen differentiation. In 1924 Guillermond reported changes in cytoplasmic "granulation" while Painter (1943) described alterations in the stainability of the "ground cytoplasm". These events have recently been shown to be associated with cycles of organelle differentiation and RNA metabolism (See figure 2a) respectively. Levels of mitochondrial DNA synthesis and replication are particularly high during microsporogenesis, and in some species these organelles become closely associated with the early post meiotic nuclear envelope (See figure 2b).

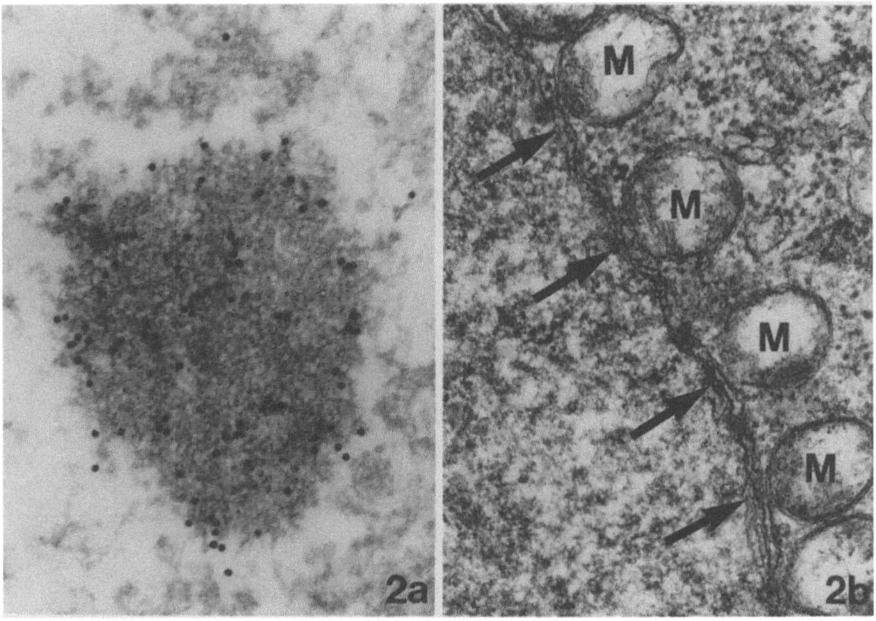

Figure 2a,b. Cytoplasmic changes during microsporogenesis. 2a. Cytoplasmic nucleoloid in the tetrad of *Lilium*, following hybridisation with *in situ* probe carrying ribosomal sequences. The nucleoloid clearly contains rDNA transcripts (see Sato *et al*, 1991 for details) 45,000X. 2b. Mitochondria (M) affixed to the envelope (arrows) of the post-meiotic nucleus in *Cosmos bipinnatus* (see Li and Dickinson, 1988 for details) 58,000X.

The relationship between pollen mother cells and their investing tapetal layer is complex, for although both cell lines share in the expression of a number of genes (Scott *et al* 1991) and in certain phases of development (Dickinson and Bell 1976a,b), the tapetum continues in a conventional mitotic cell cycle whilst the sporogenous cells enter meiosis. The genes regulating entry into meiosis in *Saccharomycetes* have been identified (McLeod and Beach, 1988) but this event is mating-type dependant in yeast, which suggests that the same genes are unlikely to be involved in plants. However, sequences controlling the "mechanical" components of meiosis such as DNA nicking, strand exchange and ligation should have some homology at the protein level. There is evidence that the delicate meiocyte/tapetal relationship can break down in some male sterile lines. Cytoplasmic male sterility (CMS) - a key tool in plant breeding (see elsewhere in this volume) - generally results from mutational rearrangements of the mitochondrial genome (Hanson *et al* 1989). Why these become phenotypically expressed only during male development remains unclear; certainly the energy requirements of the tapetum are considerable, particularly during an extended phase of DNA synthesis (See figure 3a), and there is also evidence that anther ATP/ADP ratios are affected (See figure 3b) long before structural degeneration (Liu and Dickinson 1989). Equally, it is possible that a metabolite, unique to the anther, interacts with mitochondrial gene expression or development in such a way as to result in sterility. Although "engineered" male sterility systems have been devised (Mariani *et al* 1990), an understanding of, and an ability to exploit the situation *in vivo* will be of great value in the development of systems to be used in the field.

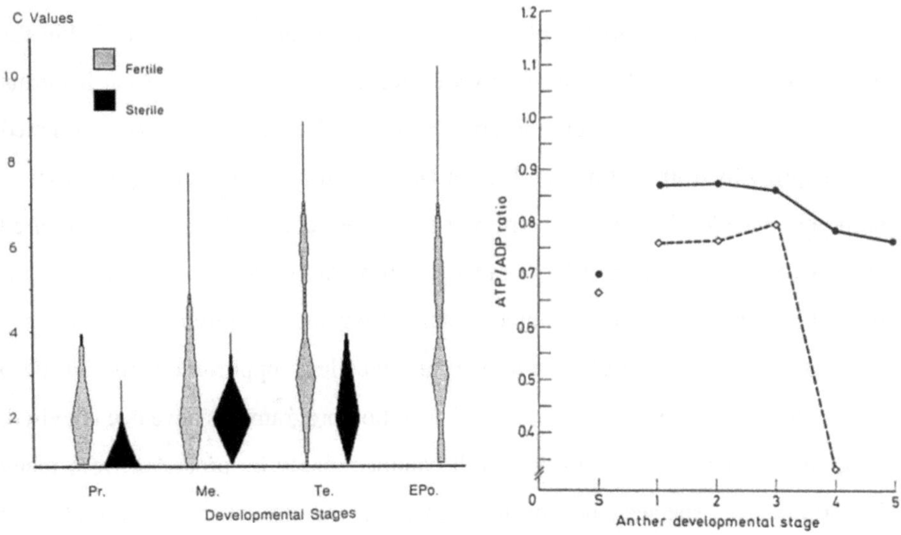

Figure 3a,b. DNA synthesis and ATP/ADP ratios in fertile and male sterile lines of *Petunia.* 3a. DNA C values of tapetal muclei of fertile and male sterile lines. The width of each histogram indicates the proportion of the total nuclear population of that particular C value. For methods see Liu and Dickinson 1989. Stages; pro: meiotic prophase; me: metaphase-I; Te: tetrad; Epo: early pollen. 3b. ATP/ADP ratios in somatic cells and in the anthers of normal and CMS lines of *Petunia hybrida* throughout male development. S Somatic (leaves), 1 premeiosis, 2 prophase, 3 metaphase, 4 tetrad, 5 early pollen.
Normal, ● CMS. ◇

Compared with the early phases of meiocyte and microspore development, large numbers of genes have been characterised following pollen mitosis I (see McCormick, 1991, for review). Function has been ascribed to many of these and some are clearly involved in pollen stigma interactions. However, the sequences directing many key developmental events, such as the assymmetic cytokinesis of pollen mitosis I, and the elimination of organellar genomes (Miyamura *et al* 1987), have yet to be identified. Interestingly, pollen expresses a very high proportion of the genes expressed in the growing plant (Mascarenhas and Willing 1984) and thus has potential value in selection strategies (Ottoviano *et al* 1988).

HAPLOID EMBRYO PRODUCTION FROM DEVELOPING POLLEN

Naturally occurring haploids have been known in the plant kingdom for many years (Blackslee, 1922) and while is was appreciated that doubled haploids would be of great use to the plant breeder, such plants were not available for most crop species, and no means existed for generating them. Such a means emerged from a chance observation by Guha and Maheshwari (1964), who noted that outgrowths from cultured anthers of *Datura*, were in fact microspore-derived embryos. Using this approach, embryos have been obtained from either cultured anthers or microspores of a large range of crop plants and, in many cases it has proved possible to regenerate haploid plants from these embryos. However, the haploid embryos themselves provide an excellent opportunity for *in vitro* selection, and a number of gametoclonal selection programmes have been devised. Despite this success, haploid lines still cannot reliably be produced from many important crop species and for this reason a number of cell and molecular biological investigations have focused on the primary induction phases of microspore embryogenesis.

The earliest stages of embryogenesis in *Brassica* have been demonstrated to involve a symmetrical first post-meiotic division, rather than the normal asymmetrical pollen mitosis I (Pechan and Keller 1988, Zaki and Dickinson 1990). This lack of asymmetry is accompanied by the formation of a sporophytic-type wall, accumulation of starch and, most interestingly, the synthesis of stress-related proteins similar to those following heat shock (Pechan pers. comm.). How these proteins can induce such comprehensive changes in metabolism and organisation of the cytoskeleton remains to be determined. There are also data from *Nicotiana* indicating that development may be directed into a sporophytic or gametophytic pathway according to the composition of the culture media (Benito Moreno *et al* 1988).

POLLEN STIGMA INTERACTIONS

A number of strategies have been developed to overcome interspecific incongruities (Hogenboom, 1975), ranging from stylar pollination to parasexual hybridisation - where protoplasts of the two species are fused *in vitro*. However, the taxonomic distance of two species is not only reflected in their pollination systems, but also in chromosomal number and structure. Thus many of these hybrids fail to succeed through abortion of either embryo or endosperm. There are no straightforward molecular approaches to this important problem but, since most wide hybrids are synthesised to transfer a small number of traits to an established crop line, direct gene transfer is often a viable alternative.

Systems of self-incompatibility (SI) occur in plants with both dry and wet stigmas. These mechanisms were described by Darwin in 1877, in connection with which he first coined the use of the word "recognition" in a biological context. While Darwin noted the various mechanical devices preventing self-pollination, work on *Nicotiana* by East and Mangelsdorf (1925) and on the Cruciferae by Bateman (1955) established the physiological regulation of SI by few multiallelic loci. The power of molecular methodology was demonstrated in 1986 when Anderson *et al* cloned the gene encoding the female S(incompatibility)-allele product from *Nicotiana*. More recently, this glycoprotein has been shown to be an active RNAse (McClure *et al* 1989), and to be internalised following a self pollination (Gray *et al* 1991). This SI mechanism may thus have evolved from RNAses already present in the stigmatic secretion - perhaps acting as a defence against pathogen attack. As yet, the male determinant of the system has not been identified and nothing is known of the recognition mechanism itself. Female S-linked glycoprotein genes have now been cloned in many members of the Solanaceae (e.g. Singh *et al* 1991), all of which encode RNAses. Interestingly the female determinant of SI in *Papaver rheoas* (Franklin Tong *et al* 1992) is not an RNAse. In an *in vitro* bioassay this polypeptide stimulates the phosphorylation of a range of novel pollen polypeptides following an "incompatible" interaction and a number of cDNAs characteristic of compatible and incompatible responses have already been cloned (Franklin *pers. comm.*).

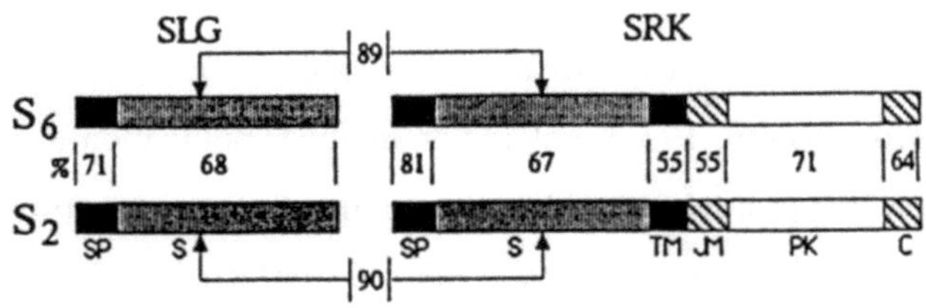

Figure 4. Concerted evolution of the SLG/SRF gene pair within S genotypes and their divergence between genotypes. The predicted SLG and SRK proteins encoded by the S_6 and S_2 genotypes are represented schematically. Numbers indicate percentage identity between putative domains. SP, signal peptide; S, S domain; TM, transmembrane domain; JM, juxtamembrane domain; PK, protein kinase domain; C, C terminus domain. From Stein *et al.* 1991.

The sporophytic SI system of *Brassica* also features female glycoproteins (Nishio and Hinata 1982), but which also have no RNAse activity. The S-locus in *Brassica* may be more complex than that of *Nicotiana* in that at least one further gene is present - termed SRK (Stein *et al* 1991). This gene (See figure 4), which is expressed in both pistil and anther, encodes a transmembrane kinase featuring an extracellular domain with >90% homology to the female S-locus glycoprotein (SLG). The physiology of the self-incompatibility response in *Brassica* is not well understood, but it is dependent upon both protein synthesis and glycosylation (See figure 5)(Sarker *et al* 1988).

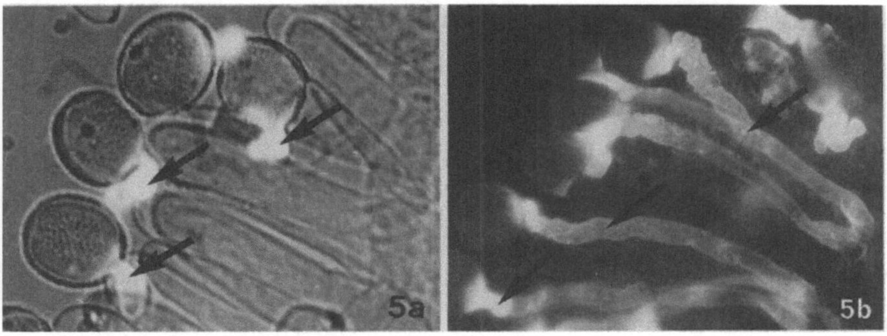

Figure 5a,b. Self incompatibility in *Brassica olearacea.* Self pollination (5a) normally results in little or no growth of the pollen tube (arrows) c. 1800X. Following tunicamycin treatment (5b), which inhibits glycosylation, long tubes (arrows) are formed 1400X. For details see Sarker *et al.* 1988.

Study of the promoter regions of the SLG gene has been revealing, for when GUS reporter constructs are transformed into self-compatible *Arabidopsis* and SI *Brassica oleracea*, the promoter drives expression in the stigmatic papillae and the anther tapetum (Toriyama *et al* 1991, Sato *et al* 1991). Since the tapetum is sporophytic and eventually coats the pollen, this finding confirms earlier suspicions that the pollen coat may be the location of the male SI determinant (Dickinson 1990). However, some GUS is also expressed in the pollen, keeping open the possibility that S-locus products survive through meiosis - or that sporophytic SI operates in addition to an existing gametophytic system.

Numerous other pistil specific genes have been cloned (e.g. Gasser *et al* 1989), but function has been assigned to very few. Recently a transmitting tissue extensin-like cDNA has been cloned from *Antirrhinum* (Baldwin *et al* 1992) suggesting a high degree of specialisation extends elsewhere in the pistil. The development of dry stigma systems was a evolutionary milestone, since it prevents the capture of pathogen spores. However, this event seems to have necessitated the development of different structure and physiology (Elleman and

Dickinson 1986, Sarker *et al* 1988), and perhaps even the evolution of a new signalling system (Doughty *et al* 1992).

A FUTURE PERSPECTIVE

While progress over the last decade has become increasingly rapid, there are a number of important areas in which we remain ignorant. In particular, very little is known of the molecular basis of female development. Certainly a number of embryo-sac specific cDNAs have been identified and some excellent ultrastructural studies have been carried out (Russell 1986), but much further work will be needed before the molecular basis of synergid function, or apomixis is understood. The manipulation of apomixis remains an important commercial research objective, but a combination of complicated genetics and bewildering cytology render it one of the most refractive of topics.

Within the next five years some of the basic research currently in progress should start to provide tangible results. In addition to the manipulation of floral genes discussed earlier, microspore embryogenesis should become routine in all but the most recalcitrant species and, if the S-locus of *Brassica* consists only of the SLG and SRK genes, SI will become transferable to new species. More tentatively, a knowledge of the epigenetic changes imposed on genes during gametogenesis (Haig and Westoby, 1989) may help in understanding endosperm failure in wide hybrids, while the discovery of sequences in *Zea mays* homologous with the SPO 11 gene, which regulates pairing in *Saccharomyces cereviseae* (Howley, *et al*, in preparation), suggests that the characterisation of plant genes involved in recombination is feasible.

In the final analysis, plant reproduction is important because it affects, directly or indirectly, the quality and quantity of all crop products. It is a process constantly in use in both the developed and developing world, and is the natural method by which new genetic combinations are generated. Increasing our understanding in this area should both provide opportunities for optimising existing processes, as well as for manipulating the central mechanisms of transmission genetics.

ACKNOWLEGEMENTS

The author wishes to thank Ann Rogers for invaluable help during the preparation of the manuscript.

REFERENCES

Anderson, M.A., Cornish, E.C., Mau, S-L., Williams, E.G., Hoggart, R., Atkinson, A., Bonig, I., Grego, B., Simpson, R., Roche, P.J., Haley, J.D., Penschow, J.D., Niall, H.D., Tregear, G.W., Coghlan, J.P., Crawford, R.J. and Clarke, A.E. 1986. Cloning of a cDNA for a stylar glycoprotein associated with the expression of self-incompatibility in *Nicotiana alata*. Nature 314, 38-44.

Baldwin,T.C., Coen, E.S. and Dickinson, H.G. 1992. The ptl1 gene expressed in the transmitting tissue of *Antirrhinum* encodes an extensin-like protein. In press.

Bateman, A.J. 1955. Cruciferae. Heredity 9, 53-68.

Benito Moreno, R.M., Macke, R., Hauser, M.T., Alken, A. and Herbele Bors, E. 1988. Sporophytes and male gametophytes from *in vitro* cultured immature tobacco pollen. In "Sexual Reproduction in Higher Plants". Eds. M. Cresti, P. Gori and E. Pacini. Springer Verlag (Heidelberg) p. 137-142.

Blackslee, A.F., Belling, J., Farnham, M.E. and Bergner, A.D. 1922. A haploid mutant in the jimson weed, *Datura shamonium*. Science 55, 1433.

Carpenter, R. and Coen, E.S. 1990. Floral homeotic mutations produced by transposon-mutagenesis in *Antirrhinum majus*. Genes and Development 4: 1483-1493

Coen, E.S. and Meyerowitz, E.M. 1991. The war of the whorls: genetic interactions controlling flower development. Nature 353, 31-37.

Darwin, C. 1877. The different forms of flowers on plants of the same species. John Murray, London. 2nd Ed.

Dickinson, H.G. 1990. Self-incompatibility in flowering plants. BioEssays 12, 155-161.

Dickinson, H.G. and Bell, P.R. 1976a. Development of the tapetum in *Pinus banksiana* preceding sporogenesis. Ann. Bot. 40, 103-114.

Dickinson, H.G. and Bell, P.R. 1976b. The changes in the tapetum of *Pinus banksiana* accompanying formation and maturation of the pollen. Ann. Bot. 40, 1101-1109.

Doughty, J., McCubbin, A., Hedderson, F., Elleman, C.J. and Dickinson, H.G. 1992. The role of the pollen grain coating in pollination and self incompatibility in *Brassica oleracea*. In press.

East, E.M. and Mangelsdorf, A.J. 1925. A new interpretation of the heredity behaviour of self-sterile plants. Proc. Natl. acad. Sci. USA 2, 166-183.

Elleman, C.J. and Dickinson, H.G. 1986. Pollen stigma interactions in *Brassica*. IV. Structural reorganisation in the pollen grains during hydration. J. Cell Sci. 80, 141-157.

Franklin-Tong, V.E. and Franklin, F.C.H. 1992. Gametophytic self-incompatibility in *Papaver rheoas*. Sex. Plant Rep. 5 (1), 1-8.

Gasser, C.S., Budelier, K.A., Smith, A.G., Shah, D.M. and Fraley, R.T. 1989. Isolation of tissue-specific cDNAs from tomato pistils. Plant Cell 1, 15-24.

Gray, J.E., McClure, B.A., Bönig, I., Anderson, M.A. and Clarke, A.E. 1991. Action of the style product of the self-incompatibility gene of *Nicotiana alata* (S-RNase) on *in vitro*-grown pollen tubes. The Plant Cell 3, 271-283.

Guha, S. and Maheshwari, S.C. 1964. *In vitro* production of embryos from anthers of *Datura*. Nature 204, 497.

Guillermond, A. 1924. Recherches sur l'evolution du chondriome pendent le development du sac embryonaire et des cellules-mères des grains de pollen dans les Liliacées et la signification des formations ergastoplasmiques. Ann Sci Nat Bot 6: 1-52

Haig, D. and Westoby, M. 1989. Parent specific gene expression and the triploid endosperm. Am. Nat. 134, 147-155.

Hanson, M.R., Pruitt, K.D. and Nivison, H.T. 1989. Male sterility loci in plant mitochondrial genomes. Oxf. Surveys. Plant Mol. Cell Biol. 6, 61-85.

Hogenboom, N.G. 1975. Incompatibility and incongruity: two different mechanisms for the non-functioning of intimate partner relationships. Proc. R. Soc. Lond. B. 188, 361-375.

Li, F.L. and Dickinson, H.G. 1988. Organelle behaviour during higher plant gametogenesis. In Division and Segregation of Organelles, eds. S.A. Boffey and D. Lloyd, Soc. Exptl. Biol. Seminar Series 35, 131-148.

Liu, X.C. and Dickinson, H.G. 1989. Cellular energy levels and their effect on male cell abortion in cytoplasmically male sterile lines of *Petunia hybrida*. Sex Pl Reproduction 2: 167-172

Mariani, C., de Beuckeleer, M., Truettner, J., Leemans, J., and Goldberg, R.G. 1990. Induction of male sterility in plants by a chimaeric ribonuclease gene. Nature 347: 737-741

Mascarenhas, J. and Willing, R.P. 1984. Analysis of the complexity and diversity of mRNAs from pollen and shoots of *Tradescantia paludosa*. Plant Physiol. 78, 887-890.

McClure, B.A., Haring, V., Ebert, P.R., Anderson, M.A., Simpson, R.J., Sakiyama, F. and Clarke, A.E. 1989. Style self-incompatibility gene products in *Nicotiana alata* are ribonucleases. Nature 342: 955-957.

McCormick, S. 1991. Molecular analysis of male gametogenesis in plants. Trends in Genetics 7 (9): 298-303

McLeod, M. and Beach, D. 1988. A specific inhibitor of the *ran1*+ protein kinase regulates entry into meiosis in *Schizosaccharomyces pombe*. Nature 322: 509-514

Miyamura, S., Kuriowa, T. and Nagata, T. 1987. Disappearance of plastid and mitochondrial nucleoloids during the formation of generative cells of higher plants revealed by fluorescence microscopy. Protoplasma 141: 149-159.

Nishio, T. and Hinata, K. 1982. Comparative studies on S-glycoproteins purified from different S-genotypes in self-incompatible *Brassica* species. I.

Purification and chemical properties. Genetics 100, 641-647.
Ottoviano, E., Sari-Gola, M., Frova, C. and Pe, E. 1988. Male gametophytic selection in higher plants. In "Sexual Reproduction in Higher Plants". Eds. M. Cresti, P. Gori and E. Pacini. Springer Verlag (Heidelberg) p. 35-42.
Painter, T.S. 1943. Cell growth and nucleic acids in the pollen of *Rheo discolor*. Bot. Gaz. 105, 58-68.
Pechan, P.M. and Keller, W.A. 1988. Identification of potentially embryogenic microspores in *Brassica napus*. Physiologia Plantarum 74, 377-384.
Russell, S.D. 1986. Dimorphic sperm cells, cytoplasmic transmission and preferential fertilization in *Plumbago zeylanica*. In The Chondriome - Chloroplast and Mitochondrial genomes. Eds. S.H. Mantell, G.P. Chapman, P.F.S. Street. Longman, Essex, p. 69-116.
Sachs, J. 1890. A history of botany. Trans. H.E.F.Garnsey. Clarendon Press, Oxford.
Sarker, R.H., Elleman, C.J. and Dickinson, H.G. 1988. Control of pollen hydration in *Brassica* requires continued protein synthesis, and glycosylation is necessary for intraspecific incompatibility. Proc. Natl. Acad. Sci. USA 85: 4340-4344.
Sato, S., Jones, K., de Los Dios Alche, J. and Dickinson, H.G. 1991. Cytoplasmic nucleoloids of *Lilium* male reproductive cells contain rDNA transcripts and share features of development with nucleoli. J. Cell Sci. 100, 109-118.
Sato, T., Thorsness, M.K., Kandasamy, M.K., Nishio, T., Hirai, M., Nasrallah, J.B. and Nasrallah, M.E. 1991. Activity of an S locus gene promoter in pistils and anthers of transgenic Brassica. The Plant Cell 3, 867-876.
Scott, R., Dagless, E., Hodge, R., Wyatt, P., Soufleri, I., and Draper, J. 1991. Patterns of gene expression in developing anthers of *Brassica napus*. Plant Mol Biol 17 (2): 195-209
Sheldon, J.M. and Dickinson, H.G. 1983. Determination of patterning in the pollen wall of *Lilium henryi*. J Cell Sci 63: 191-208
Singh, A., Ai, Y. and Kao, T. 1991. Characterization of ribonuclease activity of three S-allele-associated proteins of *Petunia inflata*. Plant Physiol. 96, 61-68.
Stein, J.C., Howlett, B., Boyes, D.C., Nasrallah, M.E. and Nasrallah, J.B. 1991. Molecular cloning of a putative receptor protein kinase gene encoded at the self-incompatibility locus of *Brassica oleracea*. Proc. Natl. Acad. Sci. USA 88, 8816-8820.
Toriyama, K., Thorsness, M.K., Nasrallah, J.B. and Nasrallah, M.E. 1991. A *Brassica* S locus gene promoter directs sporophytic expression in the anther tapetum of transgenic *Arabidopsis*. Dev. Biol. 143, 427-431.
Yanofsky, M.F., Ma, H., Bowman, J.L, Drews, G.N., Feldmann, K.A. and Meyerowitz, E.M. 1990. The protein encoded by the *Arabidopsis* homeotic gene *agamous* resembles transcription factors. Nature 346, 35-39.
Zaki, M.A.M., and Dickinson, H.G. 1990. Structural changes during the first divisions of embryos resulting from anther and free microspore culture in *Brassica napus*. Protoplasma 156, 149-162.

The Control of Flower Formation: Exogenous and Endogenous

G. BERNIER
Laboratory of Plant Physiology
University of Liège
Sart Tilman, 4000 Liège
Belgium

Introduction

Flower initiation is a paradigm of the long-distance signalling system in higher plants. In most plants, indeed, this developmental step is influenced by a variety of environmental factors. Each of these factors is most perceived by a particular plant part, either the leaves or the shoot apices or the roots. As a result, the various plant parts then exchange a number of signals. When the environmental conditions are favourable for flowering, this ultimately leads some target shoot meristems to shift from vegetative to reproductive growth and morphogenesis.

My aim here is to summarize our knowledge about this complex system of control of flower initiation involving a variety of exogenous (environmental) and endogenous (signals) factors.

Exogenous (environmental) control

Inspection of the 6 volumes of Halevy's Handbook of Flowering (1985-1989) reveals that in most plants the exogenous control of flower initiation results from an interaction between several distinct factors, most often daylength, temperature and irradiance. Other factors, like mineral nutrition, water stress, etc, might also interplay with the above factors, although their roles seem secondary (Bernier *et al.*, 1981, vol. I).

For those plants that have been the subject of detailed analyses, it is not uncommon to find that their floral transition can be caused by several different treatments (Bernier, 1988). See for example the cases of the short-day plant (SDP), *Pharbitis nil*, and the long-day plant (LDP), *Silene armeria*, in vol. IV of the above Handbook.

In our lab, we have investigated this problem in the LDP *Sinapis alba* and the SDP *Xanthium strumarium*. In *Sinapis*, we succeeded to induce flower initiation not only by exposure to a single LD, but also to a single displaced short day (DSD), to two weeks of normal SD at a higher irradiance or by growing the plants for five weeks in SD at 4 °C (Bernier, unpublished results).

In *Xanthium*, we obtained an identical flowering response in plants exposed to one long night or to two LD if the first half of these LD and the associated short nights are at 4°C and if GA_3 is applied at the midpoint of each LD (Mirolo *et al.*, 1990).

Sites of perception of exogenous factors

Each factor of the environment affecting flower initiation is perceived most effectively by a particular plant part. Clearly, daylength and irradiance are most perceived by expanded leaves; mineral nutrition and water stress by roots. Temperature is presumably perceived by all plant parts, but classical evidence supported by more recent results (Metzger, 1988; Arumuganathan *et al.*, 1991), indicates that low temperatures promoting flowering are essentially perceived by the shoot apex. In *Xanthium*, too, flower formation in LD (see above) is observed when only the shoot is cooled to 4°C while the root system is maintained at 23°C (Bernier, unpublished result). Thus, the site of low temperature perception in this last case is either the leaf, the apex or both. In pea, low temperature was found to be perceived by both the leaves and the apex (see Halevy's Handbook, vol. IV).

In other cases, other sites of perception of low or high temperatures were disclosed (Bernier, 1988). The low temperature promoting flowering in chicory is perceived by the root collar and the high temperature substituting for LD in the LDP *Silene armeria* is perceived by the roots. In *Brassica pekinensis*, high devernalizing temperature has been claimed to be perceived by the root system (Pressman and Negbi, 1981), but Rietze and Wiebe (1988) found that this is probably due to heat transfer within the plant and that the shoot apex is the real site of temperature perception.

Thus, different promotive and inhibitory exogenous factors may act at totally different sites within the plant. When these sites are not the places where flowers will be formed, i.e. the target shoot apices, there is a requirement for long-distance transmission of signals.

What is the general nature of floral signals ?

Recent electrophysiological studies in the SDP *Chenopodium rubrum* (Adamec and Krekule, 1989a) concluded that, during the photoperiodic induction of flowering, no change of membrane potential in leaf cells occurs that could be interpreted as floral signals. A similar conclusion is reached after measurements of transorgan electric potentials during photoinduction (Adamec and Krekule, 1989b). These results support the classical idea that the floral signals have a biochemical nature, not a biophysical one (Lang, 1965; Bernier, 1988).

Changes in sap composition

The movement of long-distance signals is essentially in the xylem and phloem saps. The floral signal(s), produced in the leaves as a result of exposure to favourable daylengths, are indeed known since long ago to be transported in the phloem with the bulk of assimilates (Lang, 1965; Bernier *et al.*, 1981). On the other hand, any influence exerted by roots in the flowering process is presumably transmitted in the xylem sap. Thus, analyses of changes in sap composition at critical times during the floral transition in well-defined experimental systems should be undertaken. Such a work has been initiated in various labs, including in our own.

In the SDP *Pharbitis nil*, the flowering of apices excised from noninduced plants and grown *in vitro* is markedly promoted by addition to the culture medium of the exudate (phloem sap) from induced cotyledons (Ishioka *et al.*, 1990). On the contrary, the exudate from noninduced cotyledons decreases flowering in cultured apices collected on induced plants.

The exudate from induced leaves of a SD cv of tobacco is also found to increase the number of flower buds regenerated by cultured thin cell layer explants from the same tobacco (Hatori *et al.*, 1991). The exudate from noninduced leaves has no such effect.

In these two studies, the chemical nature of the active compound(s) in the exudates is under investigation, but is still entirely unknown.

In our lab we have developed another approach of this problem using *Sinapis* plants induced to flower by either a single LD or a single DSD. These two inductive light/dark regime are known to be most perceived by the mature leaves. In this system, we took advantage that the successive steps of the floral transition are very well defined (Bernier, 1986). We also knew that part of the events normally occurring during the floral transition in the target shoot meristem can be caused either by an increase in the carbohydrate level or by an application of a cytokinin (Bernier, 1986). The events caused by carbohydrates are different from those produced by the cytokinin. These two classes of chemicals are thus good candidates for starting the chemical analysis of sap composition in *Sinapis* plants during the floral transition.

On the other hand, knowing that in other plant systems, e.g. the moss *Funaria*, part of the effects of cytokinins is apparently mediated by an increase in Ca^{2+} uptake in responsive cells (Saunders, 1990), we also analyzed the content in Ca^{2+} in the saps of *Sinapis*.

Two kinds of exudates were collected from the plants (Lejeune *et al*., 1988) : (a) the leaf exudate which represents the phloem sap flowing out of the mature leaves; (b) the root exudate which represents the xylem sap exported by the root system.

The leaf exudate - Sucrose is the predominant carbohydrate in this exudate. We observed that the level of sucrose exhibits a dramatic increase in response to induction by either one LD or one DSD (Lejeune *et al*., 1991). This rise is seen quite early, that is very soon after the start of the photoextension period of the LD or during the DSD itself.

Isopentenyladenine (iP) is the major cytokinin detected in the leaf exudate, with zeatin being present only in trace amounts. The total cytokinin activity in the leaf exudate of plants induced by one LD is markedly increased at around 8 h after start of the photoextension period of the LD, i.e. several h later than the sucrose increase (Bernier *et al.*, 1990). The cytokinin activity after induction by one DSD remains to be investigated.

There is no change at all in the level of Ca^{2+}, in the leaf exudates of plants exposed to either one LD or one DSD (Havelange and Bernier, in preparation).

<u>The Root exudate</u> - Some of the above compounds, namely cytokinins and Ca^{2+} that are found in the leaf exudate, could originate from the root system. This inference was tested by collecting and analyzing root exudates.

Contrary to the situation in the leaf exudate, trans-zeatin riboside (ZR) is the major cytokinin found in the root exudate, with isopentenyladenine riboside being a minor component. The total cytokinin activity in the root exsudate of plants induced by one LD is markedly increased and, interestingly enough, the increase is first detected very soon after the start of the photoextension period of the LD (Bernier *et al.*, 1990). In fact, the increase in cytokinin activity in the root exudate precedes that in the leaf exudate by about 7 h.

The Ca^{2+} level starts to rise in the root exudate of LD-induced plants at 4 h after start of the photoextension period and stays at a higher level for about 20 h (Havelange and Bernier, in preparation). A rather similar situation was disclosed in plants submitted to one DSD.

Discussion

It is striking that, in the leaf exudate, sucrose increases first and dramatically both in *Sinapis* plants induced by one LD, that is with an extension of the period of photosynthetic assimilation, and in those induced by a DSD, that is without such an extension. Moreover, in response to one LD or one DSD, these early increases in the sucrose level of the leaf exudate occur well before any increase in the supply from leaves of recently-synthesized assimilates (^{14}C-labelled assimilates) for the apex (Bodson *et al.*, 1977; Bodson and Remacle, 1987). All together, our data suggest that mobilization of reserve carbohydrates stored in the leaves occurs quite early as a response to induction of flowering. This is line with the well-known fact that many environmental factors promoting flower initiation, like SD, low temperature and water stress, all promote starch hydrolysis and thus increase sugar availability (Bernier *et al.*, 1981, vol. II).

In our lab, we have also started sap analyses in the SDP *Xanthium*. Despite that this plant is induced to flower by one long night, i.e. by markedly decreasing the duration of the period of photosynthetic activity, the sucrose content is also found to exhibit an early increase in the leaf exudate (Houssa *et al.*, 1991).

Given that the cytokinin level in *Sinapis* increases several h earlier in the root exudate than in the leaf exudate, one might think that the cytokinins exported by leaves originate from those which arrived previously from the roots. As against this simple idea, there is the fact that cytokinins of the zeatin family have been found to derive from an iP derivative whereas cytokinins of the iP family cannot derive from Z derivatives (McGaw, 1987; Palni *et al.*, 1990). Thus, in the present state of knowledge, the production of the iP found in the leaf exudate of *Sinapis* cannot be from the ZR of root origin. Presumably, then, iP is produced in the leaves themselves. Roots are known as the major site of cytokinin biosynthesis in vegetative plants, but leaves were shown to be additional sites of cytokinin production (Palni *et al.*, 1990). A possibility raised by our results is that iP production in leaves is somehow controlled by the level of cytokinins of the Z family.

It is usually recognized that Ca^{2+} is rather immobile in the phloem and thus is not significantly redistributed from mature leaves toward apices (Kirby and Pilbeam, 1984; Clarkson, 1984). Indeed, the Ca^{2+} level does not increase in the *Sinapis* leaf exudate during the floral transition despite the fact that the level of this cation does increase in the apex of induced plants (Havelange, 1989). Thus, it is quite possible that part of the extra supply of Ca^{2+} found in the root exudate of induced plants, is diverted via the apoplast from the axial vessel strands in the stem directly to the apex (Clarkson, 1984).

Conclusions

Clearly, the changes in the environmental conditions that are conducive to flower initiation in *Sinapis* cause early and dramatic alterations of the long-distance signalling system in the entire plant. The compositions of both the leaf and the root exudates are changed. These changes, affecting the levels of carbohydrates, cytokinins and Ca^{2+}, are well ordered in time and space. It is concluded that all plant parts participate in the exchange of signals, and that all plant parts are very rapidly instructed about the changes of the light/dark regime to which the leaves are exposed.

References

Adamec L, Krekule J (1989a) Changes in membrane potential in *Chenopodium rubrum* during the course of photoperiodic flower induction. Biol Plant 31:336-343

Adamec L, Krekule J (1989b) Changes in transorgan electric potential in *Chenopodium rubrum* during the course of photoperiodic flower induction. Biol Plant 31:344-353

Arumuganathan K, Dale PJ, Cooper JP (1991) Vernalization in *Lolium temulentum* L. : Responses of *in vitro* cultures of mature and immature embryos, shoot apices and callus. Ann Bot 67:173-179

Bernier G (1986) The flowering process as an example of plastic development. In : Jennings DH, Trewavas AJ (eds) Plasticity in Plants. Company of Biologists, Cambridge, pp 257-286

Bernier G (1988) The control of floral evocation and morphogenesis. Ann Rev Plant Physiol Plant Mol Biol 39:175-219

Bernier G, Kinet JM, Sachs RM (1981) Physiology of Flowering, Vols I and II. CRC Press, Boca Raton

Bernier G, Lejeune P, Jacqmard A, Kinet JM (1990) Cytokinins in flower initiation. In : Pharis RP, Rood SB (eds) Plant Growth Substances 1988. Springer, Berlin, pp 486-491

Bodson M, King RW, Evans LT, Bernier G (1977) The role of photosynthesis in flowering of the long-day plant *Sinapis alba*. Aust J. Plant Physiol 4:467-478

Bodson M, Remacle B (1987) Distribution of assimilates from various source-leaves during the floral transition of *Sinapis alba* L. In : Atherton JG (ed) Manipulation of Flowering. Butterworths, London, pp 341-350

Clarkson DT (1984) Calcium transport between tissues and its distribution in the plant. Plant Cell Environment 7:449-456

Halevy AH (1985-1989) Handbook of Flowering, Vols I to VI. CRC Press, Boca Raton

Havelange A (1989) Levels and ultrastructural localization of calcium in *Sinapis alba* during the floral transition. Plant Cell Physiol 30:351-358

Hatori M, Sakagami Y, Marumo S (1991) The effects of auxin and antiauxin in an *in vitro* bioassay of flower regulatory activity in leaf exudate from tobacco plants. Physiol Plant 81:1-6

Houssa P, Bernier G, Kinet JM (1991) Qualitative and quantitative analysis of carbohydrates in leaf exudate of the short-day plant, *Xanthium strumarium* L. during floral transition. J Plant Physiol 138:24-28

Ishioka N, Tanimoto S, Harada H (1990) Flower-inducing activity of phloem exudate in cultured apices from *Pharbitis* seedlings. Plant Cell Physiol 31:705-709

Kirby EA, Pilbeam DJ (1984) Calcium as a plant nutrient. Plant Cell Environment 7:397-405

Lang A (1965) Physiology of flower formation. In : Ruhlandd W (ed) Encyclopedia of Plant Physiology. Springer, Berlin, pp 1379-1536

Lejeune P, Bernier G, Kinet JM (1991) Sucrose levels in leaf exudate as a function of floral induction in the long day plant *Sinapis alba*. Plant Physiol Biochem 29:153-157

Lejeune P, Kinet JM, Bernier G (1988) Cytokinin fluxes during floral induction in the long day plant *Sinapis alba* L. Plant Physiol 86:1095-1098

McGaw BA (1987) Cytokinin biosynthesis and metabolism. In : Davies PJ (ed) Plant Hormones and their Role in Plant Growth and Development. Martinus Nijhoff Publ, Dordrecht, pp 76-93

Metzger JD (1988) Localization of the site of perception of thermoinductive temperatures in *Thlaspi arvense* L. Plant Physiol 88:424-428

Mirolo C, Bodson M, Bernier G (1990) Floral induction of *Xanthium strumarium* in long days. Ann Bot 66:475-477

Palni LMS, Nandi SK, Singh S, Letham DS (1990) An overview of cytokinin biosynthesis. In : Pharis RP, Rood SB (eds) Plant Growth Substances 1988. Springer, Berlin, pp 258-266

Pressman E, Negbi M (1981) Bolting and flowering of vernalized *Brassica pekinensis* as affected by root temperature. J Exp Bot 32:821-825

Rietze E, Wiebe HJ (1988) The influence of soil temperature on vernalization of chinese cabbage. J Hort Sci 63:83-86

Saunders MJ (1990) Cytokinin : Evidence for spatial control of signal transduction. In : Pharis RP, Rood SB (eds) Plant Growth Substances 1988. Springer, Berlin, pp 511-518

Genetic and Molecular Enrichment Steps as Cloning Strategy in the Dioecious *Melandrium album (Silene alba)*

J. Veuskens[(1)], C. Lacroix[(2)], A.T. Truong[(2)], S. Hinnisdaels[(1)], A. Mouras[(3)], I. Negrutiu[(1,2)]
[(2)]Phytotec S.A.
chaussée Romaine 77
B 5030 Gembloux
Belgium

INTRODUCTION

Despite extensive efforts to understand hermaphrodite floral developmental programmes, little is known about the mode of action of sex-determining and differentiation factors.

Sex dimorphic plant species represent less than 4 % of flowering plants. In these species, sex is most probably controlled by regulatory factors either superimposed upon, or derived from the basic pattern of hermaphrodite floral development.

Among dioecious species, *Melandrium* has long been considered a model system on sex determination (Westergaard, 1958), because its genetic control is by far the most strict among such plant species, and because it has a relatively well characterized male (Y,X) and female, (X,X) chromosome system (Ciupercescu et al, 1990).

M. album has a male chromosome complement of 2n = 24,XY (Fig. 1); the genome size is 6,2 pg, the Y chromosome representing approximately 10 %.

Sex conversion is independant of environmental or hormonal factors. Instead, Y chromosomal deletions or a non-pathogenic interaction with the fungus *Ustilago violacea* are known to result in modifications of sex expression (Ye et al, 1991).

[(1)]Plantengenetica, Instituut voor Moleculaire Biologie, Paardenstraat 65, B 1640 St Genesius Rhode, Belgium
[(3)]Université de Bordeaux II, avenue des Facultés, F 33405 Talance, France

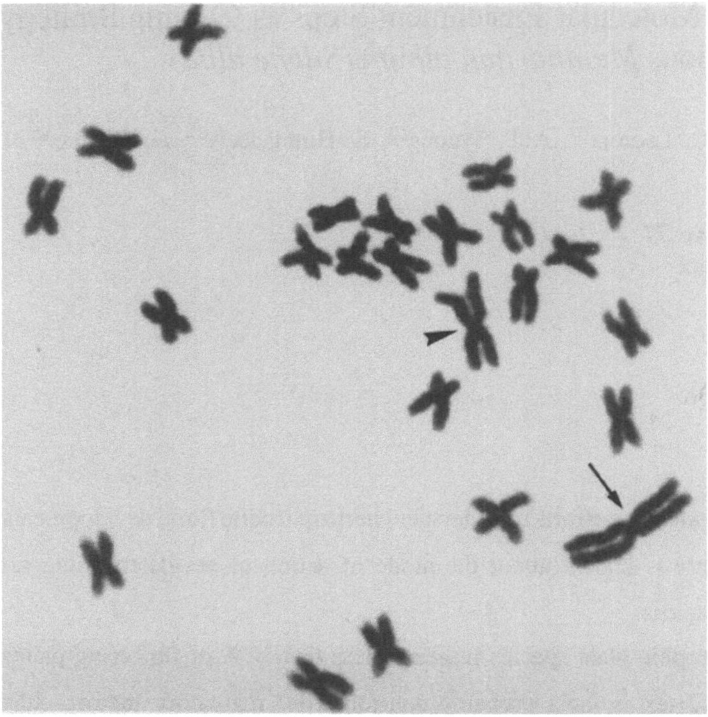

Fig. 1 : Metaphase plate from a male plant showing two distinct sex chromosomes X (arrow head) and Y (arrow) and 22 autosomes (2n = 24,XY).

Melandrium has major advantages over other plant species for molecular and genetic analysis of sex determination and differentiation :

1. The heteromorphic pair of sex chromosomes Y (maleness) and X (femaleness) contains major regulatory genes involved in sex determination. While they contain sex-linked genes which do not affect sex, it is clear that key sex differentiation genes are also located on Y and X (note that along the Y chromosome three distinct domains can be identified : one containing female suppressor gene(s), a second containing male "determinants" and a third responsible for male fertility) (Westergaard, 1958).

2. Like any flowering plant, *Melandrium* has bipotential flower primordia. Sexual dimorphism occurs by arrest of either male or female organ development at very early stages (earlier than in other dioecious species such as *Asparagus*, zkiwi). Consequently, male and female plants are a convenient source to isolate sex specific transcripts at different stages of flower development.

Both aspects represent natural systems of enrichment for sequences and functions involved in sexual reproduction.

By combining molecular and classical genetic approaches the *Melandrium* system enables to set up enrichment cascades in identifying specific functions and clonable domains of choice. This can avoid time-consuming cloning efforts for sequences of no or poor interest.

The objective of the programme is to develop a combinatorial cDNA, chromosomal and genomic library-cloning approach. As a preparatory work, we report here on chromosome sorting, cloning of male-specific sequences and isolation of sex-defective and homeotic floral mutants.

SORTING OF Y CHROMOSOME

The availability of purified individual chromosomes facilitates the study of molecular properties of eukaryotic genomes. Flow cytometric sorting offers the possibility to isolate specific chromosomes. This approach should be seen as an attempt to isolate Y-specific or Y-derived sequences to study at the molecular level factors involved in sex determination and differentiation.

A theoretical model of a flow karyotype based on the relative length of the metaphase chromosomes present in the *Melandrium* complement was previously reported (Ye et al, 1991).

The state and conditions of growth of the plant material were standardized. This was best achieved by using continuously growing *Agrobacterium rhizogenes* transformed root cultures as a source of true-to-type meristematic cells. Optimal synchronization conditions with S- and M-blocking agents (aphidicoline and oryzaline) were achieved by studying DNA synthesis inhibition/initiation and metaphase blocking. Chromosome stability was ensured by controlling the osmotic balance and avoiding bursting of protoplasts during hypotonic swelling.

A flow karyotype is depicted in Fig. 2. The histogram's peak areas suggest what sorting confirmed : autosomal clumps lie near or contaminate the

X and Y interest zones (Fig. 3). Nevertheless, Y can be sorted at 60-80 % purity. Two-color analysis might circumvent this problem if the Y-chromosome has a unique AT/GC ratio.

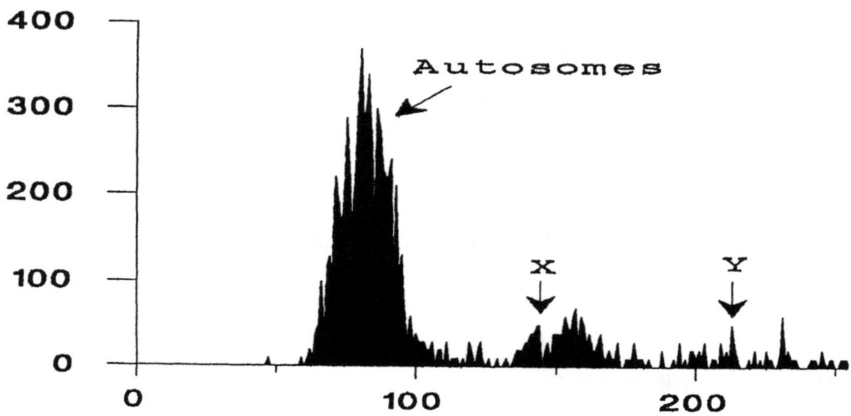

Fig. 2 : A flow karyotype (linear scale) established with DAPI (AT specific). The peak positions and their profile correspond to predicted theoretical values. The horizontal scale is fluorescence intensity (related to DNA content) and the vertical scale represents the frequency.

CLONING OF MALE-SPECIFIC SEQUENCES

Sex chromosome-specific sequences are currently studied in animals and insects. They have not been reported in plants so far.

Male specific sequences were isolated by subtractive cloning performed in total genomic libraries (essentially Hind III and Bam H1). A total of 30 male specific clones were retained, with insert size ranging from 0.1 to 6 Kb. One case is shown in Fig. 4.

Of the 30 probes investigated by Southern blotting, none hybridized with female DNA preparations. When tested on a population of 20 independant male lines, two major classes were identified :

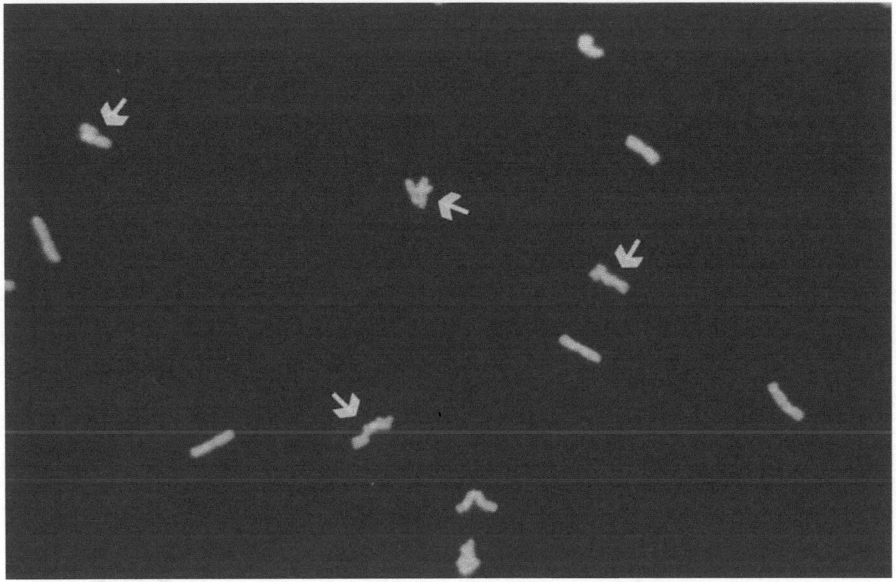

▲

Fig. 3 : Sorting onto black cellulose filters of *M. album* Y-sex chromosome. This chromosome is the largest of the complement, with virtually equal arms and represents nearly 21 % of the haploid genome (magnification : 400x). Arrows indicate contaminations by doublets and triplets of smaller autosomes.

Fig. 5 : Extent of the deletions on the Y chromosome in the two asexual ▸ mutants 5K63 and 8K40 : 12 and 20 % respectively.

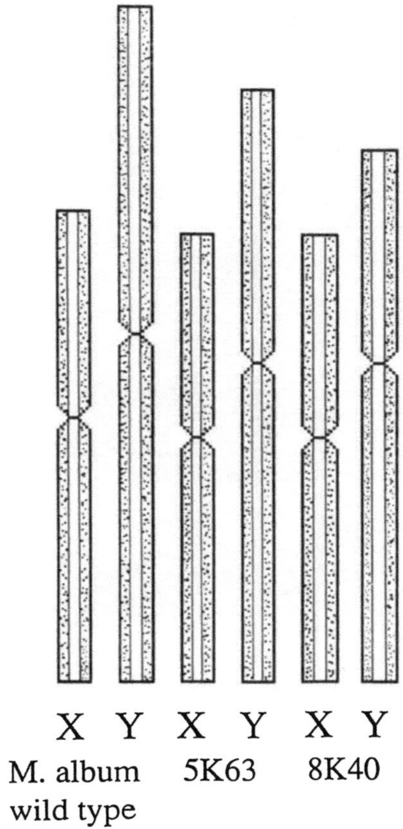

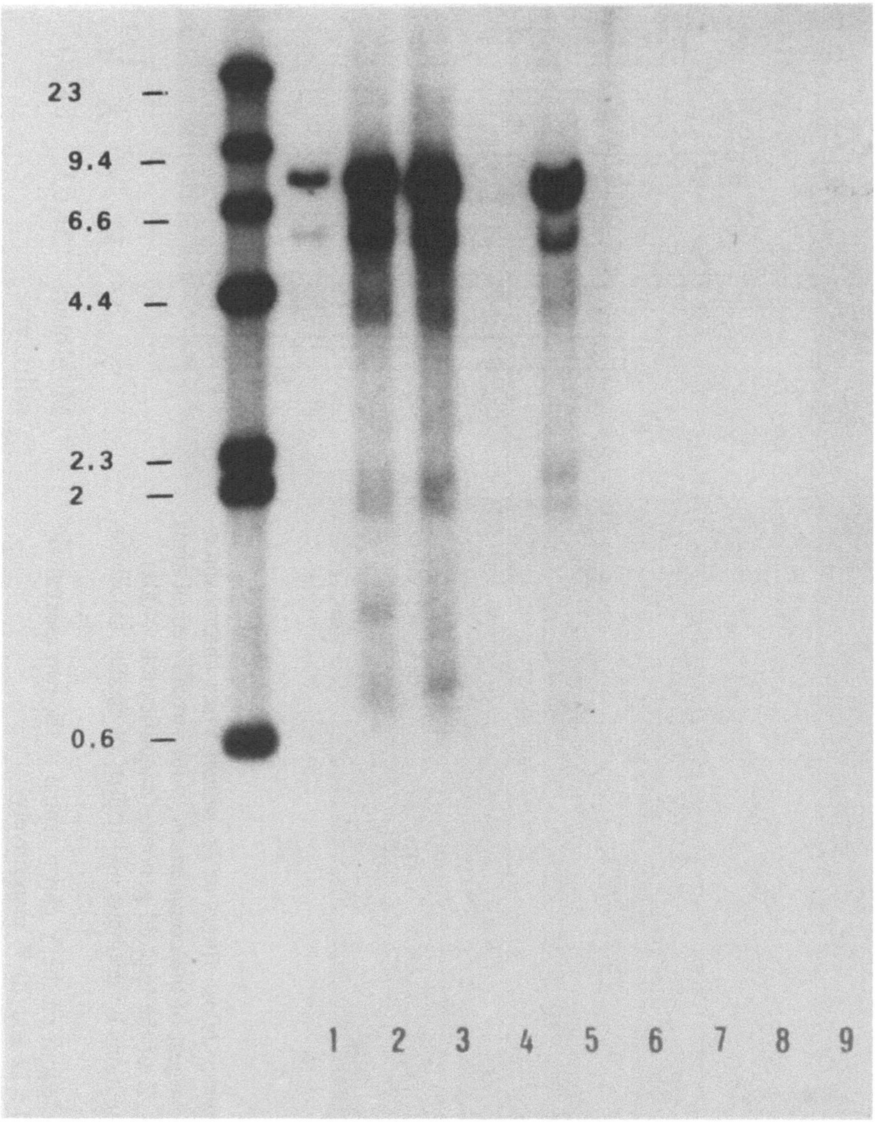

Fig. 4 : Autoradiograph of hybridization pattern of probe MaBamH_1-11 showing male specificity with BamH_1 cleaved *Melandrium* DNA from different males (lanes 1-5; very weak signal present in lane 4) and females (lanes 6-9). Size markers are indicated in kb.

class A, showing sequence homology to all or most of the probes. Certain probes exhibited polymorphism accross the responding male lines.

- class B, not recognizing any of the probe sequences used to challange those lines.

ISOLATION AND CHARACTERIZATION OF FLORAL MUTANTS.

Sex mutants have been extensively explored in *Drosophila, Caenorabdites elegans* and mammals (Truong et al, 1991). In plants, the few loci know are in maize, cucumber, *Mercurialis* (Irish and Nelson, 1989). No homeotic floral mutants have been reported so far in mono- and dioecious plants.

Several floral mutants were isolated in *M. album* via gamma-irradiated pollen. The main characteristics of two major classes, namely homeotic mutants and sexual mutants are outlined in Table 1. Of particular interest are the two asexual mutants 5K63 and 8K40. The development of both male and female organs is blocked at the same stage as the male or female ones in the wild type flowers. The mutated phenotype correlated in both mutants with defined deletions on the corresponding Y chromosome. This gives a genetic basis to the mutation, otherwise impossible to analyse by standard crosses. The deleted Y chromosomes are characterized in Fig. 5.

CONCLUSIONS

M. album appears as a suitable experimental system which enables to establish sex-, floral stage- and Y chromosome- deletion specific libraries (cDNA and/or genomic). Analysis of sex specific transcripts and sex chromosome sequences will be beneficial in identifica- tion of key sex determining and differentiation genes.

TABLE 1 : MAIN CHARACTERISTICS OF FLORAL PATTERN IN PHENOTYPIC MUTANTS OF MELANDRIUM ALBUM ISOLATED VIA IRRADIATED POLLEN. THE MUTANTS WERE SCREENED IN THE M_1 GENERATION

FLORAL PHENOTYPE	SEPALS	PETALS	STAMENS	CARPELS	DESCRIPTION
			Wild type		
Male	5	5	10	-	Carpel rudiment as a finger-like projection
Female	5	5	-	5	Anther rudiments, fused carpels
			Homeotic transformations		
5K138 (Superman)	5	5 (m.m)	> 10 m.m h.c	-	Stamina transformed into petals
7K34 (Superwoman)	5	5 (h.c)	-	h.c	Stamina transformed into carpels (fused carpelods)
			Sexual mutants		
5K63 and 8K40	5	5	-	-	Reproductive organs atrophied

Abbreviations: m.m = meristic modification
h.c = homeotic conversion

References

Ciupercescu DD , Veuskens J, Mouras A, Ye D, Briquet M and Negrutiu I (1990) Karyotyping *Melandrium album*, a dioecious plant with heteromorphic sex chromosomes. Genome 33:556-562

Irish EE and Nelson T (1989) Sex determination in monoecious and dioecious plants. The Plant Cell 1:737-744

Truong AT, Heinderyckx M, Installé P, Negrutiu I (1991) Sex determination and differentiation in organisms other than higher plants. Plant Science 80:201-211

Westergaard M (1958) The mechanism of sex determination in dioecious flowering plants. Adv. in Genetics 9:217-281

Ye D, Oliveira M, Veuskens J, Wu Y, Installé P, Truong AT, Brown S, Mouras A and Negrutiu I (1991) Sex determination in the dioecious *Melandrium* : the X/Y chromosome system allows complementary cloning strategies. Plant Science 80:93-106

Homeotic Genes Directing Flower Development in *Antirrhinum*

D.J.Bradley, R.Carpenter, E.S.Coen, L.J.Copsey, S.Doyle, R.Elliott, S.Hantke, D.Luo, P.C.M.McSteen, C.Robinson, J.M.Romero and G.W-R.Simon

Department of Genetics,
The John Innes Institute,
Colney Lane,
Norwich NR4 7UH, England

Summary

Homeotic mutants have been used to define the genetic interactions controlling flowering in *Antirrhinum*. Three categories of homeotic genes were identified by transposon mutagenesis. The first includes *floricaula* (*flo*), which is required to switch inflorescence meristems to floral. This gene has been isolated and shown to be expressed transiently in bract, sepal, petal and carpel primordia. The second group of genes controls the identity (and sometimes the number) of organs in a whorl. These genes affect overlapping whorls and their mutant phenotypes suggest a combinatorial model for gene action in determining the fate of floral primordia. Genes of the third category determine the identity of organs within one whorl and thus affect the symmetry of the flower. We propose that the interactions of these homeotic genes not only control the basic patterns of inflorescence and flower development in *Antirrhinum*, but possibly in a diverse range of plant species.

Introduction

Mutations have been isolated in numerous systems that help to define the various steps in molecular processes, such as metabolic pathways.

Homeotic mutants allow us to define developmental pathways, by revealing the genetic functions responsible for a particular mode of organ development. For example, this approach has been applied to the fruit-fly (*Drosophila*) to reveal which gene combinations control certain developmental fates. We have similarly adopted this philosophy in our dissection of the developmental pathway in plants which leads to a flower. Through the use of homeotic mutations and molecular genetics, we can now begin to reveal the molecular principles underlying the visual beauty of flowers.

In our studies we have used *Antirrhinum majus*, whose large flowers are very amenable to investigation, and whose endogenous, well-characterised transposon system provides a molecular handle on the genetic loci defined by mutation. Reference will be made to the other well-studied system, *Arabidopsis*, allowing us to see how molecular mechanisms have been conserved or have diverged in these two distant species.

Description of wild-type and mutagenesis strategy

The initial phase of *Antirrhinum* growth involves a vegetative apical meristem that produces pairs of opposite (180°) leaves at each node in a decussate phyllotaxis (i.e. each pair is at a right-angle to the previous). After a period of vegetative growth the apex changes to an inflorescence meristem, producing smaller leaves (bracts) in a spiral arrangement, generally with a single bract at each node. In the axils of these bracts, floral meristems are initiated which produce four whorls of organs separated by very short internodes, so that consecutive whorls appear adjacent to each other.

Antirrhinum flowers are zygomorphic:they can be divided into mirror-image halves by a single plane. The flower consists of four whorls of organs, numbered from the outside of the flower;thus whorl 1 is outermost and whorl 4 is central. The whorl primordia appear sequentially: whorl 1 first, 2 and 3 almost simultaneously and finally whorl 4 (Awasthi et al. 1984). Organs within a whorl are upper, if closer to the stem (adaxial), or lower if nearer the bract (abaxial). Whorl 1 consists of five sepals and whorl 2, the corolla, contains five petals which are fused for part of their length, but end in five separate lobes. In whorl 3, five stamen primordia are initiated, though the uppermost fails to develop fully. Whorl 4 is occupied by two united carpels forming a gynoecium with a bilocular ovary.

To isolate and study homeotic genes in *Antirrhinum* we have carried out an extensive transposon mutagenesis experiment (Carpenter and Coen, 1990). Plants carrying active transposons were grown at 15°C, a temperature inducing transposition in *Antirrhinum* (Carpenter et al. 1987), and self-pollinated to give 26,000 progeny, many of which were expected to carry recessive mutations in a heterozygous condition. The homozygous phenotypes were revealed by self-pollination and growing 80,000 plants. Over 15 independent homeotic mutations were obtained from this screen and here we present our conclusions derived from a study of some of the genes involved.

Genetic control of the first steps in floral development

The transition from inflorescence to floral meristems represents the first step specific to the floral developmental pathway. Mutants that are unable to carry out this transition might be expected to produce proliferating inflorescence shoots in place of flowers. One mutant obtained with this phenotype is *floricaula* (*flo*) (Carpenter and Coen, 1990). The *flo* mutant initiates vegetative growth and the transition to inflorescence meristem in a similar manner to wild-type. Instead of flowers being produced in the axils of bracts, indeterminate shoots bearing further bracts are produced, each shoot having two opposite bracts at the base followed by a spiral of single bracts. Each of these shoots in turn produce further shoots in the axils of their bracts and this sequence can repeat itself indefinitely. The wild-type *flo* product is therefore necessary for the transition between inflorescence and floral meristems so that in its absence the inflorescence programme is continually reiterated. The *flo* mutant can be considered to be homeotic because it results in one structure (the flower) being replaced by a homologous structure (indeterminate shoot). This mutant can also be viewed as heterochronic (changing the relative timing of events) because it results in an early developmental programme (inflorescence meristem) being continually reiterated.

Other mutants have been described in *Antirrhinum*, such as *squamosa*, which produce inflorescence-like shoots in the axils of bracts, although many of these shoots eventually produce flowers (Stubbe, 1966;Schwarz-Sommer et al. 1990). Interestingly, although stable *floricaula* alleles never produce flowers, under certain conditions they may eventually terminate in a carpel-like structure. A similar mutant, *leafy*, has been found in *Arabidopsis* (Haughn and Sommerville, 1988)

A second class of mutant has almost the opposite effect of those described above because it promotes the conversion of inflorescence to floral meristems in positions where this would not normally occur. Plants with indeterminate inflorescences, such as *Antirrhinum*, do not produce terminal flowers. Thus, the apical inflorescence meristem does not itself undergo the transition to floral. The *centroradialis* mutant in *Antirrhinum* produces terminal flowers, suggesting that the wild-type gene product inhibits this apical conversion (Kuckuck and Schick, 1930).

The *flo* gene of *Antirrhinum* has been isolated and extensively characterised (Coen et al. 1990). It produces a transcript of about 1.6 kb which has the potential to encode a protein whose sequence shows no extensive homologies with any other proteins in available data-banks, though it contains some motifs found in transcription factors. This suggests that FLO may be a transcriptional activator, although other roles cannot be excluded. *In situ* hybridisation shows that *flo* is expressed from a very early stage in wild-type inflorescences in a very specific temporal and spatial sequence. The earliest expression seen is in bract primordia and is followed by expression in sepal, petal and carpel primordia, but no expression is seen in stamen primordia. Expression in each organ is transient and is not observed in later stages of development. Taken together, these results suggest that *flo* acts not only as a switch between inflorescence and floral meristems, but is also involved in directing specific patterns of gene expression in the early floral meristem.

Genetic control of organ identity

Homeotic mutations were also isolated that act after the inflorescence to floral transtion has taken place. Many of the genes involved affect the identities of organs in two adjacent whorls (Carpenter and Coen, 1990;Sommer et al. 1990) and for simplicity three overlapping regions can be defined for wild-type: **A** (whorls 1 and 2); **B** (whorls 2 and 3); **C** (whorls 3 and 4). One class of mutant, *ovulata*, affects region **A** and gives carpels instead of sepals in whorl 1 and stamens in place of petals

in whorl 2, giving the overall phenotype carpel, stamen, stamen, carpel. The second class affects region **B** and gives sepals instead of petals in whorl 2 and carpels instead of stamens in whorl 3, giving the phenotype sepal, sepal, carpel, carpel (though whorl 4 does not always develop). This class includes the *deficiens* (*def*), *globosa* (*glo*) and possibly some *sepaloidea* (*sep*) mutants. A third class (e.g. *plena*) affects region **C** and gives petals in place of stamens in whorl 3 and variable structures in whorl 4. Similar classes of whorl identity genes have been found in *Arabidopsis* (Bowman et al. 1989;Haughn and Sommerville, 1988) suggesting that the mechanisms controlling whorl identity have been highly conserved in evolution. Furthermore, *def* (affecting region **B**) and *agamous* (an *Arabidopsis* gene affecting region **C**) share a highly conserved sequence of about 50 amino acids that is also found in yeast and mammalian transcription factors (Schwarz-Sommer et al. 1990).
The action of many genes functions in overlapping regions could give each whorl a unique combination of functions. This might provide sufficient information to specify each whorls identity. However, in order to model these gene interactions it is necessary to introduce some constraints as to how these gene functions are established. The simplest model allows functions active in region **B** to be established indepently of the others, while gene functions of regions **A** and **C** are antagonistic (Carpenter and Coen, 1990).

Regulation of whorl identity genes

Two general types of model can be imagined to explain the expression patterns of whorl identity genes. In a purely spatial model, concentric fields could be set up in the early floral meristem, independently of the sequence of primordium initiation. The gene functions restricted to the defined regions **A**, **B** and **C** could be activated in the appropriate field and hence specify the fate of primordia growing out from the various regions of the meristem. This type of model is formally similar to models proposed for the control of segment identity in *Drosophila* (Ingham, 1988). In the sequential type of model, which shares some features of early models of flower development (Heslop-Harrison, 1964), the consecutive growth of the primordia would be essential for the establishment of the domains of activity. The various gene functions would be activated in a manner that reflected the sequence of primordium initiation.

Some supporting evidence for the latter sequential model has come from studies on the *flo* gene of *Antirrhinum.* As described above, *flo* is expressed sequentially and appears to be activated in regions where primordia are being initiated. A further feature of *flo* expression is that it occurs transiently in bract, sepal, petal and carpel but not stamen primordia. This pattern suggests that *flo* not only acts to switch inflorescence to floral meristems, but also interacts in a sequential manner with the whorl identity genes that pattern the floral meristem (Coen et al. 1990).

Genetic control of whorl number

Some of the whorl identity mutants also affect whorl number. Extreme mutants in genes affecting region **B** often give fewer whorls than wild-type. Mutants in genes acting in region **C** usually give an increase in whorl number and a more-or-less indeterminate growth pattern. These findings reveal two functions of the whorl identity genes, namely the control of organ fate and the control of whorl number. The extent to which these separate roles are mediated by independent target genes is not fully understood.

Genetic control of differences within a single whorl

In radially symmetric flowers (actinomorphic), all organs in the same whorl have very similar or identical morphologies, while in zygomorphic flowers (e.g. *Antirrhinum*), one or more whorls have organs which are distinct. The *cycloidea* (*cyc*) mutations in *Antirrhinum* give flowers with a more radially symmetric appearance than wild-type (Stubbe, 1966) and four such mutations were isolated in our transposon mutagenesis screen. Extreme *cyc* mutations give flowers with petals all resembling the lowest petal of wild-type. Therefore, *cyc* is a homeotic mutation since organs of one type are replaced by another type. Furthermore, *cyc* appears to affect several whorls, since all stamens develop in *cyc* mutants compared to the abortion of the upper stamen in wild-type.

Does *cyc* affect organs in the same way irrespective of the whorl they occupy ? In wild-type, the abortion of the uppermost stamen depends upon *cyc* activity. When stamens are found in whorl 2, in the case of *ovu*, the upper stamens are again aborted, suggesting that *cyc* interacts with primordia in a similar way, irrespective of the whorl they occupy. Thus the fate of a primordium depends on a combination of interactions that determine whorl identity and relative upper or lower position.

Inflorescence evolution

Antirrhinum is among the species with zygomorphic flowers that also have indeterminate inflorescences and so do not normally terminate with a flower. Nevertheless, exceptional plants which produce terminal flowers have been found amongst these species and they develop terminal flowers with radial symmetry (Peyritsch, 1872). This phenotype has been studied in *Antirrhinum* and shown to be due to a recessive allele of the locus *centroradialis*. A few normal flowers are produced laterally in this homeotic mutant, but the apical inflorescence soon produces a radially symmetric terminal flower, with all petals resembling that of the lower petal of wild-type plants, similar to the conversion seen in *cyc*.
An explanation for the symmetry of this terminal flower can be found in a polar-coordinate model for zygomorphy (Carpenter and Coen, 1990). Floral meristems in axillary positions are in an asymmetrical environment with the main apex above and a bract below them, subjecting them, therefore, to a gradient of *cyc* activity. In contrast, a terminal flower is already in a radially symmetric environment and so may either experience no *cyc* gradient, or no *cyc* activity itself may occur in such an environment.
Mutations such as *centroradialis* also have implications for inflorescence evolution. Some plants may produce a large single flower at the end of its shoot, while in most species flowers are clustered to form inflorescences. In plants with determinate inflorescences, the apical meristem must convert into a floral meristem, with other flowers found in axillary psitions or terminating side-branches. An indeterminate inflorescence can produce a quite different array of flowers, since each shoot only produces flowers in axillary positions. The isolation of genes such as *flo*, that regulate the transition from inflorescence to floral

meristems, will allow us to address the question of inflorescence evolution at the molecular level. Furthermore, the interactions of *flo* and the organ identity genes should add insight as to how plants overcome the constraints imposed by meristems when trying to pattern and control sequential organ development.

References

Awasthi DK, Kumar V and Murty YS.(1984) Flower development in *Antirrhinum majus* L.(Scrophulariaceae) with a comment upon corolla tube formation.Bot.Mag.Tokoyo 97:13-22.

Bowman JL, Smyth DR and Meyerowitz EM.(1989) Genes directing flower development in *Arabidopsis*. Plant Cell 1;37-52.

Carpenter R and Coen ES.(1990) Floral homeotic mutations produced by transposon-mutagenesis in *Antirrhinum majus*. Genes Dev.4;1483-1493.

Carpenter R, Martin CR and Coen ES.(1987) Comparison of genetic behaviour of the transposable element Tam3 at two unlinked pigment loci in *Antirrhinum majus*.Mol.Gen.Genet.207;82-89.

Coen ES, Romero JM, Doyle S, Elliott R, Murphy G and Carpenter R (1990) *floricaula*: a homeotic gene required for flower development in *Antirrhinum majus*.Cell 63;1311-1322.

Haughn GW and Sommerville CR.(1988) Genetic control of morphogenesis in *Arabidopsis*.Devl.Genet.9;73-89.

Heslop-Harrison J.(1964) Sex expression in flowering plants.Brookhaven Symp.Biol.16;109-125.

Ingham PW.(1988) The molecular genetics of embryonic pattern formation in *Drosophila*.Nature 335;25-34.

Kuckuck H and Schick R.(1930) Die Erbfaktoren bei *Antirrhinum majus* und ihre Bezeichung.Z.indukt.Abstamm.u.VererbLehre.56;51-83.

Peyritsch J.(1872) Uber Pelorienbildungen.Sber.Akad.Wiss.Wien.66;1-35.

Schwarz-Sommer Z, Huijser P, Nacken W, Saedler H and Sommer H (1990) Genetic control of flower development: homeotic genes in *Antirrhinum majus*.Science 250;931-936.

Sommer H, Beltran J, Huijser P, Pape H, Lonnig W, Saedler H and Schwarz-Sommer Z.(1990) *Deficiens*, a homeotic gene involved in the control of flower morphogenesis in *Antirrhinum majus*: the protein shows homology to transcription factors.EMBO J.9;605-613.

Stubbe H.(1966) Genetik und Zytologie von Antirrhinum L. sect *Antirrhinum*. Veb.Gustav Fischer Verlag, Jena.

Gametophytic and Sporophytic Gene Expression in *Helianthus annuus* L.

A. Steinmetz, R. Baltz, C. Domon, N. Dudareva, J. L. Evrard and R. Kräuter
Institut de Biologie Moléculaire des Plantes du C. N. R. S.
12, rue du Général Zimmer
67084 Strasbourg Cedex
France

Introduction

The life cycle of plants is characterized by an alternation of haploid and diploid generations. In the haploid organism (gametophyte), specific cells differentiate into male and/or female gametes which, upon fusion, yield a diploid zygote that develops into an adult diploid organism, the sporophyte. The sporophyte, in turn, generates special cells which, through meiosis, produce haploid spores developing into haploid gametophytes. In some plants, both the haploid and diploid forms are free-living green organisms of substantial size. In ferns, for example, both the gametophyte and sporophyte forms are independent green plants. In other plants, one phase dominates and becomes the recognizable plant while the other is reduced and inconspicuous and depends on the dominant one for nutrition and physical support. In some plants, such as mosses, the haploid gametophyte is the dominant phase. In seed plants, however, the diploid sporophyte is the dominant phase, whereas the gametophyte is reduced to a few cells: two or three for the male gametophyte, also known as pollen grain, and seven cells for the female gametophyte, or embryo sac.

It is estimated that at least 60,000 diverse structural genes are expressed during the dominant phase of the life cycle of higher plants (Kamalay and Goldberg, 1980). Each organ system (leaf, root, stem, petal, anther and ovary) contains about 25,000 different mRNAs, but many of them (those coding for the so-called "housekeeping" functions) are shared by all organs. A considerable number of mRNAs (20 to 40%) appear to be organ-specific and code for proteins which determine the function of the organ or cell type.

The specificity of expression of a gene is determined by nucleotide elements found upstream of the transcription start site. These elements bind regulatory proteins (transcription factors) which assist RNA polymerase in initiating transcription of the associated gene.

DNA technology and recent progress in plant transformation provide a good experimental basis to study the molecular mechanisms responsible for or involved in the regulation of expression of tissue-specific genes. Owing to DNA technology, a specific gene can be isolated and studied. To identify regulatory regions of a gene, upstream segments can be fused to reporter genes and their ability to promote tissue-specific expression of the reporter genes can be tested in transgenic plants. Finally, and in a more applied perspective, these regulatory regions can be used to direct a tissue-specific expression of foreign genes specifying agriculturally interesting traits in genetically engineered plants.

The aim of our study is to identify the nucleotide modules and the regulatory proteins which interact with these sequences to promote the expression of a gene in sporophytic and gametophytic reproductive tissues in sunflower.

Isolation of flower-specific cDNA clones

The study of a gene and its expression requires that a molecular probe for this gene is available. Tissue-specific probes are usually isolated via differential hybridization screening of a cDNA library constructed in a cloning vector (plasmid or phage derivative) using poly(A) RNA from the tissue or organ expressing the particular gene. When replicas of the library (made on nitrocellulose or nylon membranes) are screened with radiolabeled RNA from various tissues, most of the plaques (for phage libraries) or colonies (for plasmid libraries) hybridize with each RNA probe and therefore contain cDNA sequences corresponding to mRNA species present in all these tissues; a few clones, however, produce signals exclusively with RNA from which the library has been constructed. These cDNAs are derived from RNA species not present in the RNA population from the other tissues: they are specific for that organ. The clones containing specific cDNA inserts can then be purified and amplified in bacteria for further analysis.

When the differential hybridization technique was applied to a cDNA library constructed using poly(A) RNA from a sunflower inflorescence at anthesis, we were able to isolate about 30 "flower-specific" clones (Herdenberger et al., 1990). These clones were further characterized by Northern hybridization using RNA from various vegetative and reproductive organs; the features of some of these clones are described in **Table I**. All of the clones isolated hybridized exclusively with RNA from an inflorescence, thus confirming the differential screening data. Five clones (SF1, SF2, SF5, SF6 and SF18) hybridized exclusively with RNA from anthers; the first four of these clones cross-hybridized, indicating that they have closely related nucleotide sequences [SF1 and SF2 have been shown to code for two nearly identical proline-rich proteins (Evrard et al., 1991)].

Two other clones, SF3 and SF16, were found to be pollen-specific. Clone SF21 hybridized exclusively with RNA from styles and stigmas (ovaries were removed from the pistils and tested separately), whereas clone SF28 showed a weak but specific hybridization in immature and mature ovaries.

Table I. Characterization of several flower-specific cDNA clones from sunflower

clone	size of mRNA[1]	roots	leaves	flowers	corollae	pistils[2]	ovaries	anthers	pollen
		hybridization with RNA from							
SF1	880	-	-	+	-	-	-	+	-
SF2	880	-	-	+	-	-	-	+	-
SF3	1100	-	-	+	-	-	-	-	+
SF5	N.D.[3]	-	-	+	-	-	-	+[4]	-
SF6	N.D.[3]	-	-	+	-	-	-	+[4]	-
SF9	1700	-	-	+	-	-	-	+	-
SF16	1350	-	-	+	-	-	-	-	+
SF18	900	-	-	+	-	-	-	+	-
SF21	1280	-	-	+	-	+	-	-	-
SF28	N.D.[3]	-	-	+	-	-	+	-	-

1 in nucleotides
2 styles + stigmas (not ovaries)
3 N.D. = not determined
4 Northern experiment not performed but clones cross-hybridize with clones SF1 and SF2

Expression and structure of sporophytic genes

In flowering plants, the sporophytic structures involved in reproduction are the anthers and the pistils. Anthers produce the male gametophytes or pollen grains. The development of the pollen grains is highly dependent on the surrounding sporophytic tissues. Destruction of tapetal cells, for instance by expression of a cytotoxic gene, leads to pollen abortion and therefore to male-sterile plants (Mariani et al., 1990). Sporophytic tissues are also implicated in the production of the outer pollen wall, the exine. Some of the components produced by these tissues are responsible, in several plant families such as the *Brassicaceae*, for self-incompatibility (Heslop-Harrison, 1975).

Pistils have a dual function: they produce the female gametophyte (embryo sac), but they also discriminate between compatible and incompatible pollen and hence play a decisive role in determining the genotype of the future embryo and plant.

Anther-specific genes SF2 and SF18

The two anther-specific mRNAs SF2 and SF18 are expressed in peripheral anther cells from immature and mature disk florets of an open sunflower inflorescence (Evrard et al., 1991). Their cDNAs have been sequenced and found to code for short proline-rich proteins (121 amino acids for SF2 and 176 amino acids for SF18) carrying a hydrophobic signal peptide at the N-terminus (Domon et al., 1990). The function of the two proteins has not yet been elucidated; they are believed to be cell wall proteins.

The two anther-specific mRNAs display the same temporal and spatial expression pattern. The expression of their genes is therefore similarly regulated; in other words, the two genes should contain the same or similar control elements in their upstream region. To verify this we have isolated a copy of each of the two genes from a genomic DNA library and we have studied their nucleotide sequence including more than one kbp of their upstream region.

The SF2 gene is split by a 2615 bp intron which separates the signal peptide coding sequence from that coding for the rest of the protein (Domon et al., 1991). The SF18 gene has a similar split structure: however the intron separating the two peptide domains is much smaller (212 bp). Surprisingly it is found exactly at the same location. A search for common sequences in the SF2 and SF18 genes and their neighboring regions revealed three segments displaying important (approximately 95%) sequence similarities: the largest of these segments (about 240 bp) includes the immediate upstream region, the first exon (which codes for the signal peptide), and over 100 bp on the 5' side of the intron. The two other conserved boxes, which are found upstream of the transcription start site, are much shorter (30 and 50 bp respectively). We believe that these regions contain regulatory sequences involved in the anther-specific expression of the two genes. A functional promoter analysis of the upstream region of these two genes is in progress.

Pistil-specific cDNA SF21

Pistils, the female organs of the sporophyte, comprise three functionally distinct regions: (i) the ovary (at the bottom), in which the female gametophyte and later the embryo develops; (ii) the stigma (on the top), the surface of which is covered with papillar cells that receive the pollen and on which pollen germination takes place; (iii) the style, which connects the stigma to the ovary. For fertilization of the egg cell in the embryo sac, the pollen tube (carrying the two haploid sperm cells) must grow through the stylar transmitting tissue down to the micropyle which is a specialized opening at the surface of one end of the ovule. In sunflower, the ovary can be easily separated from the pistil, leaving an incomplete pistillar structure containing the style and stigma.

The SF21 cDNA, when tested in dot blot hybridization experiments with RNA from various floral structures, showed exclusive hybridization with RNAs extracted from the incomplete pistillar structures (style + stigma). No hybridization was detected with RNA from immature or mature ovaries; Whether the gene coding for this mRNA is expressed in both style and stigma, or only in one of them, remains to be determined.

The SF21 cDNA has been entirely sequenced. It is 1380 bp long and codes for a 352 amino acid long polypeptide. In contrast to the other proteins studied (see Table II) the SF21 protein has many short hydrophobic segments, one of which is large enough (21 amino acids) to span the membrane bilayer. It lacks the signal peptide sequence at the N-terminus and is therefore not a secreted protein. The SF21 protein could be a membrane protein, whose N-terminus (approximately 100 amino acids) is directed towards the cytoplasm and the C-terminus (200 amino acids) towards the exterior of the cell. An N-glycosylation site (NCS) is found in the C-terminal region. The function of the SF21 protein is unknown, but as a putative membrane protein with an extracellular and intracellular domain it could have either a receptor or transporter function. It is not related to any of the S-Locus-Specific-Glycoproteins (SLSG) involved in self-incompatibility in *Brassica* (Nasrallah et al., 1987) and *Nicotiana alata* (Anderson et al., 1989).

Expression and structure of gametophytic genes

The best characterized of the two pollen-specific clones is the SF3 cDNA. This cDNA codes for a 219 amino acid-long polypeptide with an imperfect repeat structure. Each of the repeats (consisting of 98 amino acid residues) contains a zinc finger domain and a basic region. The carboxy-terminus of the protein consists of a sixfold pentapeptide repeat. Zinc fingers have been identified as DNA binding domains in a variety of transcription factors from higher organisms. These transcription factors have an additional domain which interacts with components of the basal transcription machinery: the activator domain. The activating function of the SF3 protein could be performed by the C-terminus which is highly acidic (net negative charge of -10 in the 44 last amino acids).

The SF3 protein could be a late pollen-specific transcription factor required for the expression of late pollen-specific genes which encode proteins or enzymes involved in a late pollen function, such as pollen tube elongation, or the migration of the gametes down the pollen tube, or even directly in fertilization. The gene encoding the SF3 protein has also been isolated and its nucleotide sequence has been determined. It is a split gene, interrupted four short intervening sequences or introns (introns are nucleotide stretches mainly in eukaryotic genes which are transcribed, but which are subsequently removed from the mRNA by a mechanism called RNA splicing) (Breathnach and Chambon, 1981). The SF3 gene of sunflower is a member of a multigene family.

The SF16 cDNA has also been sequenced and was found to code for a 331 amino acid long, very hydrophilic, basic protein (19% of the amino acids are lysines and arginines). From the absence of hydrophobic segments in the protein we infer that it is not a membrane protein, nor is it secreted. Because of its very basic nature, SF16 might be a protein with a nucleic acid binding function (as is SF3). However, no structural peptide motif typically found in transcription factors (leucine zipper, zinc finger, helix-turn-helix) have been detected in the SF16 protein. Alternatively to a potential DNA binding function it could bind to a subset of mRNAs and regulate translation of these mRNAs into proteins. It is encoded by a single-copy gene which has not yet been isolated.

Table II. Features of flower-specific proteins from sunflower

clone	organ	length (a.a.)	signal peptide	transmembrane domain	pI of protein	putative function
SF2	anther	121	yes	no	8.2	cell wall component
SF3	pollen	219	no	no	9.1	transcription factor
SF16	pollen	331	no	no	11.3	nucleic acid binding
SF18	anther	176	yes	no	5.3	cell wall component
SF21	style and/or stigma	352	no	yes	6.2	membrane receptor

Conclusions and Perspectives

We have isolated and studied five flower-specific cDNA clones from sunflower (*Helianthus annuus* L). Two of these clones are specifically expressed in the male gametophyte (pollen grain), two in the male organ (the anther) and one in the female organ (pistil) of the sporophyte. The structural features and putative functions of the proteins encoded by these cDNAs are summarized in **Table II**. None of the five proteins shows homology with protein sequences from available databases. Experiments are in progress to determine their cellular location and to identify their function. We have also initiated experiments directed towards identifying the regulatory regions which specify gene expression in these various cell types.

References

Anderson MA, McFadden GI, Bernatzky R, Atkinson A, Orpin T, Dedman H, Tregear G, Fernely R, Clarke AE (1989) Sequence variability of three alleles of the self-incompatibility gene of *Nicotiana alata*. Plant Cell 1 : 483-491

Breathnach R, Chambon P (1981) Organization and expression of eukaryotic split genes coding for proteins. Annu. Rev. Biochem. 50 : 349-383

Domon C, Evrard JL, Herdenberger F, Pillay DTN, Steinmetz A (1990) Nucleotide sequence of two anther-specific cDNAs from sunflower (*Helianthus annuus* L). Plant Mol. Biol. 15 : 643-646

Domon C, Evrard JL, Pillay DTN, Steinmetz A (1991) A 2.6 kb intron separates the signal peptide coding sequence of an anther-specific protein from the rest of the gene in sunflower. Molec. Gen. Genet. 229 : 238-244

Evrard JL, Jako C, Saint-Guily A, Weil JH, Kuntz M (1991) Anther-specific, developmentally regulated expression of genes encoding a new class of proline-rich proteins in sunflower. Plant Mol. Biol. 16 : 271-281

Herdenberger F, Evrard JL, Kuntz M, Tessier LH, Klein A, Steinmetz A, Pillay DTN (1990) Isolation of flower-specific cDNA clones from sunflower (*Helianthus annuus* L). Plant Sci. 69 : 111-122

Heslop-Harrison J (1975). Incompatibility and pollen-stigma interaction. Annu. Rev. Plant Physiol. 26 : 403-425

Kamalay JC, Goldberg RB (1980) Regulation of structural gene expression in tobacco. Cell 19 : 935-946

Mariani C, De Beuckeleer M, Truettner J, Leemans J, Goldberg RB (1990) Induction of male sterility in plants by a chimaeric ribonuclease gene. Nature 347 : 737-741

Nasrallah JB, Kao TH, Chen CH, Goldberg ML, Nasrallah ME (1987) Amino acid sequence of glycoproteins encoded by three alleles of the S-locus of *Brassica oleracea*. Nature 326 : 617-619.

Gametophytic Gene Expression

Joseph P. Mascarenhas
Department of Biological Sciences and
Center for Molecular Genetics
State University of New York at Albany
Albany, New York 12222, USA

The male gametophyte of flowering plants is a microscopic structure which during its early development is enclosed within the sporophyte tissue in the anther of a flower bud. Dramatic advances have been made recently in our understanding of gene expression in the different tissues of the anther including the tapetum (Koltunow et al, 1990: Mariani et al, 1990). Following meiosis there is a long interphase period during which the microspores enlarge greatly in size. Microspore mitosis results in an unequal partitioning of cytoplasm into the vegetative and generative cells. Following anthesis the pollen grain is deposited on the stigma of the pistil where it begins another phase of its development by germinating and extruding a tube within which the sperm cells are transported to the embryo sac. In many species studied the pollen grain at maturity contains all the proteins that are required for germination and early tube growth, since protein synthesis inhibitors do not block these events. Moreover a large number of enzymes have been reported to be present in pollen grains of various species (Mascarenhas, 1975). Protein synthesis begins early in pollen germination. Most of the studies in the literature indicate that the mature pollen grain contains a store of stable mRNAs that are translated early during germination and that play a greater or lesser role in pollen germination and tube growth depending on the plant species (reviewed in Mascarenhas, 1975, 1990a,b).

The synthesis of new RNA takes place during pollen germination. The RNAs that are synthesized during pollen germination and tube growth appear to be mRNAs. Current evidence indicates that in the pollen grain at maturity, either all the proteins that are required for germination and early tube growth are already present, or if new proteins are synthesized, the mRNAs for their synthesis already exist. Moreover the proteins synthesized on new mRNAs during germination are similar to those synthesized on premade RNAs. In other words, the genes for at least the prevalent and mid-prevalent mRNAs utilized during the latter part of maturation and those utilized during germination and tube growth appear to be the same (Mascarenhas, 1990a,b).

A large number of different genes are transcribed during post-meiosis microspore and pollen development. The mRNAs that are present in the mature pollen grains of Tradescantia and maize are the products of about 20,000 to 24,000 different genes (Willing and Mascarenhas, 1984; Willing et al, 1988). These mRNAs represent genes primarily activated late in pollen development. There is evidence that additional genes are activated after meiosis and during the early stages of microspore development although we do not currently have any estimates of the total numbers of these early genes and what fraction of them might be identical to the genes expressed later in development (Mascarenhas, 1990a,b). Male gametophyte development, at least in terms of the numbers of genes required, is thus a complex process and there is a substantial overlap between genes active in gametophytic and sporophytic tissues.

We and several other laboratories have constructed cDNA libraries made to poly(A)RNA from pollen of different stages of development and from different plant species (Stinson et al, 1987; Twell et al, 1989; Ursin et al,

1989; Breiteneder et al, 1989; Albani et al, 1990; Brown and Crouch, 1990; Mohapatra et al, 1990: Rafnar et al, 1991; Roberts et al, 1991; Scott et al, 1991; Silvanovich et al, 1991; Singh et al, 1991; Valenta et al, 1991). Using cDNA clones from our libraries we have characterized several of the pollen-expressed genes with respect to the stage in development when they are first activated and the pattern of accumulation of their mRNAs (Stinson et al, 1987; Guerrero et al, 1990). Genomic clones corresponding to several of the cDNA clones have been isolated. For one gene, the cis-elements important in promoter function for pollen-specificity and correct time of expression have been located in the 5' flanking region of the gene (Guerrero et al, 1990; Hamilton et al, 1992). The results of the experiments done thus far have identified several different types of control mechanisms, transcriptional and post-transcriptional, that are operative in the pathway of gene expression in the developing male gametophyte.

The primary regulation is at the level of transcription. In the 5' flanking DNA of the pollen-specific maize Zm13 gene, there are multiple regions that contain sequence elements involved in pollen-specific and stage-specific expression. In brief, the -100 to -54 bp region is sufficient for pollen-specific expression. Upstream of this pollen-specificity region between -260 and -100 bp are present one or more sequence elements which enhance expression. In the region between -1001 and -260 bp, a negative element or elements are present that decrease expression relative to the -260 to +1 region. The cis-elements responsible for expression at the correct stage of pollen development reside within the 260 bp upstream from the start of transcription. The information regarding the construction of the pollen-specific promoter was obtained from deletion constructs with a GUS (ß-glucuronidase) reporter gene assayed in transgenic tobacco or Arabidopsis plants and by transient assays after microprojectile bombardment into Tradescantia pollen (Guerrero et al, 1990; Hamilton et al, 1992). The Zm13 pollen-specific promoter thus contains several functional domains; a pollen-specificity domain(s), a temporal-specificity domain, up regulation and down regulation domains.

In addition to the transcriptional regulatory mechanisms, there are also control mechanisms that operate post-transcriptionally in the male gametophyte. The mRNA for the pollen-specific gene Zm13, is first detectable after microspore mitosis and continues to accumulate thereafter reaching a maximum concentration in the pollen grain at anthesis. Antibodies were made to a synthetic peptide of the Zm13 presumptive protein, and Western blot analyses of extracts of pollen from different stages of development were carried out. The data obtained show that the Zm13 protein is not detectable in developing pollen grains and is only found in mature maize pollen after anthesis. Since the mRNA is synthesized several days earlier, this would indicate the operation of some mechanism of post-transcriptional repression of translation. In transgenic tobacco plants containing a GUS coding sequence fused to the -1001 to +61 or -314 to + 61 Zm13-promoter constructs and a nopaline synthase (NOS) 3' terminator, GUS enzyme activity was detectable in the immature pollen soon after microspore mitosis (Guerrero et al, 1990). This is the correct time for activation of the Zm13 gene in vivo, but not the correct time for the translation of the Zm13 mRNA into protein. One might expect nucleotide sequences in the mRNA to be responsible for the repression of mRNA translation as has been described for the translational repression of ferritin mRNA (Aziz and Munro, 1987; Hentze et al, 1987;

Thiel, 1987). Obviously the signals for this regulation are not contained within the +1 to +61 segment of the mRNA. The normal translation start site of the Zm13 mRNA is at +128. We are presently attempting to determine which region of the mRNA specifies translational repression of the mRNA and to identify the actual nucleotide sequence elements responsible.

Studies with the Zm13 promoter fragments fused to the GUS coding sequence with either the NOS 3' terminator or the terminator from the Zm13 gene (from +611 to +1079) have shown in transient transformation assays a much greater GUS enzyme activity with the Zm13 terminator than with the NOS terminator (Hamilton et al, 1992). In the Zm13 terminator the translation stop codon is at +638 and the poly(A) addition site is at +930 (Hamilton et al, 1989). There could be several reasons for this result; the 3' untranslated region could contain an enhancer or more likely, this region could confer enhanced stability on the mRNA. We are presently attempting to determine whether the 3' region contains an enhancer or whether it has a stability determinant. If it specifies stability of the mRNA we would like to determine where the sequence elements reside that are responsible for the stability properties of the mRNA.

The expression of genetic information during male gametophyte development is thus complex and is regulated at several different levels. Identifying and understanding the mechanisms involved will help us better understand pollen development and aid in the application of this information to the biotechnology of plants.

<u>Acknowledgement:</u> Work in the author's laboratory has been supported by grants from the National Science Foundation.

Literature references

Albani D, Robert LS, Donaldson PA, Altosaar I, Arnison PG, Fabijanski SF (1990) Characterization of a pollen-specific gene family from Brassica napus which is activated during early microspore development. Plant Mol Biol 15:605-622

Aziz N, Munro HN (1987) Iron regulates ferritin mRNA translation through a segment of its 5' untranslated region. Proc Natl Acad Sci USA 84:8478-8482

Breiteneder H, Pettenburger K, Bito A, Valenta R, Kraft D, Rumpold H, Scheiner O, Breitenbach M (1989) The gene coding for the major birch pollen allergen Betv1, is highly homologous to a pea disease resistance response gene. EMBO J 8:1935-1938

Brown Sm, Crouch ML (1990) Characterization of a gene family abundantly expressed in Oenothera organensis pollen that shows sequence similarity to polygalacturonase. Plant Cell 2:263-274

Guerrero FD, Crossland L, Smutzer GS, Hamilton DA, Mascarenhas JP (1990) A maize pollen-specific promoter directs tissue-specific transcription in tobacco. Mol Gen Genet 224:161-168

Hamilton DA, Bashe DM, Stinson JR, Mascarenhas JP. 1989. Characterization of a pollen-specific genomic clone from maize. Sex. Plant Reprod. 2, 208-212.

Hamilton DA, Roy M, Rueda J, Sindhu RK, Sanford J, Mascarenhas JP (1992) Dissection of a pollen-specific promoter from maize by transient transformation assays. Plant Mol Biol 18:211-218

Hentze MW, Caughman SW, Rouault TA, Barriocanal JG, Dancis A, Harford JB, Klausner RD (1987) Identification of the iron-responsive element for the translational regulation of human ferritin mRNA. Science 238:1570-1573

Koltunow AM, Truettner J, Cox KH, Wallroth M, Goldberg RB (1991) Different temporal and spatial gene expression patterns occur during anther development. Plant Cell 2:1201-1224

Mariani C, De Beuckeler M, Truettner J, Leemans J, Goldberg RB (1990) Induction of male sterility in plants by a chimeric ribonuclease gene. Nature 347:737-741

Mascarenhas JP (1975) The biochemistry of angiosperm pollen development. Bot Rev 41:259-314

Mascarenhas JP (1990a) Gene activity during pollen development. Ann Rev Plant Physiol Plant Mol Biol. 41:317-338

Mascarenhas Jp (1990b) The male gametophyte of flowering plants. Plant Cell 1:657-664

Mohapatra S, Hill R, Astwood J, Ekramoddoulah AKM, Olson E, Silvanovich A, Hatton T, Kisil FT, Sehon A (1990) Isolation and characterization of a cDNA clone

encoding an IgE-binding protein from Kentucky bluegrass (Poa pratensis) pollen. Int Arch Allergy Appl Immunol 91:362-368
Rafner T, Griffith IJ, Kuo MC, Bond JF, Rogers BL, Klapper DG (1991) Cloning of Amb aI (antigen E), the major allergen family of short ragweed pollen. J Biol Chem 206:1229-1236
Roberts MR, Robson F, Foster GD, Draper J, Scott RJ (1991) A Brassica napus mRNA expressed in developing microspores. Plant Mol Biol 17:295-299
Scott R, Dagless E, Hodge R, Paul W, Soufleri L, Draper J (1991) Patterns of gene expression in developing anthers of Brassica napus. Plant Mol Biol 17:195-207
Silvanovich A, Astwood J, Zhang L, Olson E, Kisil F, Sehon A, Mohapatra S, Hill R (1991) Nucleotide sequence analysis of three cDNAs coding for Poa p IX isoallergens of Kentucky bluegrass pollen. J Biol Chem 266:1204-1210
Singh MB, Hough T, Theerakulpisut P, Avjioglu A, Davies S, Smith PM, Taylor P, Simpson RJ, Ward LD, McCluskey J, Puy R, Knox RB (1991) Isolation of cDNA encoding a newly identified major allergenic protein of rye-grass pollen: intracellular targetting to the amyloplast. Proc Natl Acad Sci USA 88:1384-1388
Stinson JR, Eisenberg AJ, Willing RP, Pe ME, Hanson DD, Mascarenhas JP (1987) Genes expressed in the male gametophyte of flowering plants and their isolation. Plant Physiol 83:442-447
Thiel EC (1987) Ferritin: structure, gene regulation, and cellular function in animals, plants, and microorganisms. Ann Rev Biochem 56:289-315
Twell D, Wing R, Yamaguchi J, mcCormick S (1989) Isolation and expression of an anther-specific gene from tomato. Mol Gen Genet 247:240-245
Ursin VM, Yamaguchi J, McCormick S (1989) Gametophytic and sporophytic expression of anther specific genes in developing tomato anthers. Plant cell 1:727-736
Valenta R, Duchene M, Pettenburger K, Sillaber C, Valent P, bettelheim P, Breitenbach M, Rumpold H, Kraft D, Scheiner O (1991) Identification of profilin as a novel pollen allergen; IgE autoreactivity in sensitized individuals. Science 253:557-560
Willing Rp, Bashe D, Mascarenhas Jp (1988) An analysis of the quantity and diversity of messenger RNAs from pollen and shoots of Zea mays. Theoret Appl Genet 75:751-753
Willing Rp, Mascarenhas JP (1984) Analysis of the complexity and diversity of mRNAs from pollen and shoots of Tradescantia. Plant Physiol 75:865-868

Gametophytic Competition and Plant Breeding

David L. Mulcahy
Department of Botany
University of Massachusetts
Amherst, MA 01003 USA

Mirella Sari-Gorla
Department of Genetics & Microbiology
University of Milan, Via Celoria 26
20133 Milan, ITALY

Pollen selection has been shown to effect the growth rate and competitive ability of the resulting sporophytic generation in several crop plants (reviewed by Ottaviano and Mulcahy, 1989; see also Quesada, et al., 1991). The response has, however, not been found in some tested wild species (Snow, 1990), and, although it has been reported to persist into a subsequent generation in one study (Mulcahy, et al., 1978), this was not the case in a second example (Schlicting et al., 1990). Recently, aluminum (Searcy and Mulcahy, 1990) was added to the list of cases, (cold, salt, copper, zinc, herbicide and fungal toxin) in which pollen reflects the tolerance of the sporophytic pollen source (reviewed by Searcy and Mulcahy, 1990). For each of these, pollen selection could be, and, in some cases, has been, used to select for stress tolerant sporophytes. Clearly, pollen selection shows promise for plant breeding but several issues remain to be explained. In the following paper, we report on three papers which consider a fundamental aspect of this topic : the genetic control of pollen tube growth rate. Also considered are three specific, and, we believe, potentially useful, cases in which selecting for resistance to stress in pollen resulted in sporophytes which exhibited, unexpectedly, a decrease in stress tolerance.

Genetic Control of Pollen Tube Growth Rate

The genetic control of pollen tube growth rate (here used interchangeably with pollen competitive ability) was first subject to detailed examination by Wendel, et al., (1987). Employing an F_2 population in *Zea mays*, they determined the transmission ratios of 17 segregating allozyme loci and one morphological marker locus. Significant segregation distortion was observed for 12 of these 18 loci, these 12 being distributed on 7 of the 10 pairs of chromosomes. Because all inflorescences were filled and seed germination was nealy 100%, the authors rightly concluded that the distortions observed very likely had a prezygotic origin. It should be noted that, whena comparable population was grown ina winter nursery (that is, under suboptimal conditions), no significant deviations from mendelian expectations were observed. This underscores the significance of controlling the environment during pollen selection.

The competitive ability of pollen is undoubtedly strongly influenced by processes which take place during the development of pollen, the genetic control of which has been analyzed by Kindiger, et al. (1991). Using B-A translocations to produce hypoploid plants in *Zea mays*, they obtained plants in which one half of the pollen grains exhibit a predictable chromosomal deletion. Working with twenty six B-A translocations, involving 17 of the 20 chromosomes, they concluded that, in every case, "loss of part of a chromosome arm caused abnormal microspore development, a slowing of the normal mitotic or developmental processes in the male gametophyte or a termination of development". It was suggested that cross-feeding, or the presence of duplicate loci (since maize may be an ancient amphidiploid) may have allowed the development of deficient microspores.

For a more detailed mapping of genes which have a major effect upon pollen tube growth rates, Sari-Gorla et al. (submitted) employed Recombinant Inbreds (see Burr and Burr, 1991). This method involves the use of many highly lines, each derived from a single cross, inbred for approximately 20 generations, and then characterized for a large number of RFLPs (restriction fragment length polymorphisms). Every recombinant inbred line thus presents a random series of well marked blocks within the genome. In order to determine the relative growth rate of pollen, the pollen to be tested is mixed with pollen carrying a colored aleurone marker. As pollen tubes from the mixture of pollen penetrate styles, pollen tubes which are growing faster will surpass those growing more slowly. Since the styles for basal kernels of Zea mays are longer than are those for apical kernels, the faster pollen tubes will accomplish relatively more fertilizations towards the base of the inflorescence. When pollen from each of a series of recombinant inbred lines is mixed with a marker pollen, analysis of the resulting inflorescences gives the relative competitive ability of each recombinant inbred. (Alternatively, we can say that the analysis gives the relative competitive ability of each collection of genome blocks). When the combined results are examined, it is possible to determine which blocks tend to be associated with high or low competitive ability of pollen. Sari-Gorla et al. performed this analysis with 40 different recombinant inbreds. It is important to note that neither the sources of the recombinant inbreds (T232 x CM37) nor the hybrid used as a pistillate parent (A632 x Mo17) nor the aleurone marked tester, W22, carried any known Ga/ga (gametophytic factors). The final analysis indicated that 29 loci exhibited a significant association with pollen tube growth rate (PTGR), and 26 did so with pollen grain germinability (PGG). (PGG is proportion of uncolored kernels in the apical segment of the inflorescence and indicates the relative germinability of the pollen being tested). Many significant loci are closely linked to each other. Thus it is possible that significant effects associated with more than one locus may result from these markers all being linked with one quantitative trait locus (QTL). If we take the conservative position of considering all linked markers which influence PTGR and are linked to each other, it appears that significant loci can be grouped into 5 separate chromosomal regions for PTGR and 6 for PGG. However, this is clearly overly conservative since one such region could hold many independent centers of significant influence.

Other studies (Snow and Mazer, 1988) have failed to produce evidence that pollen tube growth rate is under genetic control (or responds to selection). This may be due to the conditions under which selection occurred (recal that Wendel et al. (1987) observed differential transmission in only one of their two field studies), to the choice of organism or experimental design, or to other factors.

Unexpected or Negative Results and Their Possible Implications

Unexpected, and particularly, negative, results generally represent unwelcome challenges to an investigator. However, whenever results are statistically significant, albeit opposite to what we expected, we should consider that perhaps an opportunity is presenting itself. The following two studies, attempts to select for tolerance either to cold or to Fusarium, may represent such opportunities.

Tolerance to Cold

Zamir et al., (1982) generated plants which would be segregating for genes to cold tolerance by hybridizing *Lycopersicon esculentum* with a high altitude accession of *L. hirsutum*. Pollen from the F_1 was applied to stigmas of *L. esculentum* at either normal (24°C Light/12°C Dark) or cold (12°C Light/6°C Dark) temperatures and the resultant backcross progeny were analyzed for isozymes which marked regions of the two original parents. Pollen grains containing two separate loci, *Pgi-1*^{1} or *Adh-2*^{1} were seen to be preferentially transmitted at cold temperatures. Interestingly, pollen genotypes which carried both of these cold tolerant associated alleles were selected against in the cold. While no real explanation for this unexpected result can presently be given, we suggest that the following facts may be relevant:

1. Pollen genotypes carrying both *Pgi-1*^{1} and *Adh-2*^{1} are not selected against at normal temperatures (Zamir, et

al. 1982) so the reaction to cold must be related to their associated cold tolerance.

2. Chen and Patterson (1985) report that plants exhibiting a high degree of chill-sensitivity generally exhibit a higher level of ethylene after chilling than do chill-tolerant plants. [Could the same be true for pollen tubes?]

3. Ethylene has been reported to stimulate pollen tube growth (Search and Stanley, 1970).

The following is an admittedly speculative hypothesis. However, the data from Zamir et al. (1982) are substantial and some explanation must exist, if not the following. As one possibility, we suggest that chilling induces injury in cold sensitive pollen tubes, along with a concomitant production in ethylene. This, in turn, results in a stimulation of tube growth in these sensitive grains. Pollen grains which carry both $Pgi\text{-}1^1$ and $Adh\text{-}2^1$ are more resistant to chilling than are grains carrying only one, or neither, of the cold associated alleles. Thus the more resistant phenotypes, because they lack the injury induced stimulation of ethylene, are apparently selected against in pollen tube competition. If this hypothesis is correct, more severe chilling, or the application of antiethylene compounds, should cause the doubly resistant pollen types to outcompete (in the first case), or to equal, the sensitives (in the second case).

Tolerance to Fusaric acid or to Fusarium filtrates

Fusaric acid has been implicated as an active component of Fusarium spp. infection, thus Simon and Sanford (1986) attempted to select for resistance to this compound by applying solutions to styles being pollinated. Surprisingly, the application of fusaric acid increased susceptibility in the next generation. More recently, Rowe and Stortz-Lintz (in press) exposed both pollen and callus from 25 plants, in three cultivars of Medicago sativa L. to filtrates of two isolates of Fusarium oxysporum f. sp. medicaginis. They found a significant negative correlation of pollen germination and callus growth to the filtrates. That is, growth of callus from sensitive plants was sensitive to the filtrates while pollen of sensitive plants was not sensitive. As one hypothesis, Rowe and Stortz-Lintz suggest that, "some of the nongerminating pollen may be reacting to the filtrate by cessation of growth. This response might parallel a recognition reaction as one of the early steps in the initiation of disease resistance in the plant.. This negative correlation suggests that pollen containing genes for resistance do not germinate." Their results are thus comparable to those reported by Simon and Sanford (1986). Certainly we cannot conclude that pollen exhibits the hypersensitivity of disease resistant genotypes but the above results suggests a promising field of study.

Conclusion

In cases where analytical approaches and material are appropriate, genetic factors which influence pollen development and competitive ability appear to be widely dispersed through the genome. This confirms molecular indications (Willing, et al. 1988) that approximately 60% of the sporophytically expressed genes are expressed also in the pollen. Pollen selection can be used to modify the resultant sporophytic generation and, even in cases of unexpected results, there may be significant potential for plant breeding applications.

REFERENCES

Burr, B., and F. A. Burr. 1991. Recombinant inbreds for molecular mapping in maize: theoretical and practical considerations. Trends in Genetics 7:55-60.

Chen, Y-Z., and B. D. Patterson. 1985. Ethylene and ACC as indicators of chilling sensitivity in various plant species. Aust. Jour. Plant Physiol. 12:377-385.

Kindiger, B., J. B. Beckett, and E. H. Coe. 1991. Differential effects of specific chromosomal deficiencies on the development of the maize pollen grain. Genome 14:579-594.

Mulcahy, D. L., G. B. Mulcahy, and E. Ottaviano. 1978. Further evidence that gametophytic selection modified the genetic quality of the sporophyte. Soc. Bot. Fr. Actualites Botaniques 1978 (1-2): 57-60.

Ottaviano E., Mulcahy D.L. 1989. Genetics of Angiosperm pollen. Advances in Genetics 26: 1-64.

Quesada, M., C. Schlichting, J. Winsor, and A. Stephenson. 1991. Effects of genotype on pollen performance in Cucurbita pepo. Sexual Plant Reproduction 4:208-214.

Sari-Gorla, M., M. E. Pe, D. L. Mulcahy, and E. Ottaviano (submitted for publication) Genetic dissection of pollen competitive ability in maize.

Schlicting, C. D., A. G. Stephenson, L. E. Small, and J. A. Winsor. 1990. Pollen loads and progeny vigor in Cucurbita pepo: the next generation. Evol. 44:1358-1372.

Search, R. W., and R. G. Stanley. 1970. Stimulation of pollen growth in vitro by ethylene. Phyton 27:35-39.

Searcy K. B., and Mulcahy D. L. 1990. Comparison of the response to aluminum toxicity in gametophyte and sporophyte of four tomato (Lycopersicon esculentum Mill.) cultivars. Theor Appl Genet 80:289-295.

Simon, C. J., and J. C. Sanford. 1986. Induction of gametic selection in situ by stylar application of selective agents. Pp. 107-112 in Biotechnology and Ecology of Pollen. Ed. by D. Mulcahy, G. B. Mulcahy, and E. Ottaviano. Springer-Verlag. New York.

Snow, Allison A., and Susan J. Mazer. 1988. Gametophytic Selection in Raphanus raphanistrum: A Test for Heritable Variation in Pollen Competitive Ability. Evolution 42:1065-1075.

Snow, A. A. 1990. Effects of pollen load size and number of donors on sporophyte fitness in wild radish (Raphanus raphanistrum). Am. Nat. 136:742-758.

Wendel, J. F., M. D. Edwards, and C. W. Stuber. 1987. Evidence for multilocus control of preferential fertilization in maize. Heredity 58:297-301.

Willing, P.R., D. Bashe, and J. P. Mascarenhas. 1988. Analysis of the quantity and diversity of messenger RNAs from pollen and shoots of Zea mays. Theor. Appl. Gen. 75:75-3.

Zamir, D., S. Tanksley, and R. A. Jones. 1982. Haploid selection for low temperature tolerance of tomato pollen. Genetics 101:129-137.

Cytoplasmic Male Sterility in Higher Plants

C. J. Leaver,
Department of Plant Sciences,
University of Oxford,
South Parks Road,
Oxford, OX1 3RB
United Kingdom

The ability to produce hybrid seed has been of fundamental importance to modern agricultural practice both in terms of increasing yield and generating wealth. However the crossing of one plant with another is difficult in many agriculturally important crop species due to the intimate association of the pollen bearing anthers and the stigma, as such an association usually results in self-fertilisation. One approach which prevents self-fertilization is to prevent anther dehiscence in the seed parent either genetically, chemically or physically.

The genetic approach to the production of F_1-hybrid seed was made possible by exploitation of cytoplasmic male sterility (CMS) a common trait reported in over 140 species of higher plant. Cytoplasmic male sterility is manifested in a number of ways depending on the plant species and ranges through abnormal anther development, failure in anther dehiscence to pollen abortion at various stages during microsporogenesis. The CMS phenotype is used commercially in the production of hybrid seed by preventing self-fertilisation of the seed parent plant in such crops as maize, sorghum, rice, sugarbeet and sunflower. Crosses between a CMS plant which fails to produce viable pollen (the female parent) and a male fertile partner (the pollen parent) results in male sterile progeny, since the maternally inherited (non-Mendelian) gene(s) responsible for sterility are transferred from the female parent to the seed via the egg cell cytoplasm. The paternal parent only contributes nuclear genetic information, with the cytoplasmic genomes being eliminated by an unknown mechanism.

The commercial use of CMS lines as female parents was made possible by the discovery of specific, dominant, nuclear restorer of fertility genes (Rf genes). When these genes are introduced into a hybrid from the pollen parent the male sterile phenotype is suppressed and the F_1-hybrid progeny, which is heterozygous for the Rf genes, are male fertile and produce pollen, even though they contain the 'male sterile' cytoplasm. In maize, for example, the cytoplasms that confer sterility are classified as T, C and S and are distinguished from each other by the different nuclear genes required to restore fertility (Leaver and Gray, 1982).

The widespread use of the CMS system in plant breeding programmes, while being cost effective, is not without its drawbacks. Dependence on a single source of CMS in hybrid seed production has the inevitable consequence of conferring 'cytoplasmic genetic uniformity' on the hybrid. In maize this led to a disastrous epidemic caused by a fungal pathogen which was particularly virulent on plants with the T-cytoplasm used in the breeding programme. In 1970 over 80% of the hybrid maize lines grown in the USA carried the cms-T cytoplasm, which in contrast to the N(fertile) cytoplasm and the other male-sterile cytoplasms, was responsible for susceptibility to the fungal disease *Bipolaris maydis*, race T (Southern Corn Leaf Blight). In that year an epiphytotic outbreak of the disease destroyed in excess of 15% of the US corn crop and it was soon established that the maternally inherited disease susceptibility was due to the production of pathotoxin(T-toxin) by the fungus that specifically affected mitochondria from cms-T maize.

It is thus obviously important to increase cytoplasmic genetic diversity in crop plants by identifying or creating new sources of male sterility and also, to investigate the molecular, biochemical and physiological basis of CMS and its link in cms-T maize, to disease susceptibility.

Although both chloroplast and mitochondria contain their own genomes it is now generally accepted that in most cases CMS originated from mutations in the mitochondrial DNA (mtDNA) as a result of aberrant intra- and/or intermolecular recombination events. These rearrangements have in some cases led to modification of existing genes or, in others, generated unique genes. These chimaeric genes may be transcriptionally active and are expressed as novel or variant mitochondrial proteins which appear to be causally related to failure in mitochondrial function in the anther tapetum and microspores, leading to failure in the production of viable pollen.

STRUCTURE AND INFORMATION CONTENT OF THE PLANT MITOCHONDRIAL GENOME

The mitochondrial genomes of higher plants are large and complex, varying in size by at least 10-fold between species (200-2500kb). Physical mapping studies have shown that in most plants examined to date the mitochondrial genome can be represented as a circular master chromosome, which contains a complete set of genetic information, but probably exists as smaller sub-genomic circular molecules, each of which contains specific genetic information. This multipartite structure is due to recombination across one or more pairs of directly repeated sequences which vary in length from a few 10's of bp up to 14kb. Linear and small circular DNAs, called plasmids, are also found in the mitochondria of a number of plant species. In some cases they have sequence homology with the main genome and may recombine into the mtDNA, leading to linearisation of the mitochondrial chromosome (Newton, 1988).

The work of Palmer and his colleagues have shown that not all repeated sequences are active in high-frequency recombination and that the of structural variability observed among plant mitochondrial genomes of different species may be due to rare, low-frequency recombination events across repeated sequences dispersed throughout the genome (Palmer, 1990). Given the size of the mitochondrial genome (the majority of which is non-coding) it is likely that these rearrangements will not, in most cases, involve active genes and hence would have no phenotypic consequences, but will lead to evolutionary changes in gene order and arrangement. However in recent years a number of laboratories have demonstrated in a range of plants (maize, Petunia, sorghum, sunflower, radish, rice, etc.) that these rare recombination events have generated novel genes which have become fixed in the mitochondrial population and lead to alterations in the patterns of transcription and translation (see Hanson *et al.* 1989; Newton, 1988 for reviews). In some but not all cases these mutant genes have been implicated in CMS. There are now other reports which demonstrate that rearrangements within the 5' and 3' flanking regions of known genes occurs with relatively high frequency and may result in different termini and hence size of transcripts and transcript patterns (Makaroff, 1991). Where these are observed in mitochondria from CMS plants it is only in a relatively few cases that they have been strongly linked to the male sterile phenotype and it is very likely that many mtDNA rearrangements and consequent transcriptional differences may have no causal role in CMS.

Despite the wide variation in size of the mitochondrial genome, plant mtDNA apparently only encodes the structural ribosomal RNAs (26S, 18S and 5S rRNAs) and ca 20tRNAs (the remaining 11 tRNAs are coded by the nuclear genome and imported into the mitochondria) which are constituents of the mitochondrial translation system, and of the order of 13-20 polypeptides, which together constitute less than 10% of the mitochondrial protein mass. These polypeptides are primarily hydrophobic components of the multisubunit, electron transport, and F_1-F_0ATPase complexes of the inner membrane. They include the three largest subunits of the cytochrome c oxidase complex (*cox I, II, III*), the apocytochrome b subunit (*cob*) of the bc_1 complex and several subunits of the NADH- ubiquinone oxidoreductase (NADH dehydrogenase) complex (including *nad 1, 3, 4* and *5*). The α-subunit (*atpA*) of the F_1-ATPase complex, and subunits 6 (*atp6)* and 9 (*atp9*) of the F_0-sector, are also encoded in the plant mitochondrial genome. In addition, genes for at least three proteins of the small ribosomal subunit (*rps 12, 13* and *14)* are encoded in the mitochondrial genomes of some plants, but are apparently not synthesized in this compartment in others. Recent unpublished reports suggest that a few mitochondrial protein-coding genes (e.g. *cox II* and *rps* genes) have a different subcellular location in different plant species, this probably reflects the ongoing evolutionary migration of genes from mitochondrion to nucleus and may be associated with the variability in size of plant mtDNAs. An additional factor, which also contributes to the size and variability of plant mitochondrial genomes is the presence in some genes of single or multiple introns, varying in length from 700 to 3,400 nucleotides.

Mapping of these genes on the physical maps of several plant mtDNAs shows that they are scattered throughout the genome on both strands. Gene order varies even between closely related species, suggesting that the order and arrangement have no functional significance and are a product of extensive genome reorganization during evolution. While a few mitochondrial gene products remain to be identified, only of the order of 20 labelled polypeptides can be visualised when the products of *in organello* protein synthesis are analysed on one or two-dimensional SDS polyacrylamide gels. The remainder of the 300-400 mitochondrial proteins, namely those of the outer membrane, intermembrane space, inner membrane and the matrix, are encoded in nuclear DNA, synthesised on cytosolic ribosomes, normally as higher molecular weight precursors and imported into the mitochondria post-translationally.

CYTOPLASMIC MALE STERILITY IN MAIZE

Two main approaches have been used to identify the genetic basis and mechanisms which give rise to the CMS phenotype. The first involved a comparison of the structure and expression of the mitochondrial genome of near isonuclear lines of fertile, CMS and restored lines and the second depended upon similar studies of revertant fertile plants and their CMS progenitors (Hanson *et al.* 1989).

In the case of the three male sterile cytoplasm of maize T, C and S it was shown that they could be differentiated from each other and the normal, fertile cytoplasm on the basis of mtDNA restriction fragment length polymorphism. Further studies also showed cytoplasm specific differences in mitochondrial RNA by Northern blot analysis and variation in mitochondrial translation products (Leaver & Gray, 1982; Hanson *et al.* 1989).

Levings and his colleagues subsequently isolated a mitochondrial gene, designated T-*urf13* which is unique to the cms-T maize cytoplasm and have produced very strong evidence that the protein product of this gene (*urf13*) causes CMS and also binds T-toxin from *Bipolaris maydis*, race T. (Levings, 1990).

Urf13 is a chimaeric gene and appears to have originated by intramolecular recombination events involving other mitochondrial genes. The *urf13* sequence is 115 codons long and encodes a predicted polypeptide with a molecular weight of 12,961 D. It shares significant homology to the 3' flanking (88 codons) and coding region (18 codons) of the mitochondrial 26 rRNA gene, together with nine codons of unknown origin. The sequence immediately 5' to the open reading frame shares homology with the promoter region of the atp6 gene. Functional copies of the atp6 and 26S rRNA genes are also present elsewhere in the genome and, because of the origins and similarity of the 5' flanking sequences of T-*urf13* and *atp6* genes, it is thought that they have similar promoters. This unique open reading frame is transcribed and expressed as a 13-kD polypeptide in mitochondria from all organs of cms-T maize plants, but is not synthesized in mitochondria from N (fertile) or other CMS lines. We, and others, have shown that an antibody raised against a synthetic oligopeptide corresponding to part of the predicted amino acid sequence of T-*urf13* will immunoprecipitate the unique 13-kD polypeptide synthesized by mitochondria isolated from the T-cytoplasm. The variant 13 kD translation product is, together with the majority of the mitochondrially encoded translation products, targeted to the inner mitochondrial membrane, but its precise location or functional activity is unknown.

The role of the nucleus in regulating mitochondrial genome expression is further emphasized by our observation that when the dominant nuclear alleles (Rf1 and Rf2) that restore pollen fertility are introduced into the nuclear background of the lines carrying the T-cytoplasm, there is a marked and specific reduction in the amount of the 13-kD polypeptide synthesized by mitochondria isolated from the restored plants (Forde & Leaver, 1980; Levings, 1990). Dewey *et al* (1986) have shown that the Rf1 gene appears to affect expression of the *urf13* at the transcriptional or post-transcriptional level, while the role of the Rf2 gene is unknown; however, it does not seem to affect *urf13* expression.

The fungal toxin produced by *B. maydis*, race T, consists of a linear β-oxydioxo polyketol structure. When this toxin is applied at very low concentrations to mitochondria from cms-T maize, malate-supported state-3 respiration is inhibited and NADH-supported state-4 respiration stimulated. In addition, oxidative phosphorylation is uncoupled, the mitochondria swell, and a massive leakage of small molecules such as Ca^{2+} and NAD^+ occurs. Under similar experimental conditions, the toxin has no effect on mitochondria from other maize cytoplasms, or from other plant species (Levings, 1990).

Further evidence linking expression of T-*urf13* and the CMS phenotype has been provided by the observation that a correlation exists between reduced synthesis of the 13-kD polypeptide, reversion to the male fertile phenotype, and resistance to *B. maydis*, race T, in plants regenerated from tissue cultures of cms-T maize (Levings, 1990). This reversion to fertility is stably inherited and is associated with mitochondrial genome rearrangements which have led to deletion or mutation of the T-*urf13* gene. The high frequency of reversion in plants derived from tissue culture implies that there is a greater opportunity for selection and/or segregation of mitochondrial genotypes in tissue culture and, further, that there is heavy selection against DNA molecules containing T-*urf13*. These observations provide strong circumstantial evidence for a close link between the expression of T-*urf13*, synthesis of the 13-kD polypeptide, the CMS phenotype, and susceptibility to *B. maydis*, race T.

Compelling evidence that T-toxin sensitivity and disease susceptibility by cms-T maize is due to the interaction of the 13kD polypeptide and toxin has been provided by Levings and colleagues (1990). They showed that when T-*urf13* is cloned, transformed into *E. coli* and expressed, the 13-kD protein is localized in the bacterial membrane and that bacterial respiration and growth is inhibited by T-toxin, when compared to control, untransformed cells. The affects of the toxin on the transformed bacterial cells are analogous to those exhibited by mitochondria isolated from cms-T maize, but not normal (fertile) plants and include increased permeability to ions, swelling and inhibition of respiration. Braun *et al.* (1990) have also recently shown that the toxin binds to the 13-kD polypeptide synthesised in *E. coli* transformed with T-*urf13*. These authors propose that the 13 kD- polypeptide causes toxin sensitivity in maize by forming, in the presence of the toxin, channels in the inner mitochondial membrane, which allow ion efflux and results in loss of mitochondrial function.

What appears to be definitive proof that T-*urf13* actually causes male sterility has been reported by Chaumont *et al.* (1991). They have introduced the T-*urf13* gene linked to a mitochondrial targeting presequence into the nuclear genome of tobacco. Transgenic plants obtained with the control T-*urf13* , without presequence were sensitive to methomyl, a carbamate insecticide which causes equivalent toxic effects to T-toxin on cms-T mitochondria, but were male-fertile. Fewer transgenic plants were obtained from constructs with T-*urf13* linked to the mitochondrial presequence, they grew more slowly and they were less susceptible to methomyl. However, they were partially or completely male sterile and when fertilised with heterologous pollen gave rise to F_1 progeny which had inherited the sterility trait.

Further examples of unique chimaeric genes associated with cytoplasmic male sterility have been described in cms-C maize (Dewey *et al.*, 1991), sorghum (Bailey-Serres *et al.*, 1986), Petunia (Hanson *et al.* , 1989) Brassica (Singh & Brown, 1991) and a number of other plant species. In every case these chimaeric genes are the product of aberrant recombination events involving normal mitochondrial genes and other open reading frames of unknown origin. In many cases these rearrangements of the mitochondrial genome lead to changes in organization of the transcriptional and/or translational regulatory sequences of the progenitor genes. In all cases the chimaeric genes are expressed as variant mitochondrial polypeptides which are correlated with the male-sterile phenotype.

THE MOLECULAR AND MECHANISTIC BASIS OF CYTOPLASMIC MALE STERILITY

The results described above and those reported of studies of CMS in a number of other species provide strong circumstantial evidence that the mitochondrial genotypes characteristic of each form of CMS arose by aberrant intra- and/or inter-molecular recombination events involving short repeated DNA sequences. In specific cases these rearrangements generate unique open-reading frames, while in others they have extended open-reading frames or modified the control sequences of existing genes. Expression of these chimaeric genes as a variant polypeptide requires that they are located downstream of suitable promoter and translational control sequences. The variant polypeptides may contain completely novel amino acid sequences and/or blocks of amino acids sharing homology to the normal polypeptides encoded by the DNA sequences from which they were derived. The amino acid domains may also endow the variant polypeptides with some of the structural and/or functional characteristics of the progenitor protein(s) or confer novel characteristics. An example of the latter which has profound phenotypic consequences, is provided by the elegant studies of Levings and his colleagues who have implicated the *urf13* protein of CMS-maize with disease susceptibility.

Although the data of Chaumont *et al* (1991) appears to prove that *urf13* can cause male sterility in tobacco, the mechanism by which pollen abortion occurs is still unclear. Based upon the evidence available in the literature it can be speculated that features of the variant polypeptides (e.g. amino acid sequence, secondary structure, hydrophobicity) may target it to the inner mitochondrial membrane where it could compete with native, structurally related polypeptides and become associated with a specific enzyme complex. The presence of additional or a modified polypeptide may impose constraints on the biosynthesis, assembly or function of the multimeric respiratory enzyme complexes of the inner membrane. These may be tolerated during normal vegetative growth if the native enzyme complexes are capable of providing sufficient energy to sustain cellular development, but lead to mitochondrial insufficiency during the increased demand for mitochondrial activity associated with pollen formation.

If this model is correct, the finding that the variant polypeptides are present in mitochondria at all stages of plant development, implies that mitochondrial biogenesis and function are not a major limitation on cellular development during most of the plant life cycle. However, it is interesting to note that it is during microsporogenesis, a developmental stage associated with increased rates of respiration and very active mitochondrial biogenesis (in maize there is a 20- to 40-fold increase in mitochondrial numbers in the inner tapetal cell layers of the anther and in the developing pollen), where mitochondrial degeneration occurs as the first sign of abnormality in CMS plants.

An alternative hypothesis suggests that the promoter sequences, which are common to both the novel genes and the normal progenitor genes from which were derived, compete for limiting factors which control expression of the native and variant polypeptides genes. This could result in an imbalance in the amount of an individual subunit which is a key component of a multimeric enzyme complex, leading to impairment of mitochondrial biogenesis or function, which again only becomes limiting during pollen formation.

We can further speculate that the role of the nuclear restorer gene(s) is to suppress synthesis of the variant polypeptide below a critical threshold level by affecting transcription for example (as seen, for example, in the case of the 13-kD polypeptide in restored cms-T maize lines), or to compensate for the mitochondrial dysfunction in some way, thus restoring their respiratory potential.

The question then arises as to how a single mitochondrial aberration (mutation) could be responsible for both pollen abortion and susceptibility to T-toxin. Fortunately, a model proposed by Flavell in 1974 can be modified to accommodate the currently available facts. He proposed that the altered structure of cms-T mitochondria (the 13-kD polypeptide?) makes them susceptible to both T-toxin and an anther-specific compound produced during pollen development. The 13-kD polypeptide could interact with either T-toxin or the anther-specific compound (anther toxin) leading to channel formation and permeabilization of the inner mitochondrial membrane. Preliminary experiments from my laboratory suggests that extracts from both normal fertile and male sterile maize lines can mimic at least some of deleterious effects of T-toxin on mitochondria from cms-T but not N- cytoplasms.

The only evidence available to date linking change in mitochondrial function with a CMS associated gene locus (S-pcf) and cytoplasmic male sterility is in Petunia (Connett & Hanson, 1990). The authors demonstrated reduced alternative oxidase activity in suspension culture cells of CMS Petunia and in immature anthers from CMS plants, compared to the male fertile controls. A culture from a fertile plant carrying a nuclear fertility restorer gene and the CMS cytoplasm exhibited increased activity of the alternative oxidase pathway compared to the CMS lines. More recently they have demonstrated co-fractionation of the 25 kD protein, encoded by the S-pcf gene locus, with alternative oxidase activity.

A detailed description of all aspects of mitochondrial function and biogenesis during pollen development in fertile and CMS plants is obviously a priority. Even though the pioneering work of Leemans and his collaborators (see this volume) have demonstrated that it is possible to engineer nuclear male-sterility in a range of plant species the study of CMS will give us a greater appreciation of the molecular basis and phenotypic expression of cytoplasmically inherited variation in higher plants. This in turn should lead to the description and development of a broader base for F_1-hybrid breeding programmes. It should not be forgotten that mitochondria and chloroplasts are the major sites of energy conversion in plant cells and thus play a vital role in determining the overall peformance and productivity of crop plants. Most if not all of the major changes which occur during growth and differentiation of plants are associated with an dependent upon changes in the number, structure and metabolic activity of either one or both organelles.

REFERENCES

Bailey-Serres, J., Hanson, D.K., Fox, T.D. & Leaver, C.J. (1986). Mitochondrial genome rearrangement leads to extension and relocation of the cytochrome c oxidase subunit I gene in Sorghum. Cell, 47: 567-576.

Braun, C.J., Siedow, J.N. & Levings, C.S. III (1990). Fungal toxins bind to the *urf13* protein in maize mitochondria and *Escherichia coli*. Plant Cell 2: 153-161.

Chaumont, F., Williams, M., Levings, C.S. III, & Boutry, M. (1991). Import of the maize T-*urf13* protein into tobacco mitochondria confers male sterility. Abstract III International Congress of Plant Molecular Biology. Tucson, U.S.A.

Connett, M.B. & Hanson, M.R. (1990). Differential mitochondrial electron transport through the cyanide-sensitive and cyanide insensitive pathways in isonuclear lines of CMS, male fertile and restored Petunia. Plant Physiol. 93: 1634-1640.

Dewey, R.E., Levings, C.S. III, & Timothy, D.H. (1986). Novel recombinations in the maize mitochondrial genome produce a unique transcriptional unit in the Texas male-sterile cytoplasm. Cell 44: 439-449.

Dewey, R.E., Timothy, D.H. & Levings, C.S. III (1991). Chimeric mitochondrial genes expressed in the C male-sterile cytoplasm of maize. Curr. Genet. 20: 475-482.

Flavell, R.B. (1974). A model for the mechanism of cytoplasmic male sterility in plants, with special reference to maize. Plant Sci. Lett. 3: 59-263.

Forde, B.G. & Leaver, C.J. (1980). Nuclear and cytoplasmic genes controlling synthesis of variant mitochondrial polypeptides in male-sterile maize. Proc. Nat. Acad. Sci. USA, 77: 418-422.

Hanson, M.R., Pruitt, K.D. & Nivison, H.T. (1989). Male sterility loci in plant mitochondrial genomes. In B.J. Miflin, ed. Oxford Surveys of Plant Molecular Biology, Vol. 6, 61-85.

Leaver, C.J. & Gray, M.W. (1982). Mitochondrial genome organization and expression in higher plants. Annu. Rev. Plant Physiol. 33: 373-402.

Levings, C.S. III (1990). The Texas cytoplasm of maize. Cytoplasmic male sterility and disease susceptibility. Science 250: 942-947.

Makaroff, C.A., Apel, I.J. & Palmer, J.D. (1991). The role of coxI-associated repeated sequences in plant mitochondrial DNA rearrangements and radish cytoplasmic male sterility. Curr. Genet. 19: 183-190.

Newton, K.J. (1988). Plant mitochondrial genomes: organization, expression and variation. Annu. Rev. Plant Physiol. Plant Mol. Biol. 39: 503-532.

Palmer, J.D. (1990). Contrasting modes and tempos of genome evolution in land plant organelles. Trends in Genetics. 6: 115-120.

Singh, M. & Brown, G.G. (1991). Suppression of cytoplasmic male sterility by nuclear genes after expression of a novel mitochondrial gene region. Plant Cell 3: 13349-1362.

Genetic Engineering for Fertility Control

J. Leemans
PLANT GENETIC SYSTEMS N.V.
Jozef Plateaustraat 22
B - 9000 Gent
Belgium

Abstract

A new hybrid system based on genetically engineered genes for fertility control has been developed. Male sterility was engineered by targeting expression of chimaeric ribonuclease genes to immature anthers of transgenic plants. Ribonuclease expression destroyed the tapetal cell layer, prevented pollen formation and lead to male sterility. Male sterility was restored in progeny of crosses between male sterile plants and plants expressing a chimaeric tapetum specific ribonuclease inhibitor gene. Fertility restoration is the consequence of the suppression of cytotoxic ribonuclease activity by the formation of RNase/RNase inhibitor complexes. Fertility restored anthers have a normal tapetum, and produce normal amounts of functional pollen. The hybrid system has been successfully introduced in oilseed rape, cauliflower, chicory, lettuce, tomato, cotton and corn.

Introduction

F1 hybrids are superior in reliability of yield and quality. Their increased uniformity is a key requirement when modern growing and harvesting techniques are used. Hybrids also offer opportunities for the rapid deployment of dominant genes for resistance to diseases and pests.

Preventing self-fertilization is a key requirement in F1 hybrid seed production. To achieve this, naturally occurring phenomena such as male sterility, self-incompatibility and pollenic competition have been deployed or seed producers have used manual or mechanical emasculation. In this paper we present a new hybrid system based on genetically engineered genes for fertility control. The system can provide several advantages for F1 hybrid seed production by its increased flexibility and wider application range.

Results

A dominant gene for male sterility

Gene expression in plants is highly regulated. Unique sets of genes can be expressed in one given organ system and be silent in others. Even within an organ system, genes can be expressed uniquely in a certain cell type. To engineer male sterility, gene expression during tobacco anther development was studied (Koltunow et al, 1990). We isolated the TA29 gene which is characterized by its extreme cell specificity in the tapetal cells of immature anthers. The 5' regulatory region of TA29 (pTA29) was used to target the expression of ribonucleases such as barnase, from *Bacillus amyloliquefaciens*, specifically to the tapetum of immature anthers. Barnase expression led to the precocious degeneration of the tapetum cells, the arrest of microspore development, and male sterility (Mariani et al, 1990). These experiments were first carried out in tobacco and oilseed rape, suggesting that pTA29 maintains its cell specific expression in different plant species. In both species genetically engineered male sterile plants have been identified in which the sterility is stable under different environmental conditions. Also, the normal agronomic performance of the male sterile lines was not affected.

Maintaining male sterility

As the engineered sterility is a dominant nuclear gene, progeny of backcrosses segregate one to one for sterile and fertile plants. Two approaches have been worked out to maintain and multiply male sterile plants. First, the male sterility gene is linked to a gene encoding resistance to the herbicide glufosinate (Basta[R]) (De Block et al., 1987). The absolute linkage of the herbicide resistance to the male sterility allows the elimination of segregating fertile plants by treating F1 populations with the herbicide shortly after germination, thereby eliminating all herbicide sensitive, male fertile plants (fig. 1). This approach can be applied in the field and is most attractive when the inbred is presown for later transplanting to the field or to the greenhouse. In a second approach, the herbicide treatment can be avoided by using homozygous male sterile lines which are maintained as grandparent lines. Female parental seed is then produced by crossing homozygous male sterile lines to the untransformed line, resulting in a pure population of heterozygous male sterile plants.

Fertility restoration

We have constructed genes which can restore fertility to the genetically engineered male sterility. These chimaeric genes consist of the tapetal specific promoter of the TA29 gene linked to the intracellular inhibitor of barnase, called barstar (Hartley, 1988). When barnase and barstar are produced simultaneously in the same tapetal cells, they form a one to one complex which has no residual ribonuclease activity. Progeny of crosses between male sterile oilseed rape plants that express barnase and fertile pollinator plants that are homozygous for the barstar gene produced fully normal pollen and showed normal seed set upon selfing (Mariani et al., 1992).

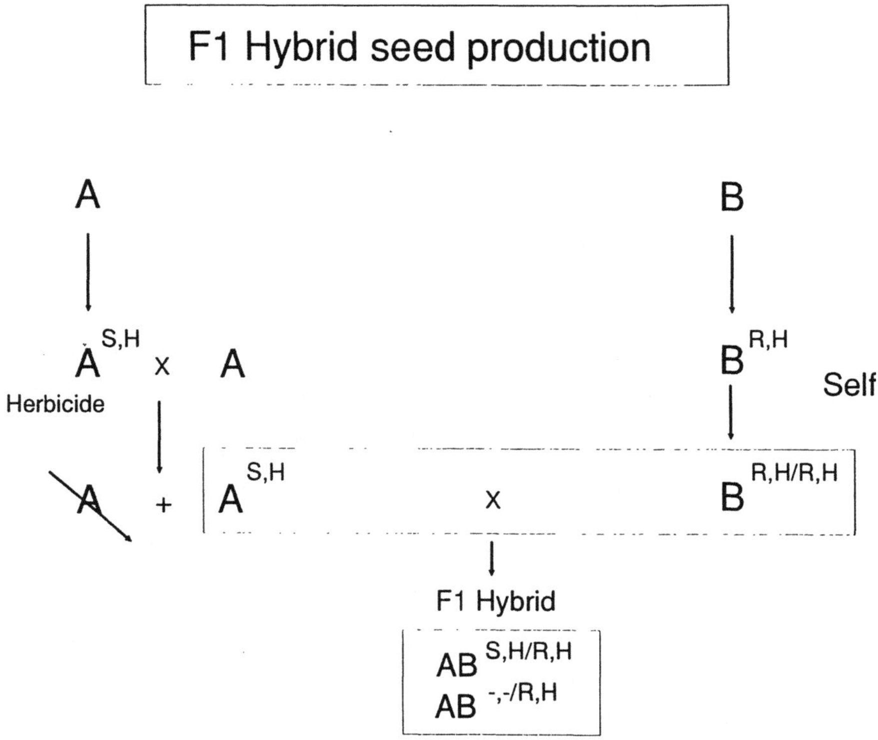

Example of the use of the genes for fertility control in hybrid seed production.

A,B : female and male parent line, respectively
S : TA29-barnase gene encoding male sterility
R : TA29-barstar gene encoding fertility restoration
H : 35S-bar gene encoding resistance to the herbicide Basta[R]

Female parent A is transformed to male sterility. Male sterility and herbicide resistance are tightly linked and inherit as a single locus. $A^{S,H}$ is multiplied by backcrossing to the untransformed line. Progeny is treated with the herbicide to select the male sterile plants.
Male parent B is transformed to fertility restoration, made homozygous and multiplied by selfing.

Applications: strategic opportunities

The new hybrid system offers the strategic opportunity to create commercial F1 hybrids in crops which today are only available as open pollinated varieties due to the lack of a practical hybrid system. A first target is oilseed rape. Male sterile and fertility restorer lines have been produced in *Brassica napus* cv. Drakkar. Field trials during 1991 on eight locations in Canada and Europe confirmed that the male sterility was stable throughout the entire flowering period. Four replicated trials in Canada and Europe confirmed that male sterile lines displayed normal agronomic performance with respect to germination, growth habit, plant vigor, flowering date and seed yield. Finally, the genotype independence of the male sterility was proven: all progeny of male sterile Drakkar backcrossed to 15 distinct spring oilseed rape lines segregated for male sterility in a Mendelian fashion.

Applications: economic opportunities

The new hybrid system can be advantageous over currently used hybrid systems in reducing the cost and increasing the reliability of hybrid seed production, in increasing the speed of hybrid breeding and in extending the available germplasm pool. A noticeable example is corn: in the US alone, each year nearly one million acres of female parent lines are mechanically and/or manually detassled. Recently, we have introduced a tapetum-specific ribonuclease gene in the corn inbred line H99. Several independent transformants have been generated which displayed shrivelled anthers with no release of pollen. Analysis of progeny of backcrosses showed that the sterility is stably inherited along with the introduced gene. A preliminary evaluation under greenhouse conditions indicated that growth and female fertility was not affected by the male sterility trait. These first results open up the possibilities for developing an economically attractive alternative for F1 hybrid corn production.

Acknowledgements

I am indebted to my colleagues who have developed, applied and evaluated the hybrid system: Prof. R.B. Goldberg (UCLA) and C. Mariani, W. De Greef, M. De Block, A. Reynaerts, J. Botterman, J. Janssen, S. Metz, B. Van Steenbrugge and P. Rüdelsheim (PGS).

References

De Block M, Botterman J, Vandewiele M, Dockx J, Thoen C, Gosselé V, Rao Mouva N, Thompson C, Van Montagu M and Leemans J (1987) Engineering herbicide resistance in plants by expression of a detoxifying enzyme. The EMBO Journal 6: 2513-2518

Hartley RW (1988) Barnase and Barstar. Expression of its cloned inhibitor permits expression of a cloned ribonuclease. Journal of Molecular Biology 202: 913-915

Koltunow AM, Truettner J, Cox KH, Wallroth M and Goldberg RG (1990) Different temporal and spatial gene expression patterns occur during anther development. Plant Cell 2: 1201-1224

Mariani C, De Beuckeleer M, Truettner J, Leemans J and Goldberg RG (1990) Induction of male sterility in plants by a chimaeric ribonuclease gene. Nature 347: 737-741

Mariani C, Gosselé V, De Beuckeleer M, De Block M, Godlberg RG, De Greef W and Leemans J (1992) A chimaeric RNase inhibitor gene restores fertility to male sterile plants. Submitted for publication.

Male Sterilities and F1 Hybrids in *Brassica*

M. RENARD[1], R. DELOURME[1], J. MESQUIDA[2], G. PELLETIER[3], C. PRIMARD[3], L. BOULIDARD[4], C. DORE[4], V. RUFFIO[1], Y. HERVE[1], J. MORICE[1]

[(1)]I.N.R.A. - Station d'Amélioration des Plantes

BP 29 - F.35650 Le Rheu - France -

Many *Brassica* species are economically important as vegetables (*B. oleracea*, *B. campestris*, *B. juncea*), condiments (*B. nigra, B. juncea*), forages (*B. oleracea, B. campestris, B. napus*) and as sources of edible or industrial oil and meal (*B. campestris, B. napus, B. juncea, B. carinata*). Reports for significant heterosis for some agronomic characters such as plant height, leaf area and yield in *B. napus* (Lefort-Buson et al., 1987) and the possibility of homogeneous production (mechanical harvesting) in *B. oleracea* have stimulated interest in the development of hybrid cultivars. But to exploit the advantages of heterosis on a commercial scale, an effective pollination control system to produce F1 hybrid seeds is essential. In diploid species (*B. oleracea* and *B. campestris*), hybrid cultivars were mostly produced by using self-incompatibility genes. The complexity of sporophytic incompatibility lead to a reduced number of lines convenient for seed production (genetic background effect). It required also bud pollination (very time-consuming) to reach homozygosity of lines and to produce isogenic lines for the establishment of the parents of hybrids. Incompatibility is scarcely complete and inbreds originated from self pollination are often present in hybrid seed lots. Numerous controls are therefore necessary to check about the good genetic quality of seeds. This leads to high seed costs to the growers. In allotetraploid species (*B. napus, B. juncea*) hybrid seed production is impeded by the preferential autogamous character of these crops. Thus an effective system of cross-pollination is necessary for hybrid rapeseed breeding.

[2] - INRA - Laboratoire de la Chaire de Zoologie - BP 29 - 35650 Le Rheu - France

[3] - INRA - Laboratoire de Biologie Cellulaire - Route de St Cyr - F.78026 Versailles Cédex - France

[4] - INRA - Station d'Amélioration des Plantes - Route de St Cyr -F.78026 Versailles Cédex - France

Beside self-incompatibility, various other systems have been used or proposed for the production of hybrid *Brassica* cultivars : treatment with male gametocides, nuclear male sterility (N.M.S.) and cytoplasmic male sterility (C.M.S.) systems. This paper mainly focuses on *B. napus* and presents information on male sterility pollination control mechanisms, hybrid seed multiplication designs, and hybrid variety types.

1 POLLINATION CONTROL MECHANISMS

Only a few papers have presented results of research on gametocides. In *B. juncea*, ethrel applications led to anther retardation (Banga and Labana, 1984). The treatments generally impaired female fertility and hybridity was low. Thus the ethrel-induced male sterility is not useful for commercial seed production. But it could be suitable for breeding. In *B. napus*, Guan and Wang (1987) induced partial to total sterile plants using gametocides like zinc-methyl arsenate. Completely sterile plants were dwarf and flowers were smaller with narrow petals. But commercial F1 hybrid varieties seem to be produced in China by this method. Gibberellic acid was applied on *B. oleracea* by Van der Meer and Vandam (1979). The sterility recovered was variable. Since the sterility induced by gametocides is generally low and variable, and phytotoxic effects are observed (dead plants, chlorosis), gametocides do not have immediate practical utility.

Many recessive NMS mutants were detected spontaneously in *Brassica* species or induced by mutagenesis (Kaul, 1988). Some dominant NMS were selected in *B. napus* (Mathias, 1985), in *B. oleracea* (Dunemann and Grunewaldt, 1991). and in *B. campestris* (Van der Meer, 1987). NMS in *Brassica* species is sometimes accompanied by a partial female fertility (bud abortion, low seed set) and is unstable. This instability can be exploited for seed multiplication of the homozygous male sterile lines in well-controlled conditions (low temperature in cauliflower). Haploidy could also be applied to fix dominant male sterile (ms) genes at the homozygous state for vegetatively propagated *Brassica* species.

Only a few marker genes linked to a ms gene were reported (Sampson, 1966). Thus the utilization of NMS is limited because male fertile plants must be discarded just before flowering in hybrid seed production fields.

N.M.S. conditioned by interaction of genes at two loci (MS and Rf) was identified in China (Li et al., 1988). This system is now applied to produce 100 % male fertile F1 hybrid rapeseed varieties (Figure 1) from a 100 % male sterile A-line (MSmsrfrf) and a restorer line (ms ms Rf Rf). More recently, Plant Genetic Systems obtained a digenic system by transferring via *Agrobacterium tumefaciens* a male sterile gene coding for a ribonuclease and a restorer gene inhibiting the RNase gene (Mariani et al., 1990 and 1991). Male sterile plants with the MS gene linked to a herbicide resistance gene can be selected in alternated strips seed production fields by treating with the herbicide.

Several male sterility-inducing cytoplasms which could potentially be developed into CMS systems for use in *Brassica* F1 hybrid production have been discovered (table 1). Programs on *B. napus* have focused on the polima system. This system was identified as a spontaneous occurence within a polish rapeseed cultivar. A restorer gene was obtained by Fang and Mc. Vetty (1989) in cultivar "Italy". But the polima system confers both a slightly temperature-sensitive male sterility (table 2) and a more or less negative effect on the yield (Mc Vetty et al., 1990). Male sterility of *B. juncea* system (Rawat and Anand, 1979) is more stable and partially restored F1 hybrids were observed in *B. juncea* and *B. napus*. The nap system identified by Thompson (1972) and Shiga (1976) in *B. napus*, is not much used due to the instability of male sterility between 20°C and 26°C (table 2) and of the large number of maintainer genes it is necessary to introduce into rapeseed material which is generally restorer. Okhawa (1984) identified a CMS in *B. campestris* closed to the nap system C.M.Ss were induced in *B. oleracea* by transferring *B. oleracea* genome into the cytoplasm of *B. napus* (Chiang and Crete, 1987) or into the cytoplasm of *B. nigra* (Pearson, 1972). Hinata and Konno (1979) reported a CMS system in *B. campestris* based on a *Diplotaxis muralis* cytoplasm. This CMS was transferred into *B. napus* by Pellan-Delourme and Renard (1987). Prakash and Chopra (1990) obtained a CMS from *B. oxyrrhina* in *B. campestris* and *B. juncea*. Alloplasmic male sterile plants are stably male sterile but midly chlorotic during initial development.

S cytoplasm derived from *Raphanus sativus* (Ogura, 1968) was transferred to *B. oleracea* and then to *B. napus* by Bannerot et al. (1974) and subsequently to *B. campestris, B. juncea* and *B. oleracea*. The Ogura CMS system was characterized by chlorophyll deficiency, low nectar secretions and lack of restorer genes in *Brassica* species. This CMS system was successively improved by protoplast fusion (Pelletier et al., 1983 ; photosynthesis and nectaries) and by interspecific crosses with a *Raphanobrassica* (Heyn, 1978 ; Pellan-Delourme and Renard, 1988 ; restorer gene) in *B.*

Table 1. Characteristics of cytoplasmic male sterility in *Brassica* species

Species	Systems	Stability	Maintener	Restorer	Nectaries	Female A-line	Fertility R-line	Photosynthesis
B. napus	.*Napus* (nap)	--	-	+	++	+	+	+
	.*Polima* (pol)	-	+/-	+	++	+	+	+
	.*Juncea*	+/-	+	+/-	++	+	+	+/-
	.*Nigra*	+/-	-	+	-	-	+	+
	.*Diplotaxis*	+/-	-	+	-	-	+	+
	.Raphanus Ogura	++	++	+/-	-	+/-	+/-	-
	.Raphanus cybrids	++	++	+	+	+	+	+
B. juncea	.*Juncea*	+/-	+	+/-	+	+	+	+/-
	.*Oxyrrhina*	+	+	-				-
	.Raphanus ogura	+	+	-	-	+/-		-
B. campestris	.*Campestris*	--	-	+	+	+	+	+
	.*Diplotaxis*	+/-	+/-					+
	.Raphanus ogura	+	+	-	-	+/-		-
B. oleracea	.*Nigra*	+	+/-		-	-		+
	.*Radish* ogura	++	++		-	+/-		-
	.*Radish* cybrids	++	++		+	+		+
	.*Napus*	-	-		+	+		+
	.*Diplotaxis*	-	-					+

napus. Restorer lines with a good female fertility were recently selected (Delourme et al., 1991). Male sterile cybrid plants of rapeseed were selected according to leaf color, nectar production, female fertility and yield. Selected cybrid plants presented cp DNA from rapeseed and a recombined mt DNA from rapeseed and radish (Vedel et al., 1986). Cybrids with a mt DNA close to *B. napus* are higher yielding and need only one restorer gene to get complete restoration. Recently, a 2.5 kb mt DNA fragment correlated to CMS was isolated from cybrids obtained by backfusion (Bonhomme et al., 1991).

At present the Ogu-INRA system appears to be the most promising C.M.S. system in *B. napus* in the short term. This system is now complete and workable since good female lines (A-lines, cybrids), maintainer lines (B- lines) and restorer lines (R-lines ; a single restorer gene linked to an isozyme marker) have been developed. S cybrid cytoplasms will be transferred to *B. napus* germplasm by backcrossing or *in situ* androgenesis. Male sterile cybrid cabbages without chlorosis, with high nectar production and very good female fertility were obtained by using the same strategy (Pelletier et al., 1989). In the case of *B. oleracea* no restoration of fertility is necessary as the plant is consumed at the vegetative stage. Cybrids of *B. napus* with combined CMS and triazin tolerance have been regenerated (Pelletier et al., 1983 ; Chuong et al., 1988).

2 HYBRID SEED PRODUCTION DESIGNS

Two important parameters in hybrid seed production are parental ratios and maximum female row to male row distances. In *B. napus*, Pinnisch and Mc Vetty (1990) established that a 10:1 A-line to R-line ratio (30 rows : 3 rows) is too high to produce economic hybrid seed yield on the polima C.M.S. due to the instability of this male sterility. There is a linear relationship between hybrid seed yield and percent hybridity *versus* distance from the pollen parent. Renard and Mesquida (1987a) proposed an alternated design of male and female strips (3 meter width). Using the non improved Ogura system with low nectar secretions, female plants are less attractive for honeybees. The honeybee population is decreasing as well as the yield when the distance from the pollen parent is increasing (table 3). The hybrid seed production is considerably improved by using ms cybrid plants with well-developed nectaries. Honeybee population is more homogeneous all over the field. The expected F1 hybrid seed yield on cybrids is about 80 % of the yield of the male parent. In french climatic conditions, the proposed ratio is 14:2 (30 m : 6 m). F1 hybrid seeds will be multiplied on about 6OOO ha for a 2 million hectare rapeseed production in

Table 2. Effect of temperature on anther type in sterile and partially sterile lines of the pol CMS and the nap CMS in *Brassica napus (Fan and Stefansson, 1986)*

	Temperature regimes (day/night °C)		
	22/16	26/20	30/24
CMS line			
pol. sterile	sterile	sterile	partially sterile
pol. partially sterile	partially sterile	partially sterile	fertile
nap. sterile	sterile	partially sterile	fertile
nap. partially sterile	partially sterile	fertile	fertile

Table 3. Effect of the distance from the pollen parent on the yield components of male sterile plants using Ogura and cybrid CMS systems in *B. napus.*

CMS line	Distance from the pollen parent (m)									
Ogura (tall male line)	0 male	3	6	9	12	15	18	21	24	27
Pod set	97,5	78,5	69,7	60,3	60,4	58,7	56,4	48,9	49,5	48,6
Seeds/pod	10,5	11,4	11,8	12,1	10,9	12,0	10,2	10,4	9,7	10,8
Yield (qx/ha)	27,0	25,0	19,6	17,7	19,0	15,9	18,8	17,0	19,3	16,4
Cybrid 27 (tall male line)										
Pod set	100,0	95,0	91,2	95,2	84,2	94,5	95,7	88.0	91,0	83,0
Seeds/pod	20,0	17,5	14,5	18,0	13,0	16,0	13,5	14,5	12,0	16,0
Yield (qx/ha)	35,0	35,5	36,2	36,8	37,3	38,0	38,7	38,7	35,2	32,0
Cybrid 58 (dwarf male line)										
Pod set	82,5	89,1	88,7	88,4	86,5	88,1	86,6			
Seeds/pod	20,6	21,3	18,9	20,8	20,1	19,7	19,0			
Yield (qx/ha)	20,3	32,2	26,6	29,7	28,7	28,9	29,6			

the E.E.C. The strip design was chosen because of the risk of plant lodging. Such a risk will be limited by using dwarf lines. Semi dwarf F1 hybrids could be produced by crossing a ms tall line with a male fertile dwarf line. Cross pollination is not disturbed by the height difference of the 2 lines (table 3).

While wind can be an effective pollen vector on short distances (Mesquida and Renard, 1982), the sticky entomophilous pollen of *Brassica* plants is better adapted to transfer by insects. Experimental F1 hybrid seeds can be produced in small cages using flies, bumble bees or honeybees as pollen vectors from the male line to the female line. Commercial F1 hybrid varieties of some cole crops (cauliflower) may be produced under plastic tunnel with the same insect pollinators. In these conditions, the proposed ratio is 2 to 3 rows of female and 1 row of male (2 plants/m^2 ; 0,80 m between rows). In France, about 2000 female plants are sufficient to provide seeds for the whole french cabbage production (1000 ha). In producing F1 hybrid seeds, the main restraints are synchronization of flowering, insect pollinator density (fields will be stocked with honeybee hives in rapeseed fields) and isolation distances to guarantee high hybridity (more than 95 %). The advised distances will be longer than for pure line seed multiplication because of the male sterility of the female line. The risk of outcrossing with fields of related species, volunteers and weedy species will be increased. Marker genes will be useful to establish isolation distances.

3 HYBRID VARIETY TYPES

The use of the different male sterility systems may lead to the creation of various varietal types such as male sterile or restored F1 hybrid, 3-way hybrid, double hybrid or n components synthetic varieties. In cole crops, F1 hybrids are produced between clones maintained by *in vivo* or *in vitro* cutting, or lines. F1 hybrid cauliflowers produced with NMS were recently registered in France. F1 hybrid cabbages on the improved ogura system are expected to be registrered in 1994. In oilseed rape (*B. napus*) parental lines are pure lines selected by pedigree breeding or haploidy, or narrow populations. The expected yield increase, with the use of rapeseed F1 hybrids, is as high as 30 % (Figure 2). The first restored F1 hybrids on the Ogura system will be in official tests in 1994.

Alloplasmic lines were selected by backcrosses with spring or winter rapeseed lines on our cybrid cytoplasms. These male sterile alloplasmic lines were tested for seed yield.

Step 1 Maintenance of the A. line

```
                 A                      A
          MS MS Rf rf (F)        MS MS rf rf (S)
          MS MS rf rf (S)   X    MS MS Rf rf (F)
```

Step 2 Multiplication of the A. line

```
                          A                    B
          50 %     MS MS Rf rf (F)
          50 %     MS MS rf rf (S)  X  ms ms rf rf (F)
```

Step 3 Production of F1 hybrid seeds

```
                                 A                    R
               100 %  Ms  ms rf rf (S)  X  ms ms Rf Rf (F)

                                   restored F1 hybrid
                                    Ms ms Rf rf (F)
                                    ms ms Rf rf (F)
```

Figure 1. Seed multiplication scheme using a digenic system

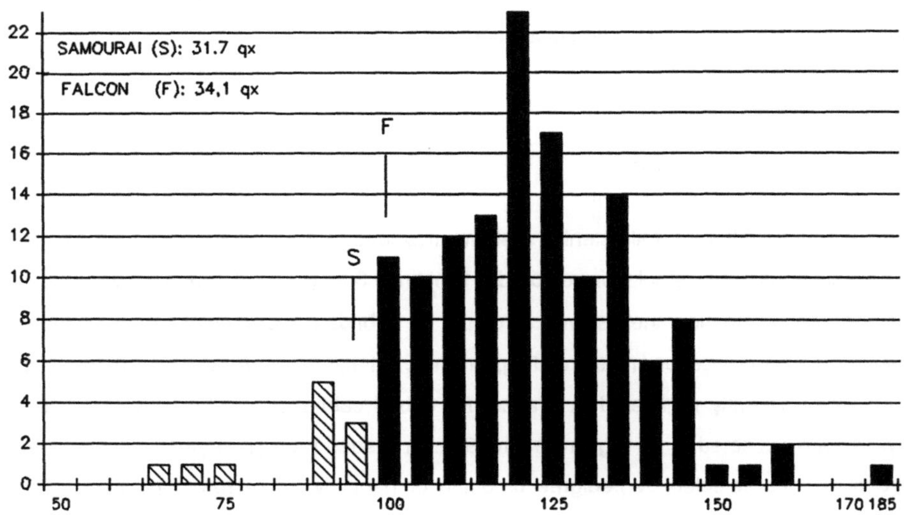

Figure 2. Distribution of the yield of male sterile rapeseed F1 hybrids (% Falcon + Samouraï) produced on a male sterile-cybrid cytoplasm (Rennes, France)

Male sterile lines on cybrids 27, 58 and 85 are generally higher yielding than the male fertile recurrent line (table 4). The yield increase could be explained by the hybridity of the cytoplasm or by the male sterility of the plants. These hypotheses will be checked from field testing of isogenic male sterile and male fertile restored lines/hybrids on a cybrid cytoplasm. To exploit this yield increase, Renard and Mesquida (1987b) imagined to grow mixed varieties obtained by blending (x) % of male sterile seeds and (1-x) % of male fertile ones. In Britanny pollination is efficient with only 5 % of pollen donor plants (table 5). A mixed Samouraï tested in different locations in 1991 (table 6) was applied in french official tests. Such a mixed stand could also simplify F1 hybrid seed multiplication using an herbicide resistance as a marker.

Many male sterility systems are proposed for the production of hybrid *Brassica* seeds. Most of these systems have been shown to possess some inherent disadvantages when used in breeding programmes. NMS is now efficiently applied in cauliflower breeding. But the improved Ogu-INRA system seems to be the most promising one in a short term in *B. napus* and *B. oleracea*. This complete CMS system was created by developing conventional breeding methods such as interspecific crosses and new technologies of cell biology. Studies are developed to characterize the products of mt DNA sequences involved in C.M.S. and of the restorer gene, and to transfer this Ogu-INRA system into other *Brassica* species (*B. juncea* and *B. campestris*). In conclusion C.M.S. represents a true revolution in *Brassica* breeding :

- in *B. napus*, it will lead to F1 hybrid varieties with higher yield,

- in *B. oleracea* (cabbage, cauliflower, kale), it brings security and allows to go towards selffertile lines as parents which means simplification and increased efficiency of breeding.

Table 4. Yield of male fertile (F) and male sterile (S) alloplasmic lines (Rennes, France)

Nucleus	Cytoplasm	Fertility	Yield
Brutor (Spring rape)	*B. napus*	F	100 (22 qx/ha)
	27	S	106
	58	S	115
	85	S	104
	118	S	101
Darmor (Winter rape)	*B. napus*	F	100 (35 qx/ha)
	27	S	118
	58	S	120
	77	S	96
	85	S	114
	118	S	103

Table 5. Characteristics of male sterile cybrid plants (Fu 85) compared to male fertile plants (%) in a mixed winter rapeseed line (Darmor) with 5 to 30 % of pollen donor plants (Rennes, France)

% of male fertile plants	pod set	seeds per pod	seed weight per plant
5 %	111	86	200
10 %	104	100	247
20 %	95	100	288
30 %	109	107	223

Table 6. Yield components of male sterile cybrid plants in a mixed rapeseed line with 10 % of pollen donor plants in different locations (% of the male fertile plants).

	Locations					
	Rennes	Prémesques	Montbartier	Dijon	Lusignan	Mean
Pod set	96,1	80,7	98,1	53,9	64,0	78,5
Seeds/pod	121,5	86,4	108,4	93,1	106,3	103,1
Seed weight/plant	237,2	161,7	191,9	214,9	237,2	208,6
Yield (qx/ha)	39,1	38,5	31,2	28,8	34,0	34,3

REFERENCES

Banga SS, Labana KS (1984) Ethrel induced male sterility in Indian mustard (Brassica juncea L. Coss). Z. Pflanzenzucht 92 (3) : 229-233.

Bannerot H, Boulidard L, Cauderon Y, Tempe J (1974) Cytoplasmic male sterility transfer from Raphanus to Brassica. Proc. Eucarpia Crop Sect. Cruciferae, 25 : 52-54.

Bonhomme S, Budar F, Ferault M , Pelletier G (1991) A 2.5 Kb NcoI fragment of Ogura radish mitochondrial DNA is correlated with cytoplasmic male-sterility in Brassica cybrids. Current Genet. 19 (2) : 121-127.

Chiang MS, Crete R (1987) Cytoplasmic male sterility in Brassica oleracea induced by B. napus cytoplasm - female fertility and restoration of male fertility. Can. J. Plant. Sci. 67 (3) : 891-897.

Chuong PV, Beversdorf WD, Powell AD, Pauls KP (1988) Somatic transfer of cytoplasmic traits in Brassica napus L. by haploid protoplast fusion. Mol. Gen. Genet. 211 : 197-201.

Delourme R, Eber F, Renard M (1991) Radish cytoplasmic male sterility in rapeseed : breeding restorer lines with a good female fertility. 8th International Rapeseed Congress : 1506-1510.

Dunemann F, Grunewaldt J (1991) Identification of a monogenic dominant male sterility mutant in Broccoli (Brassica oleracea var italica) Plant Breed. 106 (2) : 161-163.

Fan Z, Stefansson BR (1986) Influence of temperature on sterility of two cytoplasmic male sterility systems in rape (Brassica napus L.) Can. J. Plant. Sci. 66 (2) : 221-227.

Fang GH, Mc Vetty PBE (1989) Inheritance of male sterility restoration and allelism of restorer genes for the polima cytoplasmic male sterility system in oilseed rape. Genome 32 (6) 1044-1047.

Guan CY, Wang GH (1987) Studies in male sterility of rapeseed induced by chemical male gametocide. 7th International Rapeseed Congress, Poznan 243-251.

Heyn FW (1978) Introgression of restorer genes from Raphanus sativus into cytoplasmic male sterile B. napus and the genetic of fertility restoration. Proc. of the 5th Int. Rapeseed Conf. Malmö, 82-83.

Hinata K, Konno N (1979) Studies on a male sterile strain having the Brassica campestris nucleus and the Diplotaxis muralis cytoplasm. Japan. J. Breed. 29 (4) : 305-311.

Kaul MLH (1988) Male sterility in higher plants. Springer. Berlin, Heidelberg, New-York, 1005 pp.

Lefort-Buson M, Guillot-Lemoine B, Dattée Y (1987) Heterosis and genetic distance in rapeseed (Brassica napus L.) : crosses between european and asiatic selfed lines. Genome 29 : 413-418.

Li S, Qian Y, Wu Z, Stefansson BR (1988) Genetic male sterility in rape (Brassica napus L.) conditioned by interaction of genes at two loci. Can. J. Plant. Sci. 68 (4) : 1115-1118.

Mariani C, de Beuckeleer M, Truttner J, Leemans J, Goldberg RB (1990) Induction of male sterility in plants by a chimaeric ribonuclease gene. Nature 347 (6295) : 737-741.

Mariani C, De Greef W, De Block M, De Beuckeleer M, Gossele V, Leemans J (1991) Genetic analysis of engineered male sterility in oilseed rape. 8th International Rapeseed Congress Saskatoon : 352-357.

Mathias R (1985) Transfer of cytoplasmic male sterility from brown mustard (Brassica juncea L. Coss) into rapeseed (Brassica napus L.). Z. Pflanzenzucht. 95 (4) 371-374.

Mc Vetty PBE, Edie SA, Scarth R (1990) Comparaison of the effect of nap and pol cytoplasms on the performance of intercultivar summer oilseed rape hybrids. Can. J. Plant. Sci. 70 (1) 117-126.

Mesquida J, Renard M (1982) Etude de la dispersion du pollen par le vent et de l'importance de la pollinisation anémophile chez le colza (Brassica napus L.) Apidologie 13 : 353-367.

Ogura H (1968) Studies of the new male sterility in Japanese radish, with special reference to the utilization of this sterility towards the pratical raising of hybrid seeds. Mem. Fac. Agric. Kagoshima Univ. 6 : 39-78.

Okhawa Y (1984) Cytoplasmic male sterility in Brassica campestris ssp rapifera Jap. J. Breed. 34 (3) 285-294.

Pearson OH (1972) Cytoplasmically inherited male sterility characters and flavor components from the species cross B. nigra x B. oleracea. J. Amer. Soc. Hort. Sci. 97 (3) : 397-402.

Pellan-Delourme R, Renard M (1987) Identification of maintener genes in Brassica napus L. for the male sterility inducing cytoplasm of Diplotaxis muralis Plant. Breed. 99 : 89-97.

Pellan-Delourme R, Renard M (1988) Cytoplasmic male sterility in rapeseed (Brasica napus L) : female fertility of restored rapeseed with "Ogura" and cybrid cytoplasms. Genome. 30 : 234-238.

Pelletier G, Primard C, Vedel F, Chetrit P, Remy R, Rousselle P, Renard M (1983) Intergeneric cytoplasmic hybridization in cruciferae by protoplast fusion Mol. Gen. Genet. 191 : 244-250.

Pelletier F, Ferault M, Lancelin D, Boulidard L (1989) C.M.S. Brassica oleracea cybrids and their potential for hybrid seed production XII Eucarpia Congress, Göttingen, 11-7, Vorträge für Pflanzenzuchtg, 15.

Pinnisch R, Mc Vetty PBE (1990) Seed production of hybrid summer rape in the field using the pol cytoplasmic male sterility system : a first attempt. Can. J. Plant. Sci. 70 (3) 611-618.

Prakash S Chopra VL (1990) Male sterility caused by cytoplasm of Brassica oxyrrhina in B. campestris and B. juncea. Theor. Appl. Genet. 79 (2) 285-287.

Rawat DS, Anand IJ (1979) Male sterility in Indian mustard. Indian J. Genet. Plant Breed. 39 : 412-415.

Renard M, Mesquida J (1987a)Significance of nectar secretion for honeybee foraging (Apis mellifica) and consequences of pollination on oilseed rape (Brassica napus) seed production. 7th International Rapeseed Congress, Poznan : 215-221.

Renard M, Mesquida J (1987b) A preliminary study of insect pollination of male sterile oilseed rape (Brassica napus) in mixed cultures with variable proportions of pollinising plants. 7th International Rapeseed Congress, Poznan : 222-227.

Sampson DR (1966) Linkage of genetic male sterility with a seedling marker and its use in producing F1 hybrid seed of Brassica oleracea Can. J. Plant. Sci. 46 : 703.

Shiga T (1976) Studies on heterosis breeding using cytoplasmic male sterility in rapeseed (Brassica napus L.) Bull. Nat. Inst. Agric. Sci. 27 : 1-101.

Thompson KF (1972) Cytoplasmic male sterility in oilseed rape. Heredity 29 (2) : 253-257.

Van der Meer GP (1987) Chromosomal monogenic dominant male sterility in Chinese cabbage (Brassica campestris ssp pekinensis) Euphytica 36 : 927-931.

Van der Meer GP, Vandam R (1979) Giberellic acid as a gametocide for cole crops. Euphytica 28 : 717-722.

Vedel F, Chetrit P, Mathieu C, Pelletier G, Primard C (1986) Several different mitochondrial DNA regions are involved in intergenomic recombination in *Brassica napus* cybrid plants. Curr. Genet. II : 17-24.

Mechanisms of Microspore Embryogenesis

J M Dunwell
ICI Seeds
Jealott's Hill Research Station
Bracknell
Berkshire RG12 6EY
United Kingdom

INTRODUCTION

As Haldane stated in 1937 "Every species is at the mercy of its pollen grains", and although this motto is undoubtedly true in evolutionary terms, it is only relatively recently that we have been able to disrupt at will this critical phase of plant development.

This review will not deal with any of the applications of the technique (Dunwell, 1985a) whereby the typical development of a two or three celled pollen grain (gametophyte) is inhibited and instead the immature pollen (microspore) is induced (allowed?) to develop into an embryo or mass of cells capable of producing a plant (Dunwell, 1985b). Instead the review will draw parallels with some of the related processes occurring elsewhere in the plant kingdom and will focus on how molecular methods have been used recently to explore some of the underlying aspects of this fascinating developmental switch.

FACULTATIVE APOGAMY

Within the lower plant kingdom there is a phenomenon directly analogous to the induction of sporophytes from microspores; it is the apogamous production of sporophytes from gametophytes of ferns and mosses. This parallel has been drawn previously (Dunwell 1978) but it is worth devoting some more attention to the similarities of these two distantly related processes. It is well over a century ago that Farlow (1874) described the outgrowth of a shoot from a prothallus of *Pteris cretica*. Such facultative apogamy has been described subsequently in several other fern species (see review by Sheffield and Bell, 1987) and it is from some of these more recent experimental studies that the most intriguing evidence has derived. Probably the most extensively studied system is in *Pteridium aquilinum* gametophytes where Whittier and Steeves (1960) found that high concentrations of sucrose, glucose, fructose and maltose (cf Roberts-Oehlschlager *et al.*, 1990) led to an increased frequency of apogamous outgrowths. A related study (Elmore and Whittier, 1975a) showed a direct involvement of ethylene in the induction of this process but that ethylene was only active if supplied with sucrose. These same authors (Elmore and Whittier, 1975b) also demonstrated that nine gametophytic strains of this species differed in their ability to produce ethylene. However, the number of apogamous buds produced by each strain was not directly related to this ability, but rather appeared to be dependent of the ability of the gametophytes to respond to ethylene (cf Cho and Kasha, 1989).

In a contemporary study, Hirsch (1975) showed a definite nutritional rôle of sucrose in determining the balance of gametophytic to sporophytic development in *Microgramma vaccinifolia*. She concluded that increased carbohydrate supply (4% sucrose) led to increased production of sporophytes and that starvation conditions favoured gametophyte development. This concept of starvation will be considered in more detail below in relation to theories of microspore embryo development.

The most recent of these studies is that on *Equisetum arvense* by Kuriyama *et al.* (1990a) who showed that, whereas sucrose had no effect on the induction of sporophytes, large numbers of shoots could be induced by culture on medium supplemented with cytokinin, particularly benzyl amino purine. This response was also shown when gametophyte derived protoplasts were cultured (Kuriyama *et al.*, 1990b).

In summary, there are a number of studies on lower plants in which very simple stimuli have been shown to trigger sporophytic development. To date, none of these systems have been subject to analysis by molecular methods but there is no doubt that such systems would be eminently suitable for such investigations.

POLLEN DIMORPHISM, STARVATION, AND THE INDUCTION OF POLLEN EMBRYOGENESIS

It has been recognised for several decades that anthers of many, if not most, species contain, as well as the major population of starch filled mature pollen, a low frequency of small, often starch-free grains. This so-called dimorphism is found in its most extreme form in species such as *Tripogandra amplexicaulis* (Mattson, 1982) where one type of anther contains only small grains not capable of participating in fertilisation. Nutritional starvation due to breakdown of the anther filament has been implicated in this phenomenon. In some such cases the nuclei of the small grains were observed to be undergoing or to have undergone supernumerary mitosis(es). The relevance of these observations to the behaviour of immature pollen grains in cultured anthers grew out the initial finding that the protein and RNA contents of embryogenic tobacco microspores was typically at a low level one sixth that of non-embryogenic grains (Bhojwani *et al.*, 1973). In other words, the normal synthetic processes of the pollen had been inhibited in those grains destined for sporophytic development. For a long period thereafter, it was not clear whether such grains pre-existed in anthers or were induced by the culture process. Various aspects of this debate have been rehearsed over recent years (Sunderland and Huang, 1987; Roberts-Oehlschlager and Dunwell, 1991) with perhaps the most detailed recent description of the arguments being that provided by Heberle-Bors and Benito Moreno (1992). These authors are amongst the small number who have made significant progress in the analysis of embryogenesis from cultured microspores and their findings along with those of others will be described in some detail below.

There are only two experimental systems which to date have proved amenable to the combination of cytological ultrastructural, biochemical and molecular analyses required for a thorough description of the induction process. The two systems are first tobacco (*Nicotiana tabacum*, *N. rustica*) and secondly oil seed rape (*Brassica napus*). Methods have now been developed which permit the isolation and culture of relatively synchronous populations of microspores from these species and their detailed observation under of a range of cultural conditions.

The tobacco system is probably the most developed and will be considered first. It was Kyo and Harada (1985, 1986) who first developed methods which allowed biochemical investigations of embryogenesis. When highly homogeneous cultures of pollen at the mid-bicellular stage were cultured in basal medium with glutamine, most grains developed normally into mature pollen, whereas if first cultured without glutamine they underwent cytoplasmic degradation (cf Bhojwani *et al.*, 1973) and if then transferred onto medium with sucrose and glutamine, cell division occurred. In a search for biochemical markers of induction events during the "glutamine starvation" period, these authors found a series of specific phosphoproteins (Kyo and Harada, 1990a,b) which to date have not been accurately localised (Kyo and Ohkawa, 1991).

In a related study, two other pollen specific genes are being investigated as potentially useful markers of changes occurring during the induction of pollen embryogenesis. These two markers are starch synthase (Olmedilla *et al.*, 1991) and the as yet unidentified cDNA clone NTP c303 (Weterings *et al.*, 1992). The value of these two particular markers has not been proven to date.

A different, though similar approach is represented by the work of Yermakov *et al*,, (1990) who have investigated the involvement of Con A binding glycoconjugates in pollen development *in vitro*. First, an aqueous polymer two-phase partition method with PEG 4000 and Dextran 5000 was used to separate microspores and pollen grains of *Petunia hybrida*. The fraction of grains with high Con A was shown to be most embryogenic in culture. More recently, Matveyeva (pers. comm.) isolated a more purified population of tobacco pollen by utilising affinity fractionation with immobilized Con A. The data indicated that the induction of embryogenesis by cold (one of the pretreatments known to stimulate the process) produced significant changes in the surface glycoconjugates of some grains and that these changes were associated with the induction process.

Similar methods have been used by other groups who have carried out even more detailed analyses of gene expression during this induction period in tobacco. In the method developed by Benito Moreno *et al.* (1988) sucrose is the main determining component of the culture medium with induction occurring in conditions of sucrose starvation. Using this technique, Vicente *et al.* (1992) studied patterns both of proteins and RNA during the starvation period. Although they found no evidence for the synthesis of new starvation-induced proteins they did find two specific mRNAs so induced. It was noted that the storage of preformed mRNAs in particles not accessible to polysomes is a common feature of developmental systems arrested in a quiescent phase, before active cell division commences.

In an extension of this study, the authors have begun an investigation of the reactivation of the cell cycle in the vegetative cell. During normal pollen development the nucleus of this cell remains at, or close to, the 1C level of DNA - the GI phase, whereas during starvation, DNA synthesis is induced in the vegetative nucleus of a proportion of the pollen. It was concluded that the starvation induces this cell to pass through a control point in the cell cycle that may be analogous to control points in late GI in yeast known as START. It is to be hoped that the function of genes similar to cdc 10^+ from yeast (Lowndes *et al.*, 1992) can be analysed in this material.

The second of the two experimental systems used in these studies is *B. napus*. Cytological (Pechan and Keller, 1988) and flow sorting techniques (Pechan *et al.*, 1988) have been used to identify the subpopulation of microspores capable of undergoing embryogenesis. The critical period of pollen development during which induction can occur appears to be more restricted than in tobacco in being less than 8h in duration. Observations indicated that, as in tobacco, only the vegetative nucleus contributed to embryo formation, although this is not universally accepted (Hamaoka *et al.*, 1991).

More recently, Pechan *et al.* (1991) have analysed the pattern of mRNA and protein production during the inductive 8h period. It was found that, in fact, the transcription of some genes encoding high molecular weight proteins were initiated during the first 4h of culture at 32°C - the temperature required to stimulate high frequency induction. Based on two-dimensional gel analysis a number of 18-, 19-, 20- and 68-kDa proteins were identified as being unique to the "commitment phase" of microspore embryogenesis. Preliminary results have indicated that some of these proteins may indeed be heat shock proteins of the type supposed to have important functions in the regulation of cell proliferation (Pechan, 1991).

There seems little doubt that before long similar molecular studies will be reported from other species, including the cereals such as barley (Roberts-Ochlschlager and Dunwell, 1990) and maize in which significant improvements in culture technique have been made recently.

SUMMARY

Although the potential benefit of techniques for the culture of isolated microspores as a means of investigating biochemical aspects of pollen embryogenesis was obvious many years ago (Dunwell, 1976) it is only in the last few years that this potential has been realised. The studies reported in this review, particularly those on the "model" species tobacco and oil seed rape, have clearly demonstrated that the power of modern molecular methods can be brought to bear on the analysis of the gametophyte to sporophyte transition. Indeed, the whole area of reproductive biology is undergoing a startling revival (Scott *et al.*, 1991).

Although most of the detail of the mechanisms underlying the repression of normal pollen development and the induction of the "abnormal" process of embryogenesis remain to be discovered, there are already some hints about the outline of the process. The most popular model at present involves an environmental trigger leading to a "starvation" phase during which the microspores are deprived of some compound(s) required to maintain the growth of grains during the normal period of rapid biosynthesis. The evidence from studies of apogamy in Pteridophytes reported earlier is interesting in this context, since the conclusion seems to be that enrichment of medium, rather than starvation, promotes the switch towards a sporophytic mode of development. Despite this paradox, the stimulus itself, be it cold, heat, sucrose, ethylene or cytokinin, is simple and powerful. It could be argued that one would expect the control of this alternation of generations to become more rigorous during evolution of higher plants but this seems not to have been the case, at least not universally. Production of microspore - derived embryos has been described in several plant families known to have very different advancement indices (Sporne, 1974), and although it is possible that these examples only represent isolated remnants of an ability widespread amongst lower forms of plants, I consider this view to be unnecessarily pessimistic.

ACKNOWLEDGEMENT

The author is most grateful to all those who provided him with details of their work, much unpublished.

REFERENCES

Benito Moreno RM, Macke F, Hauser M-T, Alwen A, Heberle-Bors E (1988) Sporophytes and male gametophytes from *in vitro* cultured immature tobacco pollen. In: Cresti M, Gori P, Pacini E (eds) Sexual Reproduction in Higher Plants. Springer, Berlin, pp 137-142

Bhojwani SS, Dunwell JM, Sunderland N (1973) Nucleic-acid and protein contents of embryogenic tobacco pollen. J Exp Bot 24: 863-871

Cho U-H, Kasha KJ (1989) Ethylene production and embryogenesis from anther cultures of barley (*Hordeum vulgare*). Plant Cell Rep 8: 414-417

Dunwell JM (1976) A comparative study of environmental and developmental factors which influence embryo induction and growth in cultured anthers of *Nicotiana tabacum*. Env Exp Bot 16: 109-118

Dunwell JM (1978) Division and differentiation in cultured pollen. In: Thorpe TA (ed) Frontiers of Plant Tissue Culture 1978. Calgary University Press, Calgary, pp 103-112

Dunwell JM (1985a) Pollen, ovule and embryo culture as tools in plant breeding. In: Withers LA, Alderson, PG (eds) Plant Tissue Culture and its Agricultural Applications. Butterworths, London, pp 375-404

Dunwell JM (1985b) Embryogenesis from pollen *in vitro*. In: Zaitlin M, Day P, Hollaender A (eds) Biotechnology in Plant Science. Academic Press, Orlando, pp 49-76

Elmore HW, Whittier DP (1975a) The involvement of ethylene and sucrose in the inductive and developmental phases of apogamous bud formation in *Pteridium* gametophytes. Can J Bot 53: 375-381

Elmore HW, Whittier DP (1975b) Ethylene production and ethylene-induced apogamous bud formation in nine gametophytic strains of *Pteridium aquilinum* (L.) Kuhn. Ann Bot 39: 965-971

Farlow WG (1874) An asexual growth from the prothallus of *Pteris cretica*. Quart J Micros Sci 14: 266-272

Hamaoka Y, Fujita Y, Iwai S (1991) Effects of temperature on the mode of pollen development in anther culture of *Brassica campestris*. Phys Plant 82: 67-72

Heberle-Bors E, Benito Moreno RM (1992) Impact of pollen culture and pollen dimorphism on pollen plant production *in vitro*. In: Gupta SC, Babbar SB (eds) In vitro Androgenesis in Seed Plants. Kluwer, Dordrecht (in press)

Hirsch AM (1975) The effect of sucrose on the differentiation of excised fern leaf tissue into either gametotyphtes or sporophytes. Plant Phys 56: 390-393

Kuriyama A, Sugawara Y, Matsushima H, Takeuchi M (1990a) Production of sporophytic structures from gametophytes by cytokinin in *Equisetum arvense*. Naturwissenschaften 77: 31-32

Kuriyama A, Takeuchi M, Veno S, Mitsuda H (1990b) Enhancement of the division of *Equisetum arvense* protoplasts in culture by activated charcoal and their further development. Plant Cell Phys 31: 999-1004

Kyo M, Harada H (1985) Studies on condition for cell division and embryogenesis in isolated pollen culture of *Nicotiana rustica*. Plant Phys 79: 90-94

Kyo M, Harada H (1986) Control of the developmental pathway of tobacco pollen *in vitro*. Planta 168: 427-432

Kyo M, Harada H (1990a) Specific phosphoproteins in the initial period of tobacco pollen embryogenesis. Planta 182: 58-63

Kyo M, Harada H (1990b) Phosphorylation of proteins associated with embryogenic dedifferentiation of immature pollen grains of *Nicotiana rustica*. J Plant Phys 136: 716-722

Kyo M, Ohkawa T (1991) Investigations of subcellular localisation of several phosphoproteins in embryogenic pollen grains of tobacco. J Plant Phys 137: 525-529

Lowndes NF, McInerny CJ, Johnson AL, Fantes PA, Johnston LH (1992) Control of DNA synthesis genes in fission yeast by the cell-cycle gene $\underline{cdc\ 10^{+}}$. Nature 355: 449-453

Mattson O (1982) The morphogenesis of dimorphic pollen and anthers in *Tripogandra amplexicaulis*. Op Bot 66: 1-46

Olmedilla A, Schrauwen JAM, Wullens GJ (1991) Visualisation of starch-synthase expression by in situ hybridization during pollen development. Planta 184: 182-186

Pechan PM (1991) Heat shock proteins and cell proliferation. FEBS Lett 280: 1-4

Pechan PM, Keller WA (1988) Identification of potentially embryogenic microspores in *Brassica napus*. Phys Plant 74: 377-384

Pechan PM, Keller WA, Mandy F, Bergeron M (1988) Selection of

Brassica napus L embryogenic microspores by flow sorting. Plant Cell Rep 7: 396-398
Pechan PM, Bartels D, Brown DCW, Schol J (1991) Messenger - RNA and protein changes associates with induction of *Brassica* microspore embryogenesis. Planta 184: 161-165
Roberts-Oehlschlager SL, Dunwell JM (1990) Barley anther culture: pretreatment on mannitol stimulates production of microspore-derived embryos. Plant Cell Tiss Org Cult 20: 235-240
Roberts-Oehlschlager SL, Dunwell JM (1991) Barley anther culture: The effect of position on pollen development *in vivo* and *in vitro*. Plant Cell Rep 9: 631-634
Roberts-Oehlschlager SL, Dunwell JM, Faulks R (1990) Changes in the sugar content of barley anthers during culture on different carbohydrates. Plant Cell Tiss Org Cult 22: 77-85
Scott R, Hodge R, Paul W, Draper J (1991) The molecular biology of anther differentiation. Plant Sci 80: 167-191
Sheffield E, Bell PR (1987) Current studies of the Pteridophyte life cycle. Bot Rev 53: 442-490
Sporne KR (1974) The Morphology of Angiosperms, Hutchinson, London
Sunderland N, Huang B (1987) Ultrastructural aspects of pollen dimorphism. Int Rev Cytol 107: 175-220
Vicente O, Garrido D, Zarsky V, Eller N, Rihova L, Berenyi M, Tupy J, Heberle-Bors E (1992) Induction of embryogenesis in isolated pollen cultures of tobacco. In: Mulcahy DC, Ottaviano E (eds) Angiosperm Pollen and Ovules, Springer-Verlag, Berlin (in press)
Weterings K, Reijen WH, Schrawen JAM, van Herpen MMA, de Groot PFM, Oldenhof M, Wullems GJ (1992) In: Mulcahy DC, Ottaviano E (eds) Angiosperm Pollen and Ovules, Springer-Verlag, Berlin (in press)
Whittier DP, Steeves TA (1960) The induction of apogamy in the bracken fern. Can J Bot 38: 925-930
Yermakov IP, Matveyeva NP, Dubinina NP (1990) Concanavalin A binding by *Petunia* microspores. Vest Mosc Univ Ser 16, no 1: 22-26

The Use of Androgenesis in Maize Breeding

Joseph F. Petolino
DowElanco
P.O. Box 4011
Champaign, IL 61824
USA

I. Haploid Breeding of Maize

The inbred-hybrid concept, on which commercial maize breeding is based, requires the ongoing production of true-breeding or homozygous lines to evaluate as potential parents of new, productive hybrids. Typically, the production of such lines requires 5-7 generations of inbreeding to achieve the desired level of homozygosity. The main advantage of haploids is that they can be used to rapidly produce true-breeding lines. Since haploids, by definition, have only a single set of chromosomes, the doubling of the chromosome compliment of these individuals results in instantaneous line fixation. This minimizes the time associated with inbreeding while, at the same time, avoiding the sampling problems encountered if early generation testing is used.

Product development in maize is a time consuming and costly enterprise. A significant portion of the time is associated with manual self-pollination in order to fix the genetic variability released from the original cross. The use of winter nurseries is also not an uncommon cost of doing business in this highly competitive industry. Haploid breeding would certainly speed up the fixation phase of the process. However, the rapid advance to homozygosity which accompanies the doubling of haploids effectively circumvents the need for intra-progeny sampling in early generations. This can be especially important in breeding programs which regularly utilize early generation testing which lines are tested before they are completely fixed.

Intra-progeny sampling is a common concern in commercial maize breeding, since it significantly increases the number of pollinations and field evaluations that have to be made each year. For every one of the 20 or more crosses initiated each year, a breeder will typically generate 100 or more S_1 families by self-pollinating individual F_2 plants. He is then forced to make selections both among and within these families which are still on average 25% heterozygous.

At this point there is as much additive genetic variance within, as there is among, the families. Since there are practical constraints on any program, one has to limit the number of field evaluations that are carried out. The question that must constantly be addressed is 'How many within family progenies need to be evaluated in order to cover the residual variability?' Of course, more within family evaluations translate into less total families that can be looked at. Using haploid derivatives, on the other hand, one can generate completely homozygous lines directly from F_1 or F_2 plants, thus, fixing the combining ability *within* a given family thereby partitioning all of the additive genetic variance *among* the families. This allows a breeder to focus on fixed variability.

Haploid breeding in maize not only allows one to have the inbred progeny from a given cross in the field the year after a cross is made but, more importantly, eliminates the need to sample within families and, over time, evaluate more crosses more efficiently. Unfortunately, nature gives us maize haploids in exceedingly low frequency. There is a need for a method which results in the routine, low cost production of large numbers of doubled haploids before this approach gains widespread acceptance as a viable breeding strategy.

II. Maize Anther Culture

Anther culture is a means by which, theoretically, large numbers of haploid plants can be produced. Unfortunately, although rapid in some species, progress in the anther culture of maize has been slow. Although reports of pollen plant production in maize date back to 1975, only the Chinese claim to routinely use this technique for line development (Wu, 1986). The major problems have been the extremely low anther culture response frequencies as well as the difficulties associated with plant regeneration and chromosome doubling. Moreover, those few reports of successful maize anther culture outside of China involved germplasm which was not of commercial importance.

The technique known as anther culture involves isolating immature anthers and placing them only a medium which induces the cells within the anther, which would normally be destined to become pollen grains, to begin dividing and form cell cultures from which plants can be regenerated. What is so special about these cultures and the plants that are derived from them is that they are haploid and contain only a single set of chromosomes from the original plant. These individuals are sterile unless chromosome doubling occurs, either spontaneously or induced, to create what is referred to as doubled haploids which are fully fertile and completely inbred. Thus, by encouraging abnormal pollen development, culturing the resulting cells and tissues, doubling their chromosome compliment, and regenerating plants, one can go from genetically heterozygous material to essentially finished lines in a very short period of time.

Detailed protocols associated with doubled haploid line production using maize anther and microspore culture have recently been reviewed (Genovesi and Petolino, 1992). Briefly, the basic procedure starts with a cross out of which a breeder would like to generate a series of inbred lines for testing. Tassels from field- or greenhouse-grown plants are excised when the majority of the anthers contain microspores at the mid- to late-uninucleate stage of development. After a 1-4 day incubation at 8oC, anthers are excised from the florets and floated on a mannitol solution containing colchicine. After an additional 7-10 days at 8oC, the anthers are placed onto a high sucrose-containing medium which supports further divisions of the microspores. After about 3-4 weeks, structures resembling zygotic embryos become apparent. These anther-derived, embryo-like structures can be transferred to a simple mineral salt medium for germination. Doubled haploid plants can be grown to maturity and self-pollinated in the greenhouse. Seed yield on primary regenerates is generally low and of questionable quality such that second generation seed oftentimes needs to be germinated under controlled conditions for successful establishment of doubled haploid lines.

III. Genetics of Anther Culturability

To date, only certain maize genotypes have been found to undergo androgenesis to any measurable degree. These significant genotype effects suggest that genetic factors are important in determining the level of response to anther culture. Studies designed to explore the inheritance of anther culturability in maize indicated that parents which give rise to responsive hybrids can be identified and that general combining ability was highly significant (Petolino and Thompson, 1987). Indeed, progress has been made in improving the level of anther culture response. The anther culture process itself appears to represent an effective selection criterion for genes favoring androgenesis. The inter-mating of anther-derived lines was an effective means of shifting allelic frequencies toward increased responsiveness (Petolino et al., 1988; Barloy et al., 1989).

Although it would be most desirable to remove the genotype limitation altogether, a more realistic approach is to attempt to broadening the currently limited germplasm-base is to transfer androgenic capacity via genetic means. Unfortunately, labor-intensive, tissue culture-based screening, coupled with high environmental variation and low heritability, make androgenesis a difficult trait to manipulate. Recently, however, the identification of RFLP probes linked to the genes conferring high-androgenic capacity have been identified (Cowen et al., 1992), thus, making marker-based selection possible. The development of elite germplasm with enhanced levels of culturability will allow significant numbers of doubled haploid lines to be generated for agronomic evaluation.

IV. Doubled Haploid Line Production

The production of doubled haploid lines via anther culture is about a 18-month cycle from sowing of donor plant seed to harvest of second generation doubled haploid seed. It involves the interaction of the following laboratory, greenhouse, and field activities.

March-May	Plant F_1 seed @ 2-week intervals
May-June	Harvest tassels and excise anthers
June-Sep	Regenerate H_0 plants
Aug-Oct	Transfer H_0 plants to greenhouse
Sep-Jan	Self-pollinate H_0 plants
March	Have all H_1 seed harvested
April	Plant H_1 seed in greenhouse
May	Transplant H_1 plants in field
July-Aug	Self and test-cross H_1 plants
Oct	Have all H_2 seed harvested

The main bottle-necks have been the production of embro-like structures capable of regenerating plants and the ability to successfully self-pollinate primary regenerates. Nonetheless, doubled haploid lines can be generated from certain genotypes in a relatively short period of time using these procedures.

V. Performance of Anther-Derived Doubled Haploids

An important issue relative to the application of anther culture to commercial maize breeding is whether the performance of microspore-derived doubled haploids is comparable with that of lines produced by conventional inbreeding. A preliminary study designed to evaluate the hybrid performance of anther-derived doubled haploids suggests that they are comparable to conventionally-derived lines. The following results, expressed as grain yield vs. grain moisture, were obtained from doubled haploid (DH), single seed descent (SSD), and pedigree selected (PS) lines (all derived from the cross H99xPa91) in hybrid combination with a tester (LH119xB73).

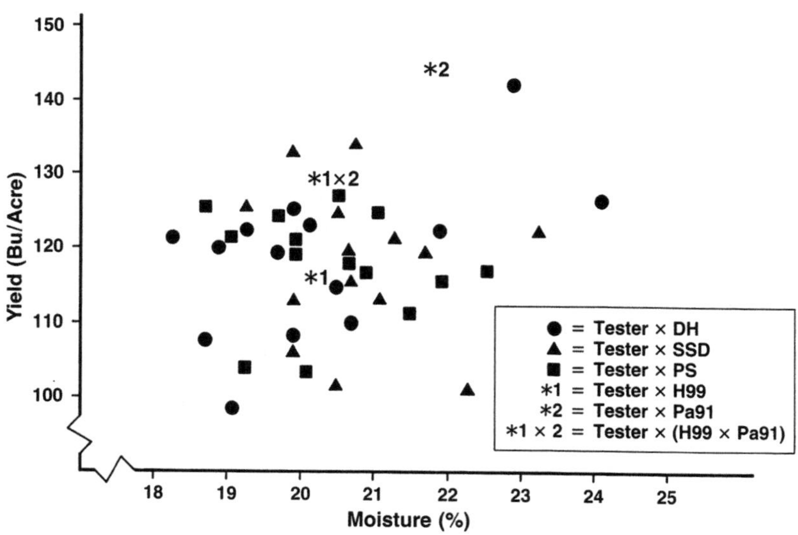

Although differences in grain yield were observed among a series of anther-derived, single seed descent, and pedigree selected lines crossed onto a tester, averaged over all lines there were no significant fixation method effects. Thus, it appears as if anther culture can be utilized to produce lines which can be used as parents of hybrids which are competitive with conventionally-derived materials.

VI. Summary

Although progress has been made in maize anther culture over the last several years, the routine production of large numbers of doubled haploid lines remains a difficult proposition. A few genotypes have been specially bred for their high-androgenic capacity, however, the germplasm-base amenable to anther culture is still quite narrow. Nonetheless, anther culturability has been associated with a set of RFLP markers corresponding to known chromosomal positions. This, together with the encouraging agronomic performance results, suggest that interest in haploid breeding of maize via anther culture will continue.

VII. Literature Cited

Barloy D, Denis L, Beckert M (1989) Comparison of the aptitude for anther culture in some androgenic doubled haploid maize lines. Maydica 34:303-308

Cowen NM, Johnson CD, Armstrong K, Miller M, Woosley A, Pescitelli SM, Skokut M, Belmar S, Petolino JF (1992) Mapping genes conditioning *in vitro* androgenesis in maize using RFLP analysis. (Submitted)

Genovesi AD, Petolino JF (1992) Anther and microspore culture. In. Freeling M, Walbot G (eds) Maize Handbook, Springer-Verlag, New York

Petolino JF, Jones AM, Thompson SA (1988) Selection for increased anther culture response in maize. Theor Appl Genet 76:157-159

Petolino JF, Thompson SA (1987) Genetic analysis of anther culture response in maize. Theor Appl Genet 74:284-286

Wu J (1986) Breeding haploid corn by anther culture. In. Hu H, Yang H (eds) Haploids of Higher Plants *in vitro*, Springer-Verlag, New York, pp 149-161

Haploidy in Barley and Wheat Improvement

P. Devaux
Florimond Desprez
P.O. Box 41
59242 Cappelle en Pévèle
France

Sufficient evidence exists now to prove that doubled haploids (DH's) represent invaluable tools for cereal improvement. In the Sixth International Barley Genetics Symposium it was noted that nearly twenty barley cultivars derived from DH's have been registrated, and most of barley genome map constructions using RFLP markers have been based on the use of DH's (Blake et *al.* 1991; Graner and Foroughi-Wehr 1991; Heun et *al.* 1991). These two examples suggest that DH production techniques have become sufficiently reliable to warrant large investment in barley breeding programmes. In wheat, although anther culture has been successful for nearly twenty years (Ouyang et *al.* 1973; Picard and de Buyser 1973), only three DH cultivars have been released (Hu Daofen 1986; Henry and de Buyser 1990). Promising improvements of anther culture and intergeneric crosses let us think that more and more DH's will be produced within the next few years. Several well documented reviews have been devoted to these topics (de Buyser and Henry 1986; Hu Han 1986; Henry and de Buyser 1990; Inagaki 1990; Laurie et *al.* 1990; Liang et *al.* 1990; Picard et *al.* 1990).

Routine production of DH's for breeders requires the manipulation of many genotypes within a year and from one year to an other. Therefore, attention must be focused on the regular production of sufficient numbers of DH's from a maximum number of hybrids over years. The use of DH's to enhance the release of competitive cultivars is the principle condition on which to base breeding programmes. But, this broad array of genetic variability also represents an invaluable tool for investigations, particularly, in verifying the improvement of techniques, initially performed on a limited number of genotypes. Also, it allows genetic studies from which strategies can be developed for a more rational use of these new techniques in breeding.

Reported herein are data on large scale DH production of barley and wheat and some experiments carried out to improve its efficiency. Also, results reported herein on the assessment of selection of gametophytes during the *in vitro* culture phase will complement preliminary results previously published (Devaux 1989).

I. STATUS OF DH PRODUCTION

Previous studies leading to the suggested use of anther culture and *Hordeum bulbosum* L. techniques in parallel for barley breeding (Devaux 1986) have been confirmed on a larger scale. From 353 F1 hybrids, 87% produced at least 10 plants (Figure 1). From 33% of the F1 hybrids, more than 50 plants per F1 were obtained. This has been reached because some of them showed such high success rates that it was difficult to manage the production and because we have focused on the production on highly potential lines. Out of the total amount of gametophyte-derived plants we produced from April 1990 to September 1991, 64% were recovered through the *H. bulbosum* method and 36% through anther culture. These data show that gametophyte-derived plant production in barley is sufficiently reliable on which to base a breeding programme.

Fig 1: Gametophyte-Derived Plant Production Efficiency from F1 Lines of Barley and Wheat

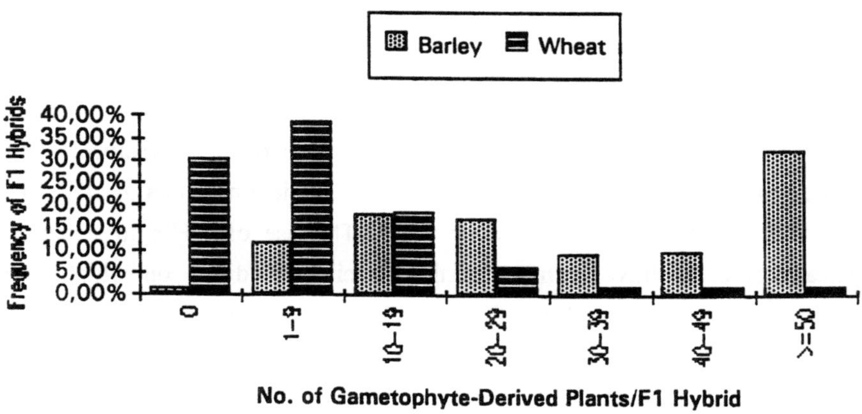

Table 1. Comparison of *H. bulbosum* method (HbM) and anther culture (AC) for haploid production efficiency in barley. Experiments carried out from April 1990 to September 1991 on F1s in our breeding programme.

Type of barley	No. of F1s having higher success rate through: HbM	AC	Total	%HbM
Winter	173	137	310	56
Spring	34	7	41	83
Total	207	144	351	59
	G (1 df) for heterogeneity = 12.18***			

***: Significant at .001 level

Comparing the relative efficiency of the two techniques, 59% of the F1 hybrids gave better success rates using *H. bulbosum vs* anther culture (Table 1). This rate is much higher for spring types (83%), and therefore, we use only the *H. bulbosum* method for DH production of spring barley. The reason of this difference is not known but has been reported by several workers (Devaux 1986; Friedt et *al.* 1987; de Kam-Smets pers. comm. 1991).

From recent contacts made with 20 Institutes and Companies working with barley throughout the world, exactly half of them have been using the *H. bulbosum* method and the other half anther culture or isolated microspore culture. It would be interesting to know the respective total number of DH's produced per year with the two techniques for further comparison.

To our knowledge all of the released DH cultivars have been obtained by the *H. bulbosum* method (Choo pers. comm. 1991), but this is certainly due to the fact that the use of anther culture for applied breeding is more recent. However, promising anther culture-derived lines have been obtained (Rossnagel pers. comm. 1991; Strube pers. comm. 1991). Agronomic performance and malting quality of our newly-released cultivar 'Vodka' obtained through the *H. bulbosum* method are indicated in Table 2. Both major characteristics i.e. grain yield and malting quality required for spring barley cultivar development in Europe appear to be associated in 'Vodka' and led us to conclude that superior cultivars can be recovered using DH production techniques.

Table 2. Agronomic performance and malting quality of the 1991-released barley cv. Vodka in French official trials (1989 + 1990).

Trait	Performance* of 'Vodka'	Range of test cultivars	Position of 'Vodka'
Grain yield	121.6**	110.75-121.6	1
Heading Precocity	5	4-5.5	2
Plant Height	4.5	3-5.5	4
Lodging Resistance	6	5-7	6
Mildew Resistance	6	4-6	1
Rust Resistance	5	3-6	2
Extract	8.0	5.0-8.0	1
Protein	4.5	4.5-6.0	1
Seed size	7.0	5.5-8.0	5
EBC Quality Index	8	0-9	2

*: Index rating except for grain yield; highest value = most desirable
**: as % of the controls cv. Triumph + Menuet + Natasha /3

As can be seen in Figure 1, gametophyte-derived plant production in wheat using anther culture is less efficient than that in barley. From a mean of 41 ± 18 spikes processed per F1 hybrid, 15 (31%) did not produce any green plants. The proportion of F1s from which at least 10 plants have been recovered was limited to 31%. However it is important to note that only one of the F1 hybrids carried a 1BL-1RS translocated chromosome. This translocated chromosome positively affects anther culture ability (Henry and de Buyser 1985) but it causes a detrimental effect on bread-making quality (Zeller et *al.* 1982). Therefore, it is less and less used in our breeding programme now.

Currently, investigations are carried out to improve gametophyte-derived plant production efficiency in wheat. Since Raquin (1983) found the beneficial effects of alternative carbohydrates other than sucrose in anther culture of *Petunia*, significant improvements have been achieved in a few crop species, especially in barley (Hunter 1987). For wheat anther culture, the stimulatory effects of glucose (Chu et *al.* 1990) and maltose (Last and Brettell 1990; Orshinsky et *al.* 1990) have been reported. We have investigated the effect of cellobiose and maltose on anther culture response of 11 F1 wheat hybrids compared to sucrose. These carbohydrate alternatives seem to influence embryoid induction greater than green plant production (Figure 2). The number of green plants/100 anthers ranged from 0.56 with sucrose up to 0.83 with maltose, while cellobiose was intermediate (0.71). Another investigation that we have recently been involved in is the effect of ficoll in wheat anther culture liquid media. The beneficial effect of liquid medium *vs* agar-solidified culture medium on embryo production has been demonstrated for several species including potato (Uhrig 1985), triticale (Charmet and Bernard 1984), barley (Kao 1981; Olsen 1987; Hou and Ullrich 1990) and wheat (Henry and de Buyser 1981; Zhou and Konzak 1989; Fadel and Wenzel 1990). In preventing embryos from sinking into liquid media and negatively affecting their regeneration ability, the addition of ficoll was recommended (Kao 1981; Olsen 1987). As can be seen from Table 3, ficoll has a stimulating effect on the number of microspore-derived embryoids for the two wheat lines, Pavon and

Fig 2. Embryoid induction and green plant production from anthers of 11 F1 wheat hybrids cultured on P2 medium with various carbohydrates. 5700 anthers per treatment.

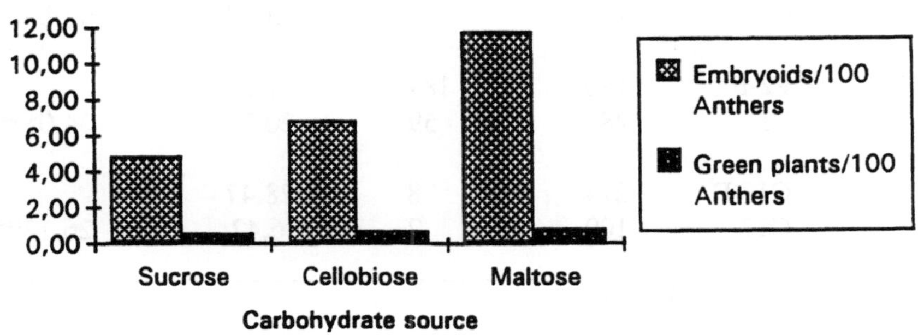

Table 3. The effect of 100g/l of Ficoll type 400 added to two anther culture liquid media, P2 and C17, on gametophyte-derived embryoid production in two wheat lines.

Genotype	Medium	No. of embryoids produced for 400 anthers plated on liquid medium:	
		With Ficoll	Without Ficoll
Pavon	P2	467	405
	C17	235	149
DH	P2	446	284
	C17	274	109

F [Genotype] = 46.37 n.s. - F [Medium] = 1581.0* - F [Ficoll] = 511,62*

Table 4. The influence of 100g/l Ficoll (-F) in two liquid anther culture media (ACM) on green plant regeneration ability of microspore-derived embryoids.

Genotype	ACM	No. embryos	No. green plants	%	G
Pavon	P2-F	467	126	26.98	
	P2	405	64	15.80	16.19***
	C17-F	235	63	26.80	
	C17	149	38	25.50	0.08 n.s.
DH	P2-F	446	184	41.25	
	P2	284	59	20.77	34.05***
	C17-F	274	78	28.47	
	C17	109	7	6.42	26.17***

***: Significant at .001 level

DH, on both Potato-2 medium (Chuang et *al.* 1978) and C17 (Wang and Chen cited by Liang et *al.* 1990) each containing 60g/l of maltose (Jeanmart and Devaux in preparation). In most cases, ficoll allows better quality of microspore-derived embryoids as the green plant regeneration ability on the 190-2 medium (Zhuang and Jia 1983) was significantly increased (Table 4).

From a practical point of view, the status of ploidy level of 4500 microspore-derived plants in barley, wheat and triticale is shown in Figure 3. Spontaneous chromosome doubling in barley (59%) is not as high as previously reported (over 90%) by Huang and Sunderland (1982) and Olsen (1987) but higher than the 27% found by Finnie et *al.* (1989). With a reliable and rapid tool such as flow cytometry, we have found it is economically valuable to check the ploidy level of these microspore-derived plants to recover self-fertility from most of the 32% haploid plants after colchicine treatment (Brown et *al.* 1991). The proportion of tetraploid plants in barley is relatively high (8%). The proportion of spontaneously DH plants in wheat (27%) and triticale (17%) is lower than in barley.

Fig 3. Ploidy Level of Microspore-Derived Plants in Some Cereals Determined by Flow Cytometry

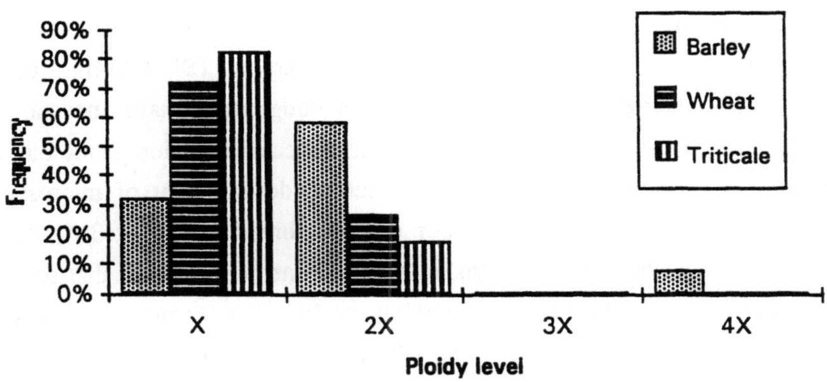

II. GAMETOPHYTE SELECTION

The possible occurence of selection during gametophyte-derived plant production may lead to distortion of segregation and should be known to those using DH's for breeding and genetic analyses. This selection can be positive in breeding if it is in favour of agronomic and/or quality inherited traits. Most studies using a range of different genetic markers have shown that probably no selection occurs through the *H. bulbosum* method (Choo et *al.* 1982; Powell et *al.* 1990; Schon et *al.* 1990). In order to investigate this further for the anther culture method, 62 DH's were derived from a single F1 hybrid of barley and subsequently analyzed by two-dimensional electrophoresis of seedling proteins. Significant preferential transmission of alleles from one parent was detected (Zivy et *al.* 1992) and allows us to mark at least one chromosome segment. This chromosome segment could also carry a gene(s) involved in haploid production. Studies are in progress to confirm this. From these data a linkage map which covers a large part of the barley genome was constructed, and it could remain valuable to locate genes encoding agronomic and malting quality traits as well.

Similar studies were conducted on wheat to assess the magnitude of *in vitro* selection among male gametophytes in their ability to produce embryoids and subsequently green plants. From 7 1BL-1RS/1BL-1BS structural heterozygotes, obtained by crossing cultivars having 1BL-1RS homozygous chromosomes with those having normal 1B chromosomes, 147 DH lines were obtained through anther culture. The presence or absence of the 1BL-1RS translocated chromosomes in the DH lines were detected by sodium dodecyl sulfate-polyacrylamide-gel electrophoresis (SDS-PAGE) of the storage proteins. One hundred and three (70.1%) of the 147 DH's analyzed had 1BL-1RS translocated chromosomes, while the 44 (29.9%) remaining ones have normal 1B chromosomes (Devaux et *al.* 1990). This significant deviation from the 1:1 expected ratio means that there is a preferential development of microspores having the 1RS chromosome arm. For practical breeding, about 30% of the derived DH's do not carry this translocation and one could expect that some of them would have sufficient bread-making quality. The strategy of crossing translocated and non-translocated wheat lines to recover sufficient numbers of bread-making quality DH lines is under investigation especially with lines recalcitrant to haploid production.

Complementary studies have been carried out on this topic but also to compare the male and female recombination frequencies between two loci, Hor 1 and Hor 2, encoding C and B hordeins, respectively, in barley. Polymorphism of the hordein fraction has been detected by reversed-phase high-performance liquid chromatography (RP-HPLC) for 49 and 69 DH's derived by *H. bulbosum*

Table 5. Numbers of DHs showing quanlitatively identical C and B hordein patterns (NR=Non-Recombinant) and DHs which have inherited C hordeins from one parent and B hordeins from the other (R=Recombinant).

Cross No.	Parents	ACDH*			HbDH**			Total
		NR	R	Total	NR	R	Total	
C.163	DH8293 x Kaskade	54	7	61	15	2	17	78
C.220	FD16 x FD21	8	0	8	27	5	32	40
Total		62	7	69	42	7	49	118
		G = 0.464 n.s.						

*ACDH = Anther Culture-Derived Doubled Haploids
**HbDH = *H. bulbosum*-Derived Doubled Haploids

and anther culture techniques, respectively (Table 5). No significant difference in recombination frequency has been found between the *H. bulbosum*- and anther-derived DH populations as far as hordein composition is concerned (Devaux et *al.* in preparation). This may also indicate that the chromosome segment we have studied by means of protein markers is probably not on the short arm of chromosome 5.

CONCLUSION

DH production in barley has been reasonably routine for several years, and practical evidence of its benefit in breeding and release of performant cultivars are currently available. Since the *H. bulbosum* and anther culture techniques show complementary advantages, these have been both used for winter types. In spring types, we prefer using the *H. bulbosum* considering the higher success rate obtained compared to anther culture. One main advantage of using both techniques for winter types is that the two techniques require differential environmental conditions for optimal donor plant growth i.e. higher temperatures for the *H. bulbosum* method than for anther culture. This allows us to save energy for heating greenhouses in winter and cooling them in summer.

DH production remains less efficient in wheat than in barley and research is needed to improve its success rate, and to limit the genotypic effect allowing the recovery of more desirable phenotypes, especially with bread-making quality. Genetic methods to improve DH techniques in wheat must be investigated too, particularly using the high responsive 1BL-1RS translocated types as one parent of F1 hybrid donor plants.

From data currently available, the anther culture-derived plant production process induces a selection pressure while the interspecific cross following by chromosome elimination, as far as barley is concerned, does not. This selection pressure is not undesirable if it is positive regarding agronomic traits, as has already been found (Foroughi-Wehr and Friedt 1984; Friedt et al. 1986). However further studies would be needed for other traits. For this reason we have been investigating wheat intergeneric crosses for haploid production.

REFERENCES

Blake T, Dahleen L, Dvorak J, Gustafson P, Hayes P, Hoffman D, Kasha K, Procunier S, Kim W, Kleinhofs A, Lapitan N, LaRoche A, Saghai Maroof MA, Molnar S, Fedak G, Scoles G, Skadsen R, Sorrells M, Tanksley S (1991) An RFLP map of barley. In: Munck L (ed) Barley Genetics VI. Proc 6th Int Barley Genet Symp, vol 1. Munksgaard Int Publish, p 245-247

Brown SC, Devaux P, Marie D, Bergounioux C, Petit PX (1991) Analyse de la ploidie par cytométrie en flux. Biofutur, Le Technoscope 105 : 1-16

Buyser de J, Henry Y (1986) Wheat: Production of haploids, performance of doubled haploids, and yield trials. In: Bajaj YPS (ed) Biotechnology in agriculture and forestry, vol 2. Springer-Verlag, Berlin Heidelberg, p 73-88

Charmet G, Bernard S (1984) Diallel analysis of androgenetic plant production in hexaploid triticale (X. *triticosecale*, Wittmack). Theor Appl Genet 69 : 55-61

Choo TM, Reinbegs E, Park SJ (1982) Comparison of frequency distributions of doubled haploid and single seed descent lines in barley. Theor Appl Genet 61 : 215-218

Chu CC, Hill RD, Brule-Babel AL (1990) High frequency of pollen embryoid formation and plant regeneration in *Triticum aestivum* L. on monosaccharide containing media. Plant Sci 66 : 255-262

Chuang CC, Ouyang J, Chia H, Chou SM, Ching CK (1978) A set of potato media for wheat anther culture. In: Proc China-Australia Plant Tissue Culture Symp. Peking 1978, p 51-66

Devaux P (1986) Comparison of anther culture and *Hordeum bulbosum* method for the production of doubled haploid in winter barley. I. Production of green plants. Plant Breeding 98 : 215-219

Devaux P (1989) Haplodiploidisation chez l'orge. In: Biotechnology for major crops. Salon International des Techniques et Energies du Futur, Toulouse, p 12-18

Devaux P, Lonnet P, Jean R (1990) Significant higher proportions of 1BL-1RS wheat-rye translocation lines among doubled haploid progenies derived from F1 hybrids between translocated and non-translocated wheat cultivars. In: Nijkamp, HJJ, LHW Van der Plas and J. Van Aartrijk (eds) Progress in plant cellular and molecular biology. Proc. 7th Int Cong Plant Tissue and Cell Cult. Amsterdam, p 422-426

Fadel F, Wenzel G (1990) Medium-genotype-interaction on androgenetic haploid production in wheat. Plant Breeding 105 : 278-282

Finnie SJ, Powell W, Dyer AK (1989) The effect of carbohydrate composition and concentration on anther culture response in barley (*Hordeum* vulgare L.). Plant Breeding 103 : 110-118

Foroughi-Wehr B, Friedt W (1984) Rapid production of recombinant barley yellow mosaic virus resistant *Hordeum* vulgare lines by anther culture. Theor Appl Genet 67 : 377-382

Friedt W, Breun J, Zuchner S, Foroughi-Wehr B (1986) Comparative value of androgenetic doubled haploid and conventionally selected spring barley lines. Plant Breeding 97 : 56-63

Friedt W, Foroughi-Wehr B, Snape JW (1987) The significance of biotechnology for the evolution of barley breeding methods. In: Yasuda S, Konishi T (eds) Barley Genetics V. Proc 5th Int Barley Genet Symp. Mazuren, Okayama, p 903-913

Graner A, Foroughi-Wehr (1991) Implementation of double-haploid techniques in RFLP-assisted breeding programs. In: Munck L (ed) Barley Genetics VI. Proc 6th Int Barley Genet Symp, vol 1. Munksgaard Int Publish, Copenhagen, p 331-334

Henry Y, Buyser de J (1981) Float culture of wheat anthers. Theor Appl Genet 60 : 77-79

Henry Y, Buyser de J (1985) Effect of the 1B/1R translocation on anther culture ability in wheat (*Triticum aestivum* L.). Plant Cell Rep 4 : 307-310

Henry Y, Buyser de J (1990) Wheat anther culture: Agronomic performance of doubled haploid lines and the release of a new variety "Florin". In: Bajaj YPS (ed) Biotechnology in agriculture and forestry, vol 13. Springer-Verlag, Berlin Heidelberg, p 285-352

Heun M, Kennedy AE, Anderson JA, Lapitan NLV, Sorrells ME, Tanksley SD (1991) Construction of a restriction fragment length polymorphism map of barley (*Hordeum* vulgare). Genome 34 : 437-447

Hou L, Ullrich SE (1990) Investigations to improve doubled haploid production of barley through anther culture. In: Use of induced mutations in connection with haploids and heterosis in cereals. Proc FAO/IAEA 3rd Res Co-ordination Meeting. IAEA, Vienna (in press)

Hu D (1986) Jinghua no. 1, a winter wheat derived from pollen sporophyte. In: Hu Han, Yang Hongyuan (eds) Haploids in higher plants *in vitro*. China Acad Publish Beijing, Springer, Berlin Heidelberg New York Tokyo, p 137-145

Hu H (1986) Wheat: Improvement through anther culture. In: Bajaj YPS (ed) Biotechnology in agriculture and forestry, vol 2. Springer-Verlag, Berlin Heidelberg, p 55-72

Huang B, Sunderland N (1982) Temperature-stress pretreatment in barley anther culture. Ann Bot 49 : 77-88

Hunter CP (1987) Plant generation method. European patent application, No 87200773.7

Inagaki MN (1990) Wheat haploids through the *bulbosum* technique. In: Bajaj YPS (ed) Biotechnology in agriculture and forestry, vol 13. Springer-Verlag, Berlin Heidelberg, p 448-459

Kao KN (1981) Plant formation from barley anther cultures with Ficoll media. Z Pflanzenphysiol 103 : 437-443

Last DI, Brettell RIS (1990) Embryo yield in wheat anther culture is influenced by the choice of sugar in the culture medium. Plant Cell Rep 9 : 14-16

Laurie DA, O'Donoughue LS, Bennett MD (1990) Wheat x maize and other wide sexual hybrids: Their potential for genetic manipulation and crop improvement. In: Gustafson JP (ed) Gene manipulation in plant improvement II. Plenum Press, New York, p 95-126

Liang GH, Qi J, Hassawi DS (1990) A direct-generation system for wheat haploid production. In: Bajaj YPS (ed) Biotechnology in agriculture and forestry, vol 13. Springer-Verlag, Berlin Heidelberg, p 425-434

Olsen FL (1987) Induction of microspore embryogenesis in cultured anthers of *Hordeum* vulgare. The effect of ammonium nitrate, glutamine and asparagine as nitrogen sources. Carlsberg Res Comm 52 : 393-404

Orshinsky BR, McGregor LJ, Johnson GIE, Hucl P, Kartha KK (1990) Improved embryoid induction and green shoot regeneration from wheat anthers cultured in medium with maltose. Plant Cell Rep 9 : 365-369

Ouyang JW, Hu H, Chuang CC, Tseng CC (1973) Induction of pollen plants of *Triticum aestivum* L. cultured *in vitro*. Sci Sin 16 : 79-95

Picard E, Buyser de J (1973) Obtention de plantules haploides de *Triticum aestivum* L. à partir de cultures d'anthères *in vitro*. CR Acad Sci Paris 277 : 1463-1466

Picard E, Rode A, Doussinault G, Rousset M, Rives M (1990) Wheat (*Triticum aestivum*): *In vitro* production and utilization of doubled haploids. In: Bajaj YPS (ed) Biotechnology in agriculture and forestry, vol 12. Springer-Verlag, Berlin Heidelberg, p 101-124

Powell W, Ellis RP, Macauley M, McNicol J, Forster BP (1990) The effect of selection for protein and isozyme loci on quantitative traits in a doubled haploid population of barley. Heredity 65 : 115-122

Raquin C (1983) Utilization of different sugars as carbon source for *in vitro* anther culture of Petunia. Z Pflanzenphysiol 111 : 453-457

Schon C, Sanchez M, Blake T, Hayes PM (1990) Segregation of mendelian markers in doubled haploid and F2 progeny of a barley cross. Hereditas 113 : 69-72

Uhrig H (1985) Genetic selection and liquid medium conditions improve the yield of androgenetic plants from diploid potatoes. Theor Appl Genet 71 : 455-460

Zhou H, Konzak CF (1989) Improvement of anther culture methods for haploid production in wheat. Crop Sci 29 : 817-821

Zeller FJ, Gunzel G, Fischbeck G, Gerstenkorn, Weipert D (1982) Veranderungen der backeigenschaften des weizens durch die weizen-roggen-chromosomen translokation 1B/1R. Getreide Mehl Brot 36 : 141-143

Zivy M, Devaux P, Blaisonneau J, Jean R, Thiellement H (1992) Segregation distortion and linkage studies in microspore derived doubled haploid lines of *Hordeum* vulgare L. Theor Appl Genet (in press)

ACKNOWLEDGEMENTS

Many thanks to Mr. Michel DESPREZ for continued encouragement and Prof. Dr. S.E. ULLRICH, Washington State University, Pullman, WA, for critically reading the manuscript.

Sporophytic Self-Incompatibility

Philippe HEIZMANN

RCAP, LA INRA 23 879, Université Claude Bernard-Lyon 1
F-69622 Villeurbanne Cedex, France.

I. INTRODUCTION

The anemogamic or entomogamic pollination processes passively sustained by land plants and the wide occurence of hermaphrodism among most plant families, convergerge to favor a natural propensity for self-pollination, inbreeding and homozigosity. In fact, many plants species are known to be able to identify and to reject their own pollen. Indeed, Darwin already described an heteromorphic form of self-incompatibility (SI) in *Primula*, where size differences between style and anthers (heterostyly) severely hamper self-pollination. Since then, a large number of species was identified to be able to recognize and to reject self-pollen exclusively by physiological reactions (homomorphic SI). Population geneticists consider that the natural selection of SI early in the rise of flowering plants has been determinant for their successful espansion and domination in the present living world (Whitehouse, 1950).

In most families, these homomorphic systems are controlled by a single genetic locus (S) possessing a large number of alleles (de Nettancourt, 1977). Graminaceae are exceptions, having two loci (S and Z) as well as *Ranunculus* and *Beta vulgaris* where three and four loci are involved respectively (Lundquist et al 1973).

Pollen is identified as self-incompatible by the stigma and rejected when both the male and female partners carry S-products coded for by the same S-allele. In the case of gametophytic SI exemplified by *Nicotiana* and the *Solanaceae*, the fate of the pollen grain is determined by its own haploid "gametophytic" genotype. The self-incompatible pollen tube is blocked in the transmitting tissues of the style by heavy callose deposition (Ebert et al., 1989 ; see also the companion paper by R.Thompson in these abstracts). In contrast, in families including the Compositae and the Cruciferae and exemplified by *Brassica oleracea* and *B.campestris*, the behavior of the pollen grain is determined by the genotype of its diploid mother plant (=sporophyte) rather than by its own haploid genome. The consequence of this situation is that pollen S-gene products (S-products) must be synthesized at a stage of microsporogenesis where diploid tissues are actively involved in pollen ontogenesis: either the pollen mother cells before meiosis, or the tapetum known to provide components that fill the crypts of the pollen grain exinic outer coat, have been proposed to provide the male S-products (Pandey, 1960; Heslop-Harrison, 1968). In sporophytic self-incompatible situations, the pollen growth is inhibited at the stigma surface by callose deposition located in the germinating tube, and less systematically on the stigma surface. This review will give some view of the present state of the molecular analysis of the sporophytic SI reported in the literature for *Brassica*.

II THE PRODUCTS OF THE S-LOCUS GENE

II-1 Identification of the S-locus specific glycoprotein (SLSGs)

S-allele specific markers were investigated in both male (pollen) and female (stigma) partners of the self-incompatible reaction. Hitherto only stigma potential S-products have been identified by means of electrophoretic (particularly IEF) and immunological methods (Nasrallah and Wallace, 1967;Hinata and Nishio, 1978). They are glycoproteins displaying all the attributes required for an involvement in pollen-stigma self-recognition:

- they are S-allele specific and are perfectly linked with the S-genotypes in F1 and F2 segregations (Nasrallah et al., 1985a);
- they are tissue specific, being detected by immuno-cytochemical (Kandasamy et al., 1990) and biochemical means only at the stigma surface expressing the self-incompatible reaction ; recently, the use of plants transformed with high sensitivity GUS reporter genes have shown that the transmitting tissues of the style and the placenta of the ovaries also express the S-gene at low level (Sato et al., 1991);
- they are developmentally regulated; their accumulation is concomitant with the acquisition of the SI competence, two to three days before anthesis (Nasrallah et al., 1985b);
- they inhibit *in vitro* the tube growth of pollen grains having the same S-type; pretreatments of pollen grains with stigma S-products of the same genotype also prevent further pollen germination on a compatible stigma (Ferrari et al., 1981)
- they are polymorphic, as expected for molecules supposed to be the support of the S-allele specificity; they are generally strongly basic, with molecular weights of about 50 to 63 kDa.

II-2 The structure and variations of the SLSGs

The high relative abundance and strict developmental regulation of the *Brassica* SLSGs allowed to isolate their cDNAs and to deduce their amino-acid primary sequences. Libraries from the S-6 stigma mRNA extracted one to two days before anthesis were differentially screened with stigma and vegetative tissue probes. Among the clones hybridizing exclusively with the stigma probe, the pBOS5 was fused with the lacZ gene in an expression plasmid and transferred into *E.coli*. The expressed polypeptides were recognized by a polyclonal antibody raised against the S6-SLSG glycoprotein (Nasrallah et al, 1985a). The corresponding sequence was the first (but only partial) of the S-locus genes (SLG) family; these data were corrected, completed and confirmed by direct amino-acid sequencing of three *B.campestris* SLSGs (S8, S9, S12) (Takayama et al., 1987). Several other stigma SLSG have since been sequenced (Nasrallah et al., 1988; Trick and Flavell, 1989; Chen and Nasrallah., 1990; Boyes et al. 1991).

The SLSG peptides deduced from cDNA sequences have common features:

- the presence of an N-terminal hydrophobic domain of about 30 amino acids, assumed to form a signal peptide which directs the SLG to the secretory system;
- a clear structural homology but also remarkable levels of divergence for various alleles of the S locus gene. The SLG variations comprise both nucleotide substitutions and insertion/deletions. Theses variations define four regions with two variable domains, B and D, being interspersed with conserved regions A and C. A highly variable domain situated between residues 182 and 274, exhibits only 40% homology, and might participate to define the S-allelic specificity;

- the C-terminal part of SLG is characterized by the presence of 12 cysteine residues, 11 of which are invariant in the sequence comparisons;

- a number of potential N-glycosylation sites (asn-X-ser and asn-X-thr), at which carbohydrate residues may be attached. Multiple glycoforms have been described for SLGs (Takayama et al., 1987), demonstrating that post-translational processing glycosylates these sites to various degree of saturation.It is indeed possible that all the potential sites for N-glycosylation may not be occupied since in *B.campestris*, only 7 out of nine potential sites are glycosylated (Takayama et al., 1987). All the SLG sequences described up to now discriminate from each other by their potential glycosylation pattern.

An interesting point is the much lower degree of homology observed between the SLSGs from the dominant S-alleles (class I including e.g. S6, S13, S14, S29 ...) and those from the recessive ones (class II including S2, S5 and S15) (review from Nasrallah et al, 1991). This dominance relationship correlates with the strength of the self-incompatible rejection: the alleles of the class I generally tolerate the growth of less than 10 pollen tubes, while those of class II allow for 10-30 tubes to penetrate the female transmitting tissues. It also correlates with large differences at the protein level, since the monoclonal antibody MAbH8 binds to class I SLSGs, but do not recognize those from the class II. This type of allelic grouping is not restricted to *B.oleracea* : indeed, in *R.raphanistrum* as well as in *B.campestris*, the MAbH8 antibody also discriminates the two classes of S-alleles. Moreover, there is a larger sequence homology between alleles of a particular class between species, than there is between classes within species; this indicates that the evolution and fixation of S-alleles preceeded speciation in these genera. These large structural differences between recessive and dominant SLSGs also address the question whether the corresponding genes are real *bone fide* alleles of the same S-locus, even if they clearly exclude each other from this single S-locus in mendelian crosses.

III THE S-COMPLEX

Used as probes on genomic Southern blots of *Brassica* DNA, the SLG cDNAs give complex restriction patterns with as many as 10 to 15 bands, indicating that the SLG locus is only one member of a multigenic family (Nasrallah et al., 1985, 1988). Many of these genes appear to be pseudogenes with open reading frames terminated with stop codons; a growing number of other S-like sequences have been identified to be expressed in various plant tissues.

III-1 Identification of S-locus related (SLR) sequences

In addition to the SLG sequences found in cDNA libraries from mature stigmas, a second SLG-like sequence was isolated (Trick and Flavell, 1989; Lalonde et al., 1989; Scutt et al., 1990). This sequence named SLR1 is co-expressed in mature stigmas; it is structurally homologous to the SLG consensus but shows substantial sequence divergence (an average of 30% at the nucleotide level and some 40% at the protein level). It is highly conserved and shows very little restriction polymorphism among various *Brassica* S homozygotes and other Crucifers. It is genetically unlinked to the S-locus and therefore not directly implicated in the determination of the S-specificity. SLR1 gene transcription was also reported in self-compatible lines of *B.campestris* and in the self-compatible amphidiploid species *B.napus*, suggesting that it may have a more general role in the physiology of the pollination event than self-compatible recognition.

Independently, Isogai et al. (1988) had previously demonstrated the existence of stigmatic glycoproteins (termed NS glycoproteins for S-allele non-specific glycoproteins) homologous with SLSGs, but which now clearly appear to be the products of the SLR1 locus on the basis of their amino-acid sequences.

Later, a second S locus related gene (SLR2) from *B.oleracea* has been idendified (Boyes et al., 1991). Sequence analysis showed that it is in fact homologous to the sequence mis-identified as the SLG from the recessive genotype S5, as a consequence of the high divergence between recessive and dominant S-alleles (Scutt et al., 1990). Their level of expression depends on the genetic background and is much higher in S-recessive lines (S5 and S15) than in S-dominant genotypes. The protein products of the SLR2 genes have not yet been identified. Like SLR1, the SLR2 sequences have the general features of the S-gene family, they seem to be highly conserved among the various S lines and they are genetically linked to the SLR1 locus; they show marked divergence with both SLG and SLR1 sequences, being more similar with the recessive S-sequences (fig. 1).

III-2 Identification of S-locus receptor kinase (SRK) sequences.

In an analysis of protein kinase sequences from maize root tissues, Walker and Zhang (1990) described a ZmPK1 clone with a potential extracellular domain showing very significant homologies with SLGs from *Brassica*. The group of Cornell (Stein et al., 1991) invested therefore extensive sequencing efforts on the flanking regions of SLG homology domains in a collection of *Brassica* genomic clones. An SRK genomic sequence and the corresponding cDNA clones were thus demonstrated to have the same organization as the ZmPK1 clone, including an extracellular SLG-type domain, linked to a transmembrane domain and followed by an intracellular potential protein kinase showing the typical conserved domains of this class of enzymes. The SRK alleles are strictly linked to the S-locus as are the SLGs. For both S-genotypes so far analyzed (S2 and S6), the sequence homology between the SLG genes and the SLG domain of SRK genes reaches about 90% in aminoacid identity; this score is much higher than than that recorded between various SLG alleles (figure 1). The protein kinase domain themselves are also very divergent. The SRK gene is expressed not only in the stigma, but also in the anthers at the stage of microspores and bicellular pollen. The various splicing intermediates are visible on northern filters, as well as apparently N-terminally and C-terminally truncated SRK mRNAs.

These results clearly confirm the expression in both male and female tissues, of SLG and/or SLG-like genes, as already demonstrated by gene transfer of GUS reporter genes controled by an SLG promoter (Sato et al., 1991) or independently by PCR detection of RNAs homologous to SLGs in microspore stage anthers and in mature stigmas (Guilluy et al., 1991). They also allow to imagine how the means by which the recognition signal might be transduced into the physiological rejection of the pollen grain, through the implication of a protein kinase activity. The main questions to answer now are those concerning the recognition process between the extracellular SLG-type domain of the SRK kinase and its S-specific ligand: what is the nature of the ligand ? is it an SLSG ? Lewis (1960) proposed that the self-incompatible reaction might be triggered by the dimerization of homologous or identical pollen and stigma S-products: the molecules likely to participate to this reaction are probably among those described above.

IV. THE EXCEPTIONAL MOLECULAR EVOLUTION OF THE S MULTIGENE FAMILY

Among the 50 to 60 allelic variants of SLG genes identified by breeders several have been sequenced allowing a significant description of the molecular evolution of the multigenic S family. The structural homology relationships between the four main subfamilies of the S-complex (SLR1, SLR2, SRK and SLG) are shown on the dendrogram of figure 3; it suggests a classical evolutionary scheme through several successive duplications, each followed by the differentiation of each subgroup. A low but significant homology is also shown between the *Brassica* S-family and the 5' extracellular domain region of the protein kinase ZmPK1 cDNA cloned from maize by Walker and Zhang (1990) (figure 1).

IV-1. Evolution of the SLR subfamilies

The SLR1 and SLR2 genes form group of sequences described as highly conserved among the *Brassica* species (Lalonde et al., 1989; Boyes et al., 1991). This conservation is shown by a high degree of identity of all the SLR sequences currently available in the literature, even if the SLR1-63 variant reported recently (Trick, 1990) is slightly more divergent from the average of other SLR1 sequences. The base or amino acid changes occurring among SLRs, are randomly scattered throughout the whole sequences, without any apparent localisation at mutational hot-spots.

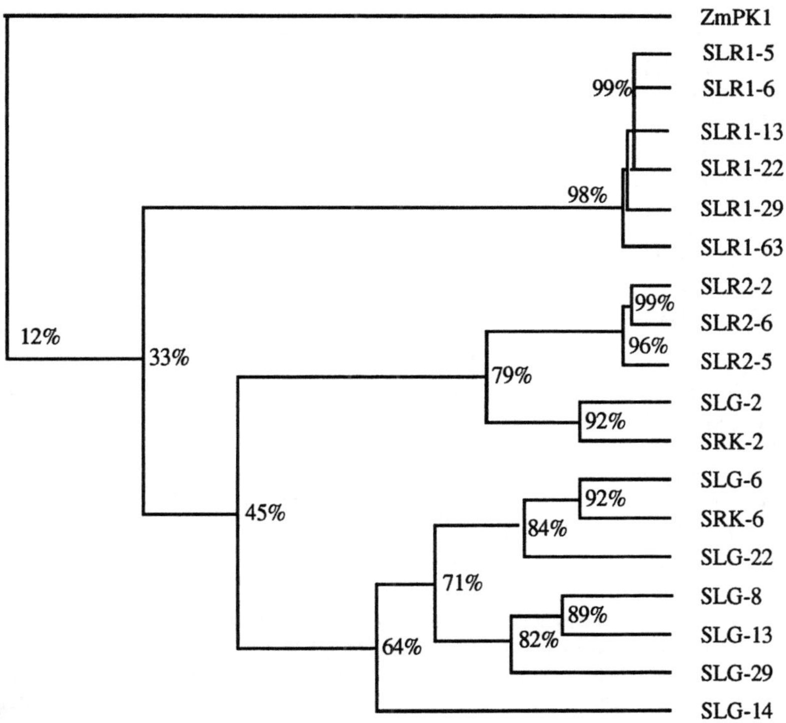

Figure 1: dendrogram of the S-type molecules. The % indicate aminoacid homologies

TABLE I
QUANTITATIVE ANALYSIS OF BASE AND AMINO ACID CHANGES

Comparison	%Syn	%Nonsyn
Active globin genes[a]:		
α1Hum/α1Chim	100%	0%
α1Hum/α1Ora	73%	27%
α1Chim/α1Ora	64%	36%
α2Hum/α2Chim	100%	0%
α2Hum/α2Ora	64%	36%
α2Chim/α2Ora	60%	40%
α2Hum/α2Mac	50%	50%
Cumulated changes	**68%**	**32%**
Globins pseudo-genes:		
βHum/ ψβHum	9%	81%
δHum/ ψη Hum	50%	50%
Cumulated changes	**25%**	**75%**
All base changes **theoretically possible**	**24.4%**	**75.6%**
Mouse V_H genes[b]	**5.9%**	**94%**
SLR1-63/SLR1-5	27%	73%
SLR1-63/SLR1-6	22%	78%
SLR1-63/SLR1-13	30%	70%
SLR1-63/SLR1-29	23%	77%
SLR1-5/SLR1-6	50%	50%
SLR1-5/SLR1-13	0%	100%
SLR1-5/SLR1-29	0%	100%
SLR1-6/SLR1-13	100%	0%
SLR1-29/SLR1-6	25%	75%
SLR1-29/SLR1-13	0%	100%
SLR2-2/SLR2-5	30%	70%
Total SLR1-2	**24%**	**76%**
SLG-6/SLG-13	28%	72%
SLG-6/SLG-29	23%	77%
SLG-8/SLG-13	38%	62%
SLG-8/SLG-29	28%	72%
SLG-13/SLG-29	26%	74%
Total SLG	**26%**	**74%**
SLG-6/SLG-29 **hypervariable**	**16%**	**84%**

[a]Globin gene sequences were extracted from EMBL and Genbank databases, and analyzed as indicated in Trick and Heizmann (1992).
Abbreviations used: α1 and α2-globin genes of human (Hum), chimpanzee (Chim), orang-outang (ora), macaque (Mac); pseudogenes of human β-globin (ψβHum) and human -globin (ψη Hum).
[b]Mouse immunoglobulin V_H gene sequences from Tanaka and Nei, 1989.

The analysis and comparison of codon changes between both SLR1 and SLR2 sequences (Table I) show an unusually high proportion of nonsynonymous versus synonymous substitutions (76% of nonsynonymous / 24% of synonymous substitutions). This is in striking contrast with the evolution observed for most genes: they change conservatively through low ratios of nonsynonymous changes (for example, 32% for αglobins), as a result of the "purifying selection" as termed by Kimura (1968), operating against deleterious mutations The particular mode of base substitutions shown for the SLR genes is prone to generate an intrinsically rapid diversification of both SLR families, in contrast with their proposed strong conservation (based on hybridization data) among *Brassica* lines or species. In fact, this proportion of 76% nonsynonymous base changes is quite close to that observed during the free evolution of pseudogenes and the theoretical nonsynonymous proportion of all possible codon changes. SLR genes thus seem to evolve through fixation of all base changes, in complete absence of purifying selection: their mode of evolution is thus a typical example of positive darwinian selection.

The rate of fixation of amino acid changes in the SLR gene families can be estimated if one assumes that the *B.oleracea* species became rapidly established and isolated within the Cruciferae, and thus is about as ancient as the family itself which appeared some 5 million years (My) ago (and, therefore, Kimura's 2T term is 10 My), according to the paleobotanical data of Muller (1981). In the case of SLR1 genes, 10 amino acid changes have been reported between the 444 residues of SLR1-29 and SLR1-63, the two most distant SLR1 sequences so far reported (Trick, 1990): the rate of amino acid substitution between these two is thus about 2.2 substitutions/100 sites/10 My or 2.2 substitutions /site/1000 My. Similar values stand for the SLR2 genes. Compared to the rates of amino acid changes reported from the literature for various proteins (Table II), this estimated maximal rate of SLR evolution must in fact be considered as rather high, allowing rapid changes of the SLR proteins.

Moreover, the SLR1 and SLR2 proteins have themselves diverged significantly; they show only 50% amino acid homology according to the dendrogram of figure 1 and their DNA sequences show very low cross-hybridization in blotting experiments (Boyes et al., 1991). These homologies are nevertheless highly indicative of a common gene lineage. Their amino acid divergence being thus around 0.5 substitutions/site, and the divergence rates between the two sequence classes being 2+2=4 substitutions/site/1000 My, they probably diverged about 120 My ago. Their common ancestor might even be much older than 120 My, indicating that genes displaying structural homologies with the SLR families should be found in most land plants if the angiosperms appeared some 100-120 My ago, as proposed by Muller (1981).

IV-2. Evolution of the SLG and SRK subfamilies

Examination of SLG sequences illustrates the very high polymorphism of the SLG alleles in *Brassica*. As for the other members of the S-complex, there is a very high proportion of nonsynonymous base changes (for instance, 76.5% of the changes between SLG-6 and SLG-29 are nonsynonymous). For the hypervariable regions the proportion of nonsynonymous base changes can be as high as 84% (for instance, in the region spanning from amino acid positions 258 to 344 for SLG-6 and SLG-29).

The divergence rate of SLG proteins in *B.oleracea* ,was estimated with the same assumptions as above for SLR genes. This yields the average divergence rates of 30-35 substitutions/site/1000 My, and the overwhelming rates of 60 substitutions/site/1000 My for the hypervariable regions. The combination of rapid

TABLE II
AMINO ACID SUBSTITUTION RATES FOR VARIOUS PROTEINS

Protein	Amino acid changes/site/1000 MYr
Histone H4	0.01
Haemoglobin a/b chains	1.2
Ribonucleases, Lysozymes	2-3
Immunoglobulin V genes:	7-8
Pseudogenes	12-20
Brassica SLR1 and SLR2	2
Brassica SLG	30-35
SLG hypervariable region	60
Nicotiana S-glycoproteins	Significantly higher values than for Brassica SLG

Data for non-S proteins from Nei (1987) and Kimura (1983).

base substitution and of apparent absence of purifying selection provides a mode of positive diversifying evolution, in perfect contrast with the neutral selection observed for most genes These changes are so rapid that they probably cannot simply result from stochastic accumulation and fixation of point mutations; they are better explained by radical changes like those occurring during gene conversions, as demonstrated for the variable regions of immunoglobulins; up to now, this unique combination of rapid aminoacid or nucleotide changes with fixation of high proportions of nonsynonymous substitutions had been described only for immunoglobulins (Hughes et al. 1990). Such hypotheses have already been proposed by Ebert et al. (1989) to explain the same unusual divergence rates of SLG alleles of the gametophytic SI system in *Nicotiana*.

The SLG genes and their SRK counterparts are structurally more closely related than the various SLG alleles. The occurence of SRK genes in several S lines indicates that the duplication of the SLG/SRK segments must be general and probably occured before the divergence of the S alleles themselves. Some mechanism of concerted evolution maintained a higher degree of relatedness between SLG and SRK than between SLGs. However, the percentage of nonsynonymous changes observed between the SLG and SRK pairs (63 to 70%) is as high as that observed between the various SLG sequences (62 to 77%), indicating that the constrainsts for coevolution might have been recently released.

IV-3 Mutagenesis of S-alleles.

Experimental mutagenesis using ionising radiation has never produced new S-alleles. Only loss of SI has been observed in pollen, style or both tissues; these data have been integrated into a model of a tripartite structure for the S-locus (Lewis,1960). It has been reported, however, that new alleles would have been created at high ratios through forced inbreeding of self-incompatible populations of plants.

Several reports of such events concerning the gametophytic systems of *Trifolium pratense* , *Nicotiana* and *Lycopersicon* have been reviewed by de Nettancourt (1977). In fact, the "new" alleles following forced inbreeding were generally identified as alleles already existing among S-genotype collections: for instance, inbreeding of an S1S2 genotype in *Lycopersicon* produced a stable allele identified as an S3 genotype (de Nettancourt et al., 1971). In a sporophytic system, Zuberi and Lewis (1988) reported the creation of S5, S6 and Sx alleles from an S1 allele in an S1S3 genotype of *Raphanus*; several cases of reversion to the original S1 allele were subsequently observed. Thus a limited number of S-structures is involved in the formation of these mutants, instead of an infinite number of possible variations: a reversible exchange of some basic building blocks forming the different possible S-genes would explain these observations. The large S multigene family in *Brassica* could provide the pool of S-related sequences for such recombinations, occuring, probably through unequal crossing-over or gene conversion (Zuberi and Lewis, 1988; de Nettancourt et al., 1971). In addition, an important question about these allelic changes is whether a given S-allele corresponds to a unique S-sequence expressed in the stigma and in the anthers. In the case of the S2 recessive allele, Chen and Nasrallah (1990) described two signicantly different S-cDNA sequences from two different *B.oleracea* cultivars *alboglabra* and *italica*, both belonging to the same class of S2 alleles. It seems thus that the mechanisms of recombination possibly involved in these allelic changes are not necessarily very accurate at the molecular level; this point should be answered by the analysis of such mutants at the molecular level.

Acknowledgements: this work was supported by INRA and EC BRIDGE grants.

REFERENCES

BOYES, D.C., CHEN, C.H., TANTIKANJANA, T., ESCH, J.J. and NASRALLAH, J.B. 1991. Isolation of a second S-locus related cDNA from Brassica oleracea: genetic relationships between the S-locus and two related loci. Genetics 127: 221-228.

CHEN, C.H. and NASRALLAH, J.B. 1990. A new class of Ssequences defined by a pollen recessive self incompatibility allele of Brassica olearacea. Mol. Gen. Genet. 222 : 241-248.

De NETTANCOURT, D., ECOCHARD, R., PERQUIN, M.D.G., van der DRIFT, T., and WESTERHOF, M. 1971. The generation of new S alleles at the incompatibility locus of L. peruvianum Mill. Theor. Appl. Genet. 41: 120-129.

De NETTANCOURT, D. 1977. Incompatibility in angiosperms. In: Monographs on theoretical and applied genetics. No. 3. Frankel, R., Gall, G.A.E. and Linskens, H.F. (eds.). Berlin, SpringerVerlag.

EBERT, P.R., ANDERSON, M.A., BERNATZKY, R., ALTSCHULER, M. and CLARKE, A.E. 1989 Genetic polymorphism of self incompatibility in flowering plants. Cell 56: 255-262.

FERRARI, T.E., BRUNS, D. and WALLACE, D.H. 1981. Isolation of a plant glycoprotein involved with control of intercellular recognition. Plant Physiol. 67: 270-277.

GUILLUY,C.M., TRICK, M., HEIZMANN, P. and DUMAS, C. 1991. PCR detection of transcripts homologous to the self incompatibility gene in anthers of *Brassica*. Theor. Appl. Genet. 82: 466-472.

HESLOP-HARRISON, J. 1968. Pollen wall development. Science, 161: 230-237.

HINATA, K. and NISHIO, T. 1978. S-allele specificity of stigma proteins in Brassica oleracea and B. campestris. Heredity 41: 93-100.

HUGHES, A.L., OHTA, T.and NEI, M. 1990. Positive darwinian selection promotes charge profile diversity in the antigen binding cleft of class I major histocompatibility molecules. Mol. Biol. Evolution 7, 515.

ISOGAI, A., TAKAYAMA, S., SHIOZAWA, H., TSUKAMOTO, C., KANBARA, T., HINATA, K., OKAZAKI, K. and SUZUKI, A. 1988. Existence of a common glycoprotein homologous to S-glycoproteins in two self-incompatible homozygotes of Brassica campestris. Plant Cell Physiol. 29: 1331-1336.

KANDASAMY, M.K., DWYER, K.G., PAOLLILO, D.J., DONEY, R.C., NASRALLAH, J.B. and NASRALLAH, M.E. 1990. Brassica S-proteins accumulate in the intercellular matrix along the path of pollen tubes in transgenic tobacco pistils. Plant Cell 2: 39-49.

KIMURA, M. 1968. Evolutionary rate at the molecular level. Nature 217: 624-626.

KIMURA, M. 1983. In "Evolution of genes and proteins" M.NEI and R.K. KOEHN, eds. pp. 208-233. SINAUER Associates, SUNDERLAND, Mass.

LALONDE, B.A., NASRALLAH, M.E., DWYER, K.G., CHEN, C-H., BARLOW, B. and NASRALLAH, J.B. 1989. A highly conserved Brassica gene with homology to the S-locus specific glycoprotein structural gene. Plant Cell 1: 249-258.

LEWIS, D. 1960 Genetic control of specificity and activity of the S antigen in plants. Proc. Roy. Soc. (Lond.) B., 151: 468-477.

LUNDQVIST, A., OSTERBYE, U., LARSEN, K. and LINDELAURSEN, I.B. 1973. Complex selfincompatibility systems in Ranunculus acris L. and Beta vulgaris L. Hereditas 74 : 161-168.

MULLER, J. 1981. Fossil pollen records of extant angiosperms. Bot. Rev. 47: 1-142.

NASRALLAH, M.E. and WALLACE, D.H. 1967. Immunogenetics of self- incompatibility in Brassica oleracea L. Heredity 22: 519-527.

NASRALLAH, J.B., KAO, T.H., GOLDBERG, M.L. and NASRALLAH, M.E. 1985a. A cDNA clone encoding an S-locus-specific glycoprotein from Brassica oleracea. Nature 318: 263-267.

NASRALLAH, J.B., DONEY, R.C. and NASRALLAH, M.E. 1985b. Biosynthesis of glycoproteins involved in the pollen-stigma interaction of incompatibility in developping flowers of Brassica oleracea L. Planta, 165: 100-107.

NASRALLAH, J.B., YU, S.M. and NASRALLAH, M.E. 1988. Self-incompatibility genes of Brassica oleracea: Expression, isolation and structure. Proc. Natl. Acad. Sci. USA 85: 5551-5555.

NASRALLAH, J.B., NISHIO, T. and NASRALLAH, M.E. 1991. The self-incompatibility genes of Brassica: expression and use in genetic ablation of floral tissues. Ann. Rev. Plant Physiol. Plant Mol. Biol. 42: 393-422.

NEI, M. 1987. Molecular Evolutionary Genetics. Columbia University Press, New York.

PANDEY, K.K. 1960. Evolution of gametophytic and sporophytic systems of self-incompatibility in angiosperms. Evolution, 14: 98-115.

SATO, T., THORNESS, M.K., KANDASAMY, M.K.., NISHIO, M., NASRALLAH, J.B. and NASRALLAH, M.E. 1991. Activity of an S locus gene promoter in pistils and anthers of transgenic Brassica. Plant Cell 3: 867-875.

SCUTT, L.P., GATES, P.J., GATEHOUSE, J.A., BOULTER, D. and CROY, R.R.D. 1990. A cDNA encoding an S-locus specific glycoprotein from Brassica oleracea plants containing the S5 self-incompatibility allele. Mol. Gen. Genet. 220: 409-413.

STEIN, J.C., HOWLETT, B., BOYES, D.C. NASRALLAH, M.E. and NASRALLAH, J.B. 1991. Molecular cloning of a putative receptor protein kinase gene encoded at the self-incompatibility locus of Brassica oleracea. Proc. Natl. Acad. Sci. USA 88: 8816-8820.

TANAKA, T. and NEI, M. 1989. Positive darwinian selection observed at the variable-region genes of immunoglobulins. Mol. Biol. Evol. 6: 447-459.

TAKAYAMA, S., ISOGAI, A., TSUKAMOTO, C., UEDA, Y., HINATA, K., OKAZAKI, K. and SUZUKI, A. 1987. Sequences of S-glycoproteins, products of the Brassica campestris self-incompatibility locus. Nature 326: 102-105.

TRICK, M. and FLAVELL, R.B. 1989. A homozygous S genotype of Brassica oleracea expresses two S-like genes. Mol. Gen. Genet. 218: 112-117.

TRICK, M. 1990. Genomic sequence of a Brassica S-locus related gene. Plant Mol. Biol. 15: 203-205.

TRICK, M. and HEIZMANN, P. 1992. Sporophytic self-incompatibility systems: Brassica S-gene family. Int. Rev. Cytol. in press.

WALKER, J.C. and ZHANG, R. 1990. Relationship of a putative kinase from maize to the S-locus glycoproteins of Brassica. Nature (Lond.) 345: 743-746.

WHITEHOUSE, H.L.K. 1950. Multiple allelomorph incompatibility of pollen and style in the evolution of angiosperms. Ann. Bot. N.S. 14: 198-216.

ZUBERI, M.I. and LEWIS, D. 1988. Gametophytic-sporophytic incompatibility in the Cruciferae - Brassica campestris. Heredity 61: 367-377.

Molecular Analysis of Gametophytic Self-Incompatibility

Hans-Hubert Kirch, Helgard Kaufmann, Thomas Wemmer,
Junqi Liu and Richard D. Thompson
Max-Planck-Institut für Züchtungsforschung,
Carl-von-Linné-Weg 10,
5000 Köln-30, Germany

Introduction

The control of gametophytic self-incompatibility as exhibited in most plant species is exerted through the action of a single locus S. When haploid pollen grains genotype Sx alight on a style expressing the same S allele, fertilization is prevented, whereas pollen grains whose S-alleles are not matched by those expressed in the style grow, and potentially effect fertilization (de Nettancourt, 1977). The mechanism by which pollen cells are distinguished on the basis of their S-alleles has been studied by physical, biochemical and more recently molecular techniques, without a complete picture having so far emerged. The best-studied system is that of the Solanaceae and this will form the focus of this paper.

Characterization of the gene products of the S-locus

In 1986, the first molecular characterization of an S-associated protein was reported in the Solanaceae, from styles of the ornamental tobacco Nicotiana alata (Anderson et al., 1986). The presence of related S-linked stylar proteins has since been confirmed for self-incompatible Solanum, Petunia and Lycopersicon species (reviewed in Thompson et al., 1991a). In contrast, self-compatible species in some cases either lack these proteins or have very much reduced amounts (McClure et al., 1989). The S-linked proteins accumulate to high (mM) concentrations around anthesis, and are deposited in the intercellular matrix of the transmitting tract and placental epidermis of the ovary i.e., precisely the tissues encountered by the pollen tube en route to fusion with an ovum (Cornish et al., 1987).

The peak rate of accumulation of the S-linked proteins around anthesis, coincides with the acquisition of a self-incompatible phenotype by the style; at earlier stages, bud pollination can give seed set. Thus there is circumstantial evidence implicating the S-linked proteins in self-incompatibility.

In Figure 1, a typical stylar protein complement for an SI Solanaceous species is presented (Kirch et al., 1989). In this example, two potato lines differing in S-allele constitution are compared. A cluster of basic polypeptides is apparent including two invariant spots, (SK1 and SK2), and the S-linked proteins S1 through S4. The function of SK1 is currently unknown. SK2 is a pistil-specific endochitinase (Wemmer and Thompson, manuscript in preparation) restricted to potato and tomato. Pistil extracts from other Solanaceous species also contain high levels of other pathogenesis-related proteins, such as 1,3 ß-glucanase in the case of Nicotiana tabacum (Ori et al., 1990). These proteins probably represent different strategies for dealing with the general problem of fungal and/or insect ingress in the style. Indeed it is conceivable that the S-locus itself has developed from such a role. The analogies between the Brassica S-locus and loci for race-specific resistance have been developed by others. (Hodgkin, Lyon and Dickinson, 1988).

Figure 1. Two-dimensional gel electrophoresis of style extracts of potato clones 191(S1S2) and 194 (S3S4). Stylar proteins were separated by NEPHGE/SDS-PAGE. Polypeptides were detected by silver stain. Each extract shows in the high pI-range two specific S-allele-associated proteins indicated by arrows, as well as two constant polypeptides (circled, O = Sk1, o = Sk2).

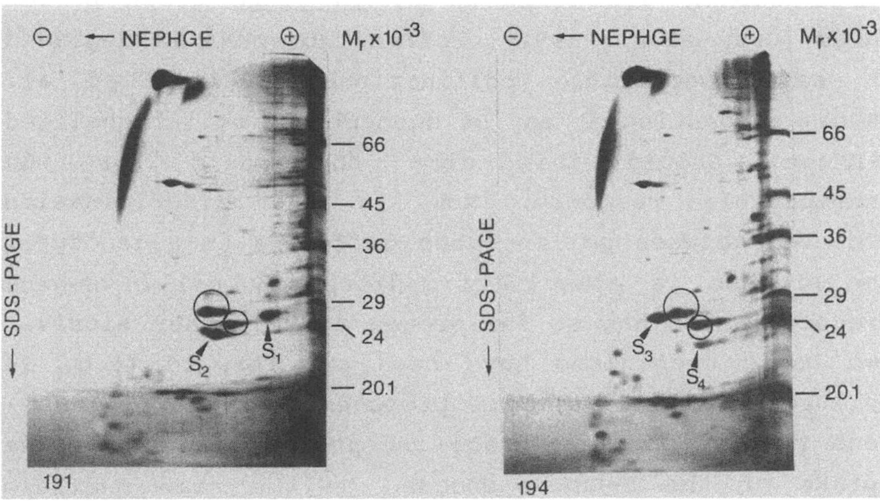

The S-linked stylar proteins have been purified from several Solanaceous species and the corresponding genes isolated. Sequence comparisons indicate the presence in the coding sequence of a mosaic structure of highly conserved stretches insterspersed with "hypervariable" regions (Ebert et al., 1989). The conserved stretches have been shown to have homology to a family of fungal ribonucleases and the S-linked proteins themselves exhibit ribonuclease activity. (McClure et al., 1989). Over 20 S-alleles have now been sequenced and in all of them the RNAse active site residues are conserved, thus strongly implicating RNAse activity in the function of these genes. The hypervariable domains are thought to confer S-allele-specific secondary structure important in the recognition of pollen tubes according to their S-genotype.

In a series of elegant experiments, Clarke and colleagues have demonstrated that the S-linked proteins (or S-RNAses) are taken up by in vitro grown pollen tubes (Gray et al. 1991). Pollen tube rRNA is degraded in self-incompatible pollinations (McClure et al. 1990), suggesting it may be susceptible to internalised S-RNAse. Should this form the basis for the incompatible response, how is the discrimination achieved between pollen tubes differing in S-genotype? The process is presumably mediated by pollen S-gene product, which has so far proved to be rather elusive. Two basic mechanisms have been put forward (e.g. in McClure et al., 1990). One proposes that the pollen S-gene product is a membrane receptor which allows the uptake of the S-RNAse encoded by the same S-allele only. In the second model, the pollen S-gene product interacts with and inactivates the S-RNAse action intracellularly. In this case pollen S-gene product from allele Sx must inactivate all S-RNAses except that encoded by Sx.

Previous attempts to identify pollen proteins linked to the S-locus have been unsuccessful; it seems that the protein(s) involved is expressed at very low levels. An alternative approach is to look for pollen proteins capable of specifically binding to the S-RNAses, on the basis of mobility shifts in native gel electrophoresis. We have been unable to identify such complexes using cross-linking reagents (Kaufmann, 1991) suggesting that if these complexes are formed they do not survive our extraction or assay conditions.

The structure of the gametophytic S-locus as derived from genetic analysis suggests one structural gene ("specificity" component) is driven by two separate pollen and stylar "activity" components or promoters. Attempts to identify S-RNAse protein or transcripts in pollen have so far failed, and it is now generally assumed that the pollen S-gene product is derived from sequences distinct from the S-RNAse sequence, although located nearby on the chromosome. The search has therefore begun for pollen-expressed ORFs in the vicinity of the S-RNAse, and the first reports of such a sequence have appeared (cited in Chasan, 1991). At the time of writing this sequence has not been further characterized. In line with the situation at the Brassica S-locus (Stein et al. 1991), a complex locus containing more than one ORF, now seems likely.

In view of the probable complexity of the S-locus, it is perhaps not surprising that its operation can be comprised by a number of mutations, including a mutant which discriminates S-alleles in its action. In the best-studied examples the mutants are non-allelic to S (Thompson et al. 1991b) and may either correspond to translocations of components of the S-locus, or to interactions with non-S-genes.

The regulation of S-locus expression

The isolation of S-RNAse genes (Kaufmann, Salamini and Thompson, 1991) has permitted us to study the sequences necessary for style-specific expression. Three clones for S1 and one clone for S2 were isolated from a genomic library using DNA from a diploid potato clone of S-genotype S1S2. To illustrate the structure of the S-locus of potato, the restriction pattern of the genomic S1-clone G193-8 and the S2-clone G193-1 is shown in Fig. 2. The maps of both S-clones with several restriction enzymes show no common features, indicating the highly polymorphic character of the S-locus. There is a single small intron located between amino acid residues 81 and 82, showing small variations in size and position between the alleles. The S1 intron is 113 bp in length, and the S2 intron 117 bp, the positions of insertion within the open reading frame apparently differing from one another by 3 bp. A TATA sequence motif exists in both S-promoter regions, but a typical TCAAT-box, located at position -61, is found solely in the S2-promoter.

Figure 2. Restriction maps of the genomic clones G193-8 (S1) and G193-1 (S2). The coding region is boxed and enlarged to delineate a more precise gene structure. ATG and TAG define start and stop sites of the translated sequence. Putative regulation signals such as TATA- and CAAT-box are indicated. The intron is depicted as a thin line. Poly A designates the final points of the cDNA-clones. The hatched boxes indicate the border sequences of the λ-DNA; L = left arm, R = right arm. B = BglII, H = HindIII, R = EcoRI, S = SalI, Xb = XbaI, Xh = XhoI.

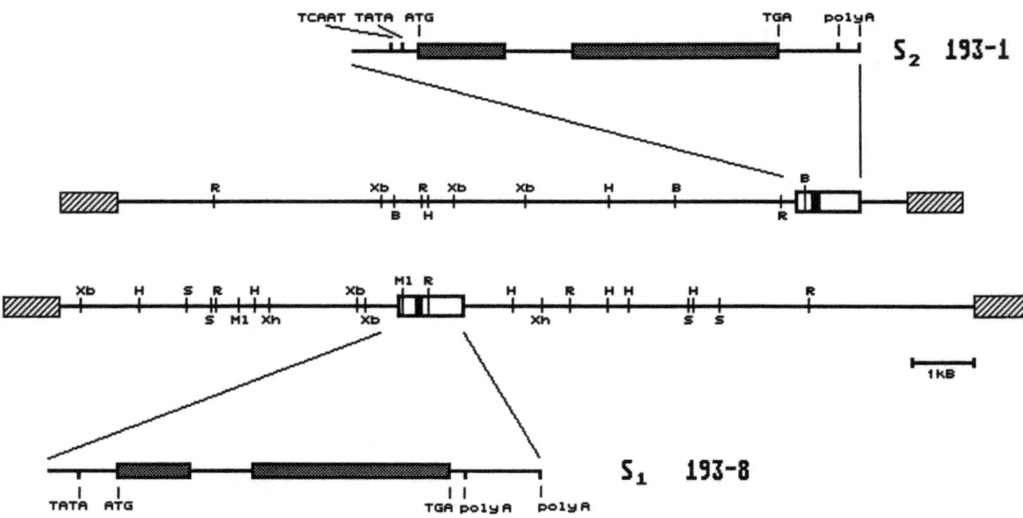

As a first step in our analysis of sequence elements required for the regulated expression of the S1- and S2-gene, their 5' flanking DNA sequences were compared to identify conserved motifs.

A computer analysis of the S2-clone revealed a combination of conserved elements in the region from -310 to -200 (Fig. 3). A sequence element of 6 bp (CAAGAA) appears several times, in part as 1.5x or 2x repeats and is framed by inverted repeats. Moreover, part of this sequence, from position -301 to -269, is repeated from position -242 to -211 with 80% identity. Another repetition of the CAAGAA-motif, again flanked by inverted repeats, is found at position -470.

A 20 bp sequence motif, common to both the S1 and the S2-promoter, is located at position -340 in the S1- and position -522 in the S2-promoter, with an overall homology of nearly 90% (Fig. 4a). Interestingly this motif shows significant homology to a regulatory sequence element of two pollen-specific promoters of tomato (Fig. 4b). It has been shown that this 56/59 box has a regulatory function and is suggested that it plays a determinative role in turning on expression of the LAT56 and LAT59 genes in pollen (Twell et al., 1991).

In a first attempt to investigate the expression of the S-promoter in floral tissues, the technique of microprojectile bombardment was used. Expression of chimeric S-promoter GUS constructs (Fig. 5) was monitored either by histochemical staining (data not shown) or by fluorometric assay of GUS-activity. Significant levels of GUS-activity were directed by the S2-promoter after bombardment of tobacco SR1 pollen. Blue-staining pollen was not observed following

Figure 3. Internal sequence motifs in the S2-promoter. The region shown with direct and inverted motif-repeats was identified by UWGCG-computer-program analysis. Motif repetitions are boxed and inverted repeats indicated by arrows.

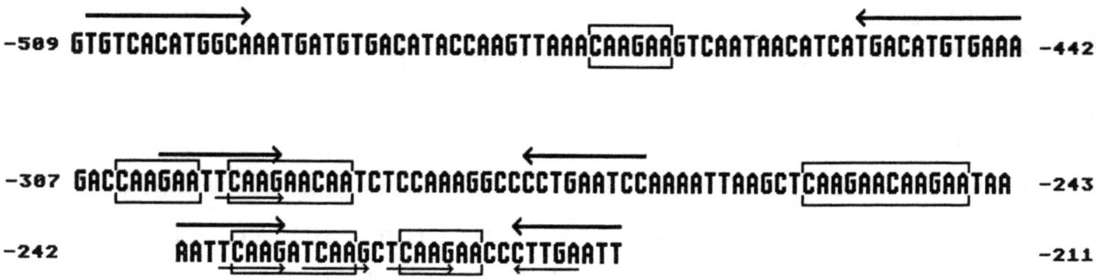

Figure 4. Sequence motifs conserved between the S1- and S2-promoter of potato and two LAT-gene promoters of tomato. Identities between pairs of sequences are shown as vertical lines unshaded.

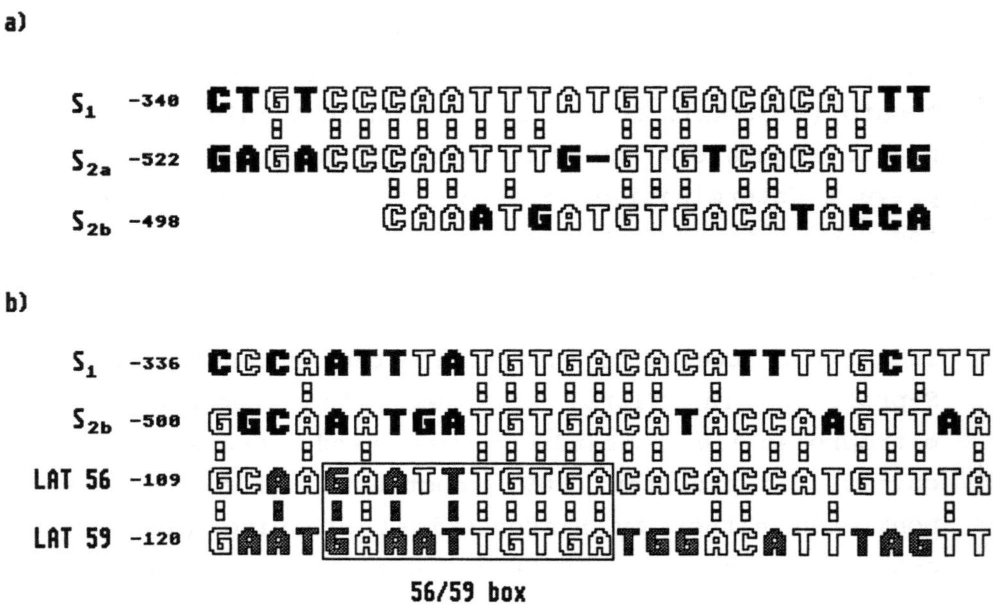

bombardment with the "promoterless" GUS-nos3' plasmid pUCGUSC indicating that the results could not be attributed to endogenous enzyme activities reported by others (Plegt and Bino, 1989). A quantitative analysis shows that the GUS-activity of the S2-promoter constructs is approximately 50% of that of control plasmid pLAT52-7 (Fig. 6). In contrast, the corresponding 0.5 kb S1-promoter construct is apparently expressed only very weakly in tobacco pollen. Tobacco leaves bombarded with the S-promoter reporter fusions show no GUS-activity above that measured in leaves bombarded the negative control plasmid, pUCGUSC, whereas leaves bombarded with pRT103GUS have GUS-activity in the range of the control plasmid pLAT52-7.

Figure 5. Chimeric gene fusions used for bombardment. An NcoI restriction site was introduced into the S-promoter fragments at the ATG start codon by <u>in vitro</u> mutagenesis or PCR amplification and transcriptionally fused to GUS-nos3'. The plasmid pLAT52-7 (*) served as a positive control and was kindly provided by S. McCormick (Twell et al. 1991).

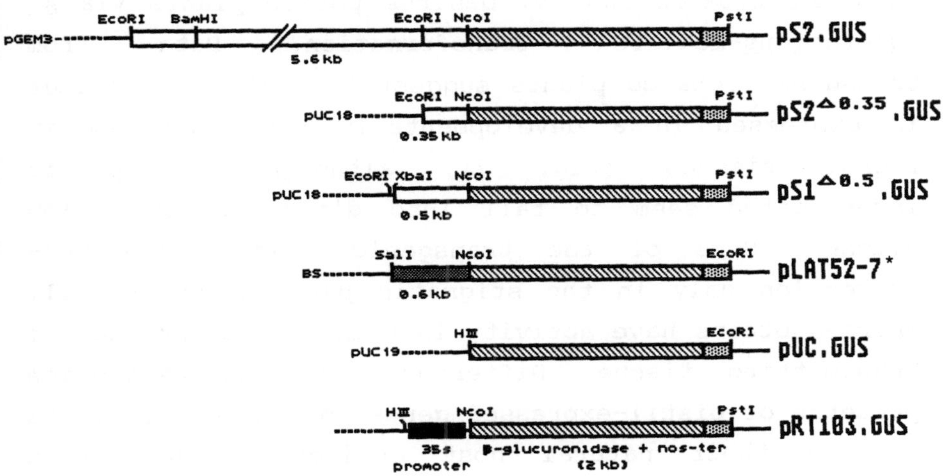

Figure 6. Transient expression of GUS activity in tobacco pollen and leaves. Approximately 6 x 10^5 mature tobacco pollen grains were bombarded with tungsten microprojectiles coated with 7μg plasmid DNA.

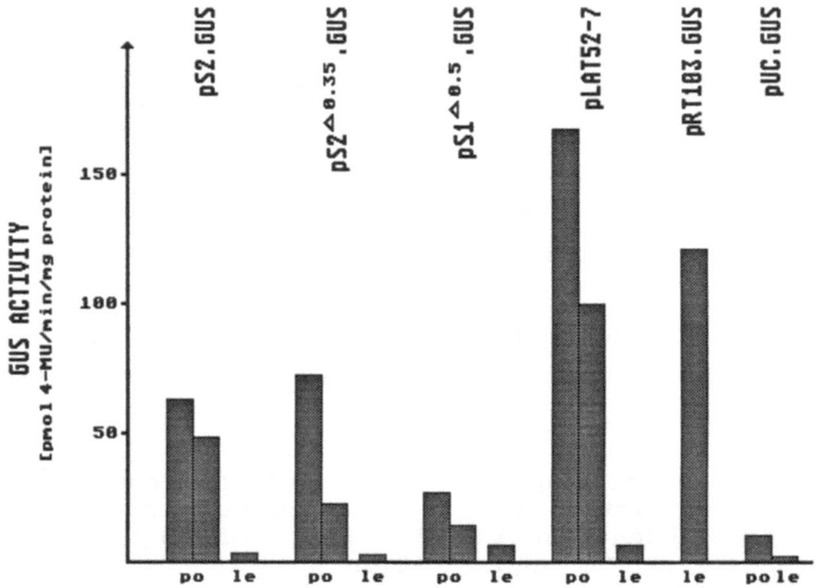

The 5.6 kb S2-promoter GUS fusion (Fig. 5) was subcloned into pBin19 and has been stably integrated into SR1 tobacco and cv. Desireé potato plants via A. tumefaciens-mediated transformation. Data from transgenic tobacco plants suggest that the S2-promoter is expressed in a developmentally regulated way in floral tissues (Fig. 7), although the plants investigated seem to fall into different expression classes. Some of the transgenic plants show GUS-expression only in the stigmatic part of the pistil, whereas others have activity in both the stigma and the transmitting tissue. Differences between expression patterns of pistil-expressed genes in transgenic plants and in their natural host environment have been reported by others (Budelier, Smith and Gasser, 1989).

Figure 7 Localization of GUS activity in pistil (a) and pollen (b) of an S2-GUS transgenic tobacco plant.

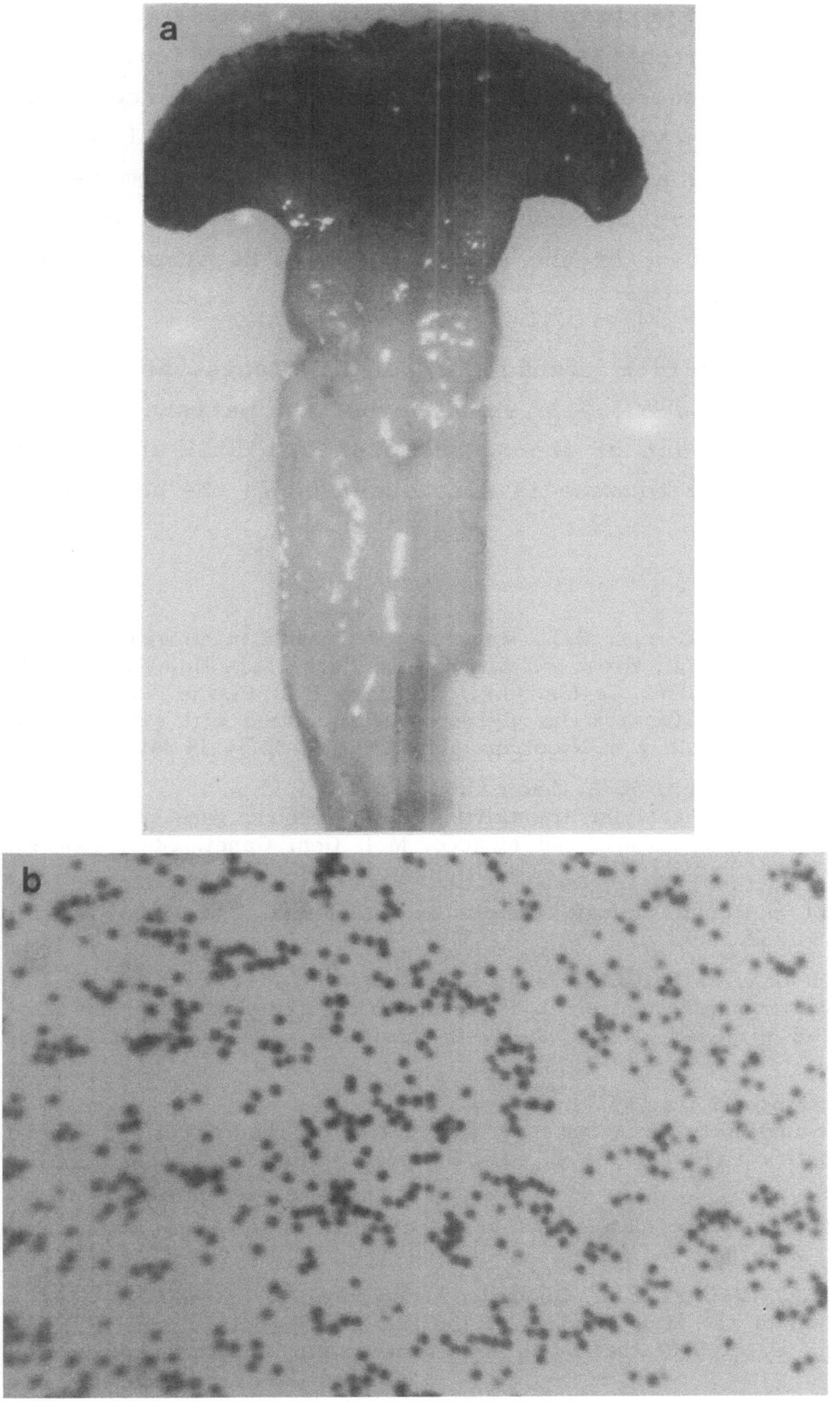

All of the analysed plants nevertheless have a very high GUS-activity in pollen, which is in correspondance with the transient expression data but contrasts with the failure to detect S-RNAse transcripts in pollen. It is, therefore, unclear what gene product arises from this expression in vivo. The fact that virtually all of the pollen grains stain blue indicates that the promoter is probably active sporophytically, as expression arising from segregation of the transgene would be otherwise detected.

Data arising from transgenic potato plants, which will be available soon, will show whether the pattern of regulated expression of the S2-promoter in pistil and pollen of transgenic tobacco is also observed in the homologous host.

Literature

Anderson M.A., Cornish E.C., Mau S.L., Williams E.G., Hoggart R., Atkinson A., Bonig I., Grego B., Simpson R., Roche P.J., Haley J.P., Niale H.D., Tregear G.W., Coghlan J.P., Crawford R.J., Clarke A.E., (1986). Cloning cDNA for a stylar glycoprotein associated with expression of self-incompatibility in Nicotiana alata. Nature 321 : 38-44.

Budelier J.A., Smith A.G., Gaser C.S., (1990). Regulation of a stylar transmitting tissue-specific gene in wild type and transgenic tomato and tobacco. Mol. Gen. Genet. 224 : 183-192.

Chasan R., (1991). Plant molecular biology blossoms in the desert. The Plant Cell 3 : 1255-1262.

Cornish E.C., Pettitt J.M., Bonig I., Clarke A.E., (1987). Developmentally-controlled, tissue specific expression of a gene associated with self-incompatibility in Nicotiana alata. Nature 326 : 99-102.

de Nettancourt D., (1977). Incompatibility in angiosperms. Monographs on Theoretical and Applied Genetics 3 (Frankel et al. ed.), Springer Berlin.

Ebert P.R., Anderson M.A., Bernatzky M., Clarke A.E., (1989). Genetic polymorphism of self-incompatibility in flowering plants. Cell 56 : 255-262.

Gray J.E., McClure B.A., Bönig I., Anderson M.A., Clarke A.E., (1991). Action of the style product of the self-incompatibility gene of Nicotiana alata (S-RNAse) on in vitro grown pollen tubes. The Plant Cell

3 : 271-282.

Hodgkin T., Lyon G.D., Dickinson H.G., (1988).
Recognition in flowering plants : A comparison of the Brassica self-incompatibility system and plant pathogen interactions. New Phytol 110 : 557-569.

Kaufmann H., (1991).
Sequenzvariabilität und Genstruktur des Selbstinkompatibilitäts-Lokus in Solanum tuberosum L.. PhD-Thesis.

Kaufmann H., Salamini F., Thompson R.D., (1991).
Sequence variability and gene structure at the self-incompatibility locus of Solanum tuberosum L.. Mol. Gen. Genet. 226 : 457-466.

Kirch H.H., Uhrig H., Lottspeich F., Salamini F., Thompson R.D., (1989).
Characterization of proteins associated with self-incompatibility in Solanum tuberosum. Theor. Appl. Genet. 78 : 581-588.

McClure B.A., Haring V., Ebert P.R., Anderson M.A., Simpson R.J., Sukiyama F., Clarke A.E., (1989).
Style self-incompatibility gene products of Nicotiana alata are ribonucleases. Nature 342 : 955-957.

McClure B.A., Gray K.E., Anderson M.A., Clarke A.E., (1990).
Self-incompatibility in Nicotiana alata involves degradation of pollen rRNA. Nature 347 : 757-760.

Ori N., Dessa G., Lotan T., Himmelhoch S., Fluhr R., (1990).
A major stylar matrix polypeptide (Sp41) is a member of the pathogenesis-related proteins superclass. EMBO J. 9 : 3429-3436.

Plegt L., Bino R.J., (1989).
β-glucuronidase activity during development of the male gametophyte from transgenic and non-transgenic plants. Mol. Gen. Genet. 216 : 321-327.

Stein J.C., Howlett B., Boyes D.C., Nasrallah M.E., Nasrallah J.B., (1991).
Molecular cloning of a putative receptor protein kinase gene encoded at the self-incompatibility locus of Brassica oleracea. Proc. Natl Acad. Sci. USA 88 : 8816-8820.

Thompson R.D., Kaufmann H., Kirch H.H., Salamini F., (1991).
Self-incompatibility in flowering plants. In : Genetics and Breeding of Ornamental species (Harding J., Singh F., Mol J.N.M. (eds)). Current Plant Sci. and Biotechnol. in Agricult. 11 : 367-385.

Thompson R.D., Uhrig H., Hermsen J.G.T., Salamini F., Kaufmann H., (1991).
Investigation of a self-compatible mutation in Solanum tuberosum clones inhibiting S-allele activity in pollen differentially. Mol. Gen. Genet. 226 : 282-288.

Twell D., Yamaguchi J., Wing R.A., Ushiba J., McCormick S., (1991).
Promoter analysis of genes that are coordinately exressed during pollen development reveals pollen-specific enhancer sequences and shared regulatory elements. Genes & Development 5 : 496-507.

The Use of Incompatibility in Breeding Programmes

P. Olesen, L. Bruun (a) and **P. Steen** (b)

Maribo Biotechnology
Langebrogade 1
P.O. Box 17
DK-1001 Copenhagen K
Denmark

INTRODUCTION

In the course of evolution nature has developed a multitude of solutions to prevent inbreeding and promote outcrossing within the plant kingdom. Apart from the mechanisms that are based on spatial or chronological separation of sexes (Van Marrewijk 1989) there is a physiological mechanism of inbreeding prevention defined as **incompatibility.** Although most authors define incompatibility as a prefertilization barrier and consider post-fertilization failures in the development of the zygote as a separate phenomenon, much confusion exists on the specificity of incompatibility. Hogenboom (1975) advocated a clear distinction between **intraspecific incompatibility** which is under the control of one or a few specific S-loci and **interspecific incompatibility** based on genetic inability of the crossing partners to set the process of pollen germination and style penetration going (**incongruity**). Most authors, however, doubt whether the two mechanisms are completely independent and always distinguishable and rather believe that interspecific incompatibility involves alleles of the S-locus in combination with other loci than the S-gene (e.g., de Nettancourt 1977).

In simplistic terms **self-incompatibility** in higher plants is defined as the inability of a fertile hermaphrodite seed-plant to produce zygotes after self-pollination (de Nettancourt 1977). However, it is important to note that, unlike in animal and algal systems, the incompatibility reaction is not between the gametes themselves, but between a male determinant (gametophytic or sporophytic by origin) carried by the pollen grain and the somatic pistil tissue of the female sporophyte.

a)
University of Copenhagen
Botanical Laboratory
Gothersgade 140
DK-1123 Copenhagen K
Denmark

b)
Maribo Seed
Højbygaardvej 14
P.O. Box 29
DK-4960 Holeby
Denmark

For simplicity, reflecting also its practical use in breeding programmes, breeders are using the terms self-incompatibility (SI) for intraspecific incompatibility (although not correct in strict terms) and **cross-incompatibility** (CI) for interspecific incompatibility. Generally CI is much less understood and explained in genetic and biochemical terms than SI.

GOALS AND DEFINITIONS

Self-incompatibility, constituting either a nuisance or a benefit to the plant breeder, occurs in most plant families and is common among cultivated species (forage crops, cereals, root crops, fruit trees, vegetables and ornamentals, cf. de Nettancourt 1977). However, weakening of SI during the process of inbreeding is often observed and may result from adverse recombinations of the genetic background (van Marrewijk 1989).

In general, practical use of SI in breeding is associated with either methods to overcome the incompatibility, methods to induce or transfer incompatibility, or methods to analyse and detect S-allele patterns (Table 1).

Table 1: Areas of main interest for plant breeders to utilize incompatibility in breeding programmes. For detailed procedures and reference to examples, see de Nettancourt (1977) and Van Marrewijk (1989).

Goal	Target	Result
Overcome incomp.	SI	F_1 hybrid seed Inbred homozygotes
	CI	Intraspecific hybrids Interspecific hybrids Intergeneric hybrids
Induce/transfer incomp.	SI	Heterotic F_1 hybrids Suppressed fructification
Diagnostics of incomp.	SI	Identified S-alleles Predicted S-allele behaviour

OVERCOMING INCOMPATIBILITY AND THE USE OF PSEUDO-COMPABILITY

Some of the most essential features of plant breeding programmes, such as homozygotes, intraspecific, interspecific and intergeneric hybrids, are all related to and somehow depending on incompatibility systems. For the production of inbred homozygotic lines either a multiscale production of double haploids or a breakdown of the self-incompatibility barriers is absolutely necessary. In self-incompatible plants as well as in the production of intra- and interspecific and intergeneric hybrids, the crossing barriers are generally caused by reactions occurring during the process of pollination, fertilization and embryogenesis (Zenkteler 1980).

Many attempts have been made to overcome these barriers, ranging from changes in the physiological environmental conditions during growth of the plants to a whole array of different in vitro pollination/fertilization manipulations. Common to all these attempts are that the functional incompatibility system is transiently broken down but continues to be inherited in the offspring (Richards 1986), a phenomenon which has been defined as **pseudocompatibility** (Pandey 1959). Most of these aspects have been carefully reviewed by de Nettancourt (1977), Richards (1986) and Van Marrewijk (1989).

The practical applications of pseudocompatibility can be divided into two groups, one of which is based on the natural existence of pseudocompatibility, e.g. delayed pollination, end of season effects, mentor effects, and high temperature. The use of manipulations such as stylar and pollen irradiation, biological inhibitors, chemicals, bud pollination and in vitro pollination/fertilization forms another group of practical applications of pseudocompatibility.

A list of effective techniques to overcome incompatibility was published by Shivanna (1980, Table 2). Although many methods have been developed, the general effectiveness is still much higher in intraspecific than in interspecific crosses (Shivanna 1980). The in vitro techniques include in vitro pollination of flowers developed in culture (test tube fertilization), in vitro pollination on dissected cultured pistils, and intra-ovarian and placental pollinations (Shivanna 1980 and references herein) and fusion of isolated gametes (Kranz et al. 1991 a,b).

In a recently published study on the effect of *in vitro* culture on self-incompability in *Trifolium pratense* and *T. repens*, *in vitro* pollination was compared to *in situ* pollination regarding pollen tube growth and seed set. Their data clearly illustrate the problems in using *in vitro* culture for overcoming SI, e.g. problems with different species and even genotypes having different ability to overcome incompatibility. The SI response was strongly weakened *in vitro* for *Trifolium repens* but not for *T. pratense*, and it was concluded that *in vitro* pollination is a good tool to produce inbred lines of *T. repens*, but not of *T. pratense* (Douglas and Connolly 1989, Leduc et al.1990).

In vitro fusion of single isolated gametes which has now been done successfully in maize (Kranz et al. 1991 a,b) provides a new and promising application for overcoming incompatibility, at least if the fusion products can be cultured into viable plants.

INDUCTION AND DIAGNOSIS OF INCOMPATIBILITY

For the production of F_1 hybrid seed it must be ensured that fertilization only occurs between two defined groups of plants and that selfing does not take place. For this purpose several hybrid breeding programmes are exploiting male sterility systems or is regulated through the use of gametocides. But since SI is a natural system of breeding regulation, and since both gametes are fertile in every plant, the use of SI is obviously to be preferred.

However, full use of SI still creates problems for the breeders. The degree of SI is often not fully reliable, especially under field conditions (weak S-alleles?), and dominance relationships between S-alleles are poorly understood (Thompson and Taylor 1966). In other crops, reliable SI is very hard to find and, even when available, it would be very useful to be able to turn SI off and on at will, to cheapen the cost of seed production. In conclusion, there seems to exist an almost desperate need to develop techniques for the transfer of SI to a number of key crop families where male sterility is currently the only option (Dickinson 1989).

Very recently, studies on the development of an SI-based production system for high-yielding hybrid alfalfa cultivars (Bauchan et al. 1991) has elegantly demonstrated the high potential being present in self-incompatible clones with the ability to produce somatic embryos. When coupled with the advancing technology for the use of artificial seeds, such system would allow for the production of elite germplasm in production fields.

Since the determination of S-genotypes often is a complex and difficult task, especially in perennials with long reproduction cycles, it is desirable to replace the tedious procedures of test pollinations and progeny analysis by more direct methods allowing the immediate identification of the S-allele pattern in each individual plant considered. For this purpose it would certainly be beneficial if the analyses could be made on vegetative tissues. For assays based upon gene products such as S-locus specific glycoproteins (e.g., using allele-specific antibodies) this would probably not be feasible since the expression of these compounds occurs in strictly tissue-specific patterns in reproductive structures only (see below). For this and other reasons (e.g., improved specificity and sensitivity) assays based on nucleic acid probes (various hybridization tests) appear to have much larger potentials for the identification and characterization of S-alleles.

MOLECULAR GENETICS OF INCOMPATIBILITY

The most frequent form of SI in the angiosperms, and occurring in more than half of the plant families, is the one-locus multi-allelic gametophytic system. Here, between 3 and over 400 alleles appear to occur at a single locus termed the S-locus. It is generally considered that all other mechanisms of SI are secondary, having arisen from the one-locus gametophytic system (Richards 1986).

In both gametophytic and sporophytic SI-systems a family of highly charged glycoproteins have been discovered in the pistil and which segregate with S-phenotypes (Clarke et al. 1989). Now both cDNA and genomic clones exist for these S-glycoproteins from several plant species. Although the functional role of the S-glycoproteins in SI-interactions has not yet been demonstrated _in vivo_, many lines of evidence strongly suggest their involvement in the recognition and rejection of self-pollen (Haring et al. 1990 and data in the preceding two chapters in the present volume).

In gametophytic SI-systems sequence homology between the S-glycoproteins and fungal ribonucleases together with the demonstrated ribonuclease activity of isolated S-glycoproteins now indicate that the inhibitory action of S-glycoproteins may be exerted through their ribonuclease activity in a pollen-specific way (McClure et al. 1989). For sporophytic S-glycoproteins no sequence similarities to ribonucleases have been found. However, a cDNA clone for a receptor protein kinase from maize has been reported to show significant homology to the Brassica S-glycoprotein sequence (Walker and Zhang 1990) indicating that S-locus glycoproteins may be involved in transmembrane signalling processes related to recognition and rejection.

Final confirmation of the function of the S-glycoproteins and that their corresponding genes represent S-alleles must await functional analyses in transgenic plants - to analyse their effect on the SI phenotypes. In a recent study on the expression of a sporophytic S-glycoprotein gene from Brassica in the transmitting tissue of transgenic tobacco plants (Moore and Nashrallah 1990), SI was not conferred by the introduced gene - but the expression data clearly imply similarities in the promoter regions of sporophytic and gametophytic putative SI-genes.

Still, a major barrier to progress in understanding the molecular basis of SI, is the fact that the male protagonist has yet to be identified in either sporophytic or gametophytic systems (Dickinson 1990).

Compared to SI, virtually nothing is known about the control of interspecific incompatibility systems (cross-incompatibility, CI). Although crossing studies have encouraged speculation about some genetic principles, the basis of CI remains obscure (Pandey 1962, Fritz and Hanneman 1989). However, in relation to the above mentioned possible involvement of a protein kinase in sporophytic SI, a change in activity of adenylate cyclase (another molecule of a putative secondary messenger signalling pathway) has been reported in association with compatible versus incompatible pollinations in interspecific crosses between Populus species (Rougier et al. 1988).

POTENTIALS FOR THE FUTURE

Since putative S-allele genes normally expressed in female pistil tissues still have to be analyzed carefully in transgenic combinations, and male SI-components have not been identified at all, we are still pretty far away from using genetic manipulation to transfer naturally occurring SI-systems or parts thereof to functional heterologous combinations. As pointed out earlier (de Nettancourt 1977) it must also be remembered that the extreme sensitivity of the S-locus to its genetic background may constitute, in this respect, some constraint on the heterologous function of SI systems.

However, with the advent of PCR and other techniques for rapid production of gene probes and RFLP/RAPD genome mapping being developed in most crop species, it can be foreseen that methods for S-allele identification and predictability will become available in the near future. Along these lines there are also good reasons to believe that the molecular structure of different S-loci will be characterized in enough detail to determine which domains confer specificity and, even, "weakness" and "strength" of the S-alleles.

It is to be anticipated that in vitro bioassays for the SI-response mechanism will play an important role in future work, as indicated by recent developments in gametophytic (Franklin-Tong et al. 1988) as well as sporophytic (Singh and Paolillo 1989) systems - although, as pointed out by Jackson and Linskens (1990), several pitfalls are to be considered.

For many decades plant breeders have been looking for methods to break crossing barriers, i.e. to manipulate interspecific incompatibility in such a way that crossings not being possible today would be facilitated (wide crossings, introgression etc.). Since the molecular basis of cross-incompatibility still remains obscure and the relationship between CI and SI has not been further understood, the possible manipulation with CI seems even further away than for SI.

In conclusion, the fulfilment of the plant breeders' and biotechnologists' dreams on modulating incompatibility systems to allow for the breakdown of crossing barriers on one hand and to fully control crossing behaviour on the other hand, has come some steps closer through the progress made in the last decade. This progress has mainly been in the molecular understanding of SI and the possibilities for S-allele diagnosis. However, much work remains to be done, not only on continued molecular characterization of the S-locus, but also on many other aspects that are needed in the multidisciplinary approach which is a prerequisite for a better understanding of the complex incompatibility processes - and its utilization in plant breeding.

REFERENCES

Bauchan GR, Campbell TA, O'Neill NR, Elgin JH Jr (1991) Self-incompatibility in two alfalfa populations. Crop Sci 30: 1205-1210

Clarke AE, Anderson MA, Atkinson A, Bacic A, Ebert PR, Jahnen W, Lush WM, Mau SL, Woodward JR (1989) Recent developments in the molecular genetics and biology of self-incompatibility. Plant Molecular Biology

Dickinson HG (1989) Core biology of plant breeding. CEC, Bruxelles

Dickinson HG (1990) Self-incompatibility in flowering plants. Bioassays 12: 155-161

Douglas GC, Connolly V (1989) Self-fertilization and seed set in Trifolium repens L. by in situ and in vitro pollination. Theor Appl Genet 77: 71-75

Franklin-Tong VE, Lawrence MJ, Franklin FCH (1988) An in vitro bioassay for the stigmatic product of the self-incompatibility gene in Papaver rhoeas L. New Phytol 110: 109-118

Fritz NK, Hannemann RE Jr (1989) Interspecific incompability due to stylar barriers in tuber-bearing and closely related non-tuber-bearing Solanums. Sex Plant Reprod 2: 184-192

Haring V, Gray JE, McClure BA, Anderson MA, Clarke AE (1990) Self-incompatibility: A self-recognition system in plants. Science 250: 937-941

Hogenboom NG (1975) Incompatibility and incongruity: Two different mechnisms for the non-functioning of intimate partner relationships. Proc R Soc London Ser B 188: 361-375

Jackson JF, Linskens HF (1990) Bioassays for incompatibility. Sex Plant Reprod 3: 207-212

Kranz E, Bautor J, Lörz H (1991) In vitro fertilization of single, isolated gametes of maize mediated by electrofusion. Sex Plant Reprod 4: 12-16

Kranz E, Bautor J, Lörz H (1991) Electrofusion-mediated transmission of cytoplasmic organelles through the in vitro fertilization process, fusion of sperm cells with synergids and central cells, and the reconstitution in maize. Sex Plant Reprod 4: 17-21

Leduc N, Douglas GC, Monnier M, Connolly V (1990) Pollination in vitro. Effects on the growth of pollen tubes, seed set and gametophytic self-incompatibility in Trifolium pratense L. and T. repens L. Theor Appl Genet 80: 657-664

McClure BA, Haring V, Ebert PR, Anderson MA, Simpson RJ, Sakiyama F, Clarke AE (1989) Style self-incompatibility gene products of Nicotiana alata are ribonucleases. Nature 342: 955-957

Moore HM, Nashrallah JB (1990) A Brassica self-incompatibility gene is expressed in the stylar transmitting tissue of transgenic tobacco. Plant Cell 2: 29-38

Nettancourt D de (1977) Incompatibility in angiosperms. Springer, Berlin Heidelberg New York

Pandey KK (1959) Mutations of the self-incompatibility gene (S) and pseudo-compatibility in angiosperms. Lloydia 22: 222-234

Pandey KK (1962) Interspecific incompatibility in Solanum species. Am J Bot 49: 874-882

Richards AJ (1986) Plant Breeding Systems. George Allen Unwin, London

Rougier M, Inoud N, Dumas C (1988) Localization of adenylate cyclase activity in Populus: its relation to pollen-pistil recognition and incompatibility. Sex Plant Reprod 1: 140-149

Singh A, Paolillo DJ Jr (1989) Towards an in vitro bioassay for the self-incompatibility response in Brassica oleracea. Sex Plant Reprod 2: 277-280

Shivanna KR (1980) Pollen-pistil interaction and control of fertilization. In: Johri BM (ed.) Experimental embryology of vascular plants, pp 131-165

Thompson K, Taylor J (1966) Non-linear dominance relationships between S-alleles. Heridity 21: 345-362

Van Marrewijk GAM (1989) Overcoming incompatibility. In: Wright CJ (ed) Manipulation and fruiting. London, pp 173-191

Walker JC, Zhang R (1990) Relationship of a putative receptor protein kinase from maize to the S-locus glycoproteins of Brassica. Nature 345: 743-746

Zenkteler M (1980) Intraovarian and in vitro pollination. Int. Rev. Cytol. Supp. IIB: 137-155.

Fertilization in Angiosperm Plants

J.L. van Went
Department of Plant Cytology and Morphology
Wageningen Agricultural University
Arboretumlaan 4
6703 BD Wageningen
The Netherlands

In angiosperm plants the interaction between male and female gametophytes, the pollen tube and the embryo sac respectively, results in double fertilization. The pollen tube delivers two male gametes, the sperm cells, into the embryo sac. One sperm cell fuses with the egg cell to form the embryo, and the second sperm cell fuses with the central cell to form the endosperm. Most of our knowledge of the fertilization process is based on microscopical observations of fixed and sectioned materials (Jensen 1972; Van Went and Willemse 1984; Willemse and Van Went 1984). The embryo sac and the fertilization events taking place within it, cannot be observed directly, because it is positioned deeply inside the ovule, surrounded by a large quantity of untransparant sporophytic tissues.

Since a few years it has become possible to isolate male and female gametes in living condition (Theunis et al. 1991). Embryo sacs and composing cells can be separated from the surrounding sporophytic cells and accordingly isolated, by the use of enzymatic maceration techniques or combinations of enzymatic maceration and microdissection (Hu et al. 1985; Zhou and Yang 1986; Mol 1986; Huang and Russell 1989; Wagner et al. 1989; Van Went and Kwee

1990; Kranz et al. 1991). Living sperm cells can be isolated in large quantities by osmotic bursting of pollen (Russell 1986), pH-mediated pollen bursting (Dupuis et al. 1987), and grinding or squashing of pollen (Southworth and Knox 1989; Theunis and Van Went 1989). Presently, in a number of laboratories molecular and cell biological research is in progress focused on the developmental and functional aspects of male and female gametophytes and gametes. Isolated and living embryo sacs are used to analyze the fertilization process (Zhou 1987), and in-vitro fusion experiments with isolated male and female gametes are used to clarify the basic principles of gamete fusion in angiosperm plants (Kranz et al. 1991).

Sperm cell delivery in the embryo sac

Since the 1960's, a large number of articles has been published on the ultrastructural aspects of the entrance of the pollen tube into the embryo sac and the discharge of the tube content, including the delivery of the sperm cells (for review of literature see Van Went and Willemse 1984). Invariably, it has been observed that the pollen tube enters the embryo sac at the micropylar side, where it penetrates through the filiform apparatus into one of the synergids. The pollen tube never enters the egg cell or the central cell. Even in species that do not have synergids, like *Plumbago zeylanica*, the pollen tube approaches the embryo sac at the micropylar side and it delivers the sperm cells without distortion of the egg cell and the central cell (Russell 1982). In the latter species the portion of the embryo sac wall at the micropylar apex of the embryo sac and the adjacent portion of the wall in between the egg cell and the central cell develops into a structure that highly resembles the filiform apparatus formed by the synergids in other species. It is into this specialized cell wall part that the pollen tube penetrates, as it does into the filiform apparatus of the synergids. Evidently, the filiform apparatus or a comparable structure plays a key role in the interaction between the male and the female gametophytes.

After the pollen tube has passed the filiform apparatus it enters into the cytoplasm of one of the synergids. In many species the latter degenerates before the arrival of the pollen tube. In other species, however, either both synergids degenerate before the pollen tube arrives, or both synergids remain intact until the moment of actual penetration of the pollen tube. Soon after the pollen tube has reached the synergid cytoplasm, its growth ceases and the tube opens. In some species the pollen tube simply bursts at its tip. In other species, however, a distinct subterminal pore is formed in the pollen tube wall. Subsequently, a considerable portion of the tube content, including the vegetative nucleus and the two sperm cells, is injected into the penetrated synergid. After the entrance of the pollen tube content, the vacuole of the synergid has always disappeared, the volume of the synergid has strongly enlarged, and the synergid content shows various signs of degeneration. Frequently, a degeneration or rupture of the synergid plasma membrane is observed in the chalazal region of the cell. Degenerated cytoplasm of the synergid is seen in between the adjacent cells, in close contact to the plasma membranes of the egg cell and the central cell. At the mature embryo sac stage in this region the cells of the egg apparatus (synergids and egg cell) and the central cell are separated from each other by their plasma membranes only, and not by solid pecto-cellulosic cell walls. Apparently, the plasma membranes of the egg cell and central cell are protected in some way from being affected by the contact with the degenerating synergid cytoplasm. Degeneration also occurs in the penetrating pollen tube material, except in the sperm cells. Inside the pollen tube, the sperm cells are enveloped by two plasma membranes (one of the sperm cell itself and one of the surrounding vegetative cell) and a very thin intercellular layer. After the discharge of the tube content in the synergid, the outer one of those membranes (the plasma membrane of the vegetative cell) degenerates, but apparently the sperm cells themselves are protected for degeneration, as are the egg cell and the central cell. The sperm cells, however, remain not unaltered. Inside the pollen tube they have a clear spindle shape, and they contain numerous bundles of cortical microtubules. Inside the penetrated synergid, their shape changes to spherical, and the cortical microtubules disappear (Mogensen 1982). In these respects, they strongly resemble sperm cells after artificial isolation from pollen grains or tubes (Theunis 1990). In species like *Plumbago zeylanica* which lack synergids, the pollen tube injects its content into the filiform apparatus, from which it then invades in between the egg cell and the central cell. The injected material shows similar degenerative changes as are observed in penetrated synergids.

As result of the events described above, at the time of actual fertilization both male and female gametes have been transformed into complete or partial naked protoplasts, and the plasma membranes of the sperm cells become directly exposed to the plasma membranes of the egg cell and the central cell.

Gamete fusion

It is not known by which mechanism the male and female gametes are brought together. It might be that the mechanical forces of the injection process cause the sperm cells to move to their final positions for the subsequent fusion. Likely, in the pollen tube the sperm cells are subjected to a transport mechanism based on an actin-myosin system located in the vegetative cell (Heslop-Harrison and Heslop-Harrison 1989). But it is unlikely that such a mechanism is operating in the penetrated and degenerating synergid. Furthermore, it is unlikely that angiosperm sperm cells move autonomously, since independent and active movement of in situ or isolated sperm cells has never been observed. The mechanism of sperm cell transfer and positioning in the synergid becomes even more intriguing since it has been established that, at least in some species mating of the gametes is not at random. In these species preferential fertilization occurs, in which one sperm cell is predetermined to fuse with the egg cell, while the second one always fuses with the central cell. For *Zea mays* it has been established that the sperm cell containing the B-chromosomes routinely fuses with the egg cell (Roman 1948). In *Plumbago zeylanica*, a species with dimorphic sperm cells, the sperm cell containing the plastids generally fuses with the egg cell (Russell 1985). Preferential fertilization likely requires an adequate recognition system, by which the female gametes can recognize their respective fusion partners. Another possibility is that the predetermined fusion partners are brought together by a very accurate deposition mechanism.

Gametic fusion may occur very rapidly. In *Plumbago zeylanica* it has been established that the presence of unfused sperm cells in the female gametophyte is lasting less than 5 minutes, while the progamic phase takes a period of 8.5 hours (Russell 1983). Fusion of the male and female gametes starts with local contact and fusion of their respective membranes. In *Plumbago zeylanica*, gametic fusion is initiated by a single fusion event between the membranes of the male and female gametes, followed by secondary fusions at different locations along their surfaces. Evidently, gametic fusion in angiosperms is a membrane-based event, for which the characteristics, exposure and contact of the gametic membranes are essential. As is pointed out before, exposure and contact of the gametic membranes is facilitated by the degeneration of the synergid and pollen tube cytoplasm and plasma membranes. The degenerative changes in the penetrated synergid are not just a result of the pollen tube entrance and the interaction between the pollen tube and synergid content, but they appear to be vital prerequisites for successful fertilization. In addition, the gametic membranes must have specific characteristics to enable the membrane recognition and fusion. Such specific characteristics have been observed in sperm cell plasma membranes by freeze-fracture analysis (Van Aelst et al. 1990). With this technique it is possible to establish the number, position and arrangement of intramembrane protein particles (IMPs). Generally, in somatic plant cells the cytoplasmic half of a plasma membrane (PF face) contains more IMPs than the peripheral half (EF face). In plasma membranes of sperm cells the EF face appears to contain three times the number IMPs as present in the PF face. Wether this high number of IMPs in the EF face of the sperm cell plasma membrane is related to specific recognition and fusion capacities of the gametic plasma membranes remains to be solved.

The membrane-based gamete fusion concept implicates that not only the male gamete nucleus is transmitted, but that there may also be co-transmission of sperm cell cytoplasm. Co-transmission of sperm cell cytoplasm may have important implications for cytoplasmic inheritance (Keijzer et al. 1988). Co-transmission of sperm cell cytoplasm has been established by genetical studies using marker genes located in male cytoplasmic organelles (Kirk and Tilney-Bassett 1978). Ultrastructural evidence for male cytoplasmic organelle transmission has been presented by Meyer and Stubbe (1974) for *Oenothera erythrosepala*, and by Russell (1983) for *Plumbago zeylanica*. In the latter species, the sperm cell plastids are morphologically different from the plastids of the egg cell. Plastids of sperm cell origin have been identified inside the fertilized egg cell, in numbers matching the number of plastids in the unfused sperm cell (Russell 1985). However, co-transmission of sperm cell cytoplasm is not a rule. In barley, a male cytoplasmic body is found in the degenerated synergid just outside the egg cell at the point where the male nucleus has entered (Mogensen 1982). This structure is identified as the entire male cytoplasm that was left outside the egg cell during the fusion between egg cell and sperm cell (Mogensen 1988). In addition, no identifiable male cytoplasmic organelles were found in the fertilized egg cell. These observations strongly suggest that in barley the male cytoplasm does not enter the egg cell during fertilization. Also for barley it has been established that gamete fusion is based on interaction and fusion of the plasma membranes of the gametes. The way in which the male cytoplasm is excluded from being transmitted to the egg cell remains to be solved. However, only one cytoplasmic body per fertilized embryo sac is found, located near the egg cell. It is likely, that the cytoplasmic content of the second sperm cell is transmitted to the central cell. This indicates that there might be fundamentally different mechanisms of fertilization between the first sperm cell and the egg cell, and the second sperm cell and the central cell, respectively.

The next step in the process of double fertilization is the fusion of the sperm cell nucleus with the egg cell nucleus, and the fusion of the second sperm cell nucleus with the (fused) polar nuclei. The nuclear fusions start with local contact and fusion of subsequently the outer nuclear membranes and the inner nuclear membranes (Jensen 1964; Mogensen 1982). Frequently, the nuclei to be fused are seen connected to each other by cisterns of endoplasmic reticulum, which is continuous with the outer nuclear membranes. The endoplasmic reticulum could be involved in initiating the nuclear fusion, or in the movement of the nuclei towards each other. The fusion of the outer and inner nuclear membranes results in the formation of nucleoplasmic bridges between the fusing nuclei, which enlarge and coalesce. Finally, the contents of the male and female nuclei intermingle, which marks the completion of fertilization.

References

Dupuis I, Roeckel P, Matthys-Rochon E, Dumas C (1987) Procedure to isolate viable sperm cells from corn (*Zea mays* L.) pollen grains. Plant Physiol 85:876-878

Heslop-Harrison J, Heslop-Harrison Y (1989) Myosin associated with the surfaces of organelles, vegetative nuclei and generative cells in angiosperm pollen grains and tubes. J Cell Sci 94:319-325

Hu S-Y, Li L-G, Zhou C (1985) Isolation of viable embryo sacs and their protoplasts of *Nicotiana tabacum*. Acta Bot Sin 27:337-344

Huang B-Q, Russell SD (1989) Isolation of fixed and viable eggs, central cells and embryo sacs from ovules of *Plumbago zeylanica*. Plant Physiol 90:9-12

Jensen WA (1964) Observations on the fusion of nuclei in plants. J Cell Biol 23:669-672

Jensen WA (1972) The embryo sac and fertilization in angiosperms. Harold L. Lyon Arbor Lect 3:1-32

Keijzer CJ, Wilms HJ, Mogensen HL (1988) Sperm cell research: the current status and applications for plant breeding. In: Wilms HJ, Keijzer CJ (eds) Plant sperm cells as tools for biotechnology. Pudoc, Wageningen, pp 3-8

Kirk JTO, Tilney-Bassett RAE (1978) The plastids: their chemistry, structure, growth and inheritance. Elsevier/North Holland Biomed Press, Amsterdam

Kranz E, Bautor J, Lörz H (1991) In vitro fertilization of single, isolated gametes of maize mediated by electrofusion. Sex Plant Reprod 4:12-16

Meyer B, Stubbe W (1974) Das Zahlenverhaltnis von mutterlichen und vaterlichen Plastiden in den Zygoten von *Oenothera erythrosepala* Borbas (syn. *Oe. Lamarckiana*). Ber Dtsch Bot Ges 87:29-38

Mogensen HL (1982) Double fertilization in barley and the cytological explanation for haploid embryo formation, embryoless caryopses, and ovule abortion. Carlsberg Res Commun 47:313-354

Mogensen HL (1988) Exclusion of male mitochondria and plastids during syngamy in barley as a basis for maternal inheritance. Proc Natl Acad Sci, USA 85:2594-2597

Mol R (1986) Isolation of protoplasts from female gametophytes of *Torenia fournieri*. Plant Cell Rep 3:202-206

Roman H (1948) Directed fertilization in maize. Proc Natl Acad Sci, USA 34:46-52

Russell SD (1982) Fertilization in *Plumbago zeylanica*: entry and discharge of the pollen tube in the embryo sac. Can J Bot 60:2219-2230

Russell SD (1983) Fertilization in *Plumbago zeylanica*: gametic fusion and fate of the male cytoplasm. Amer J Bot 70:416-434

Russell SD (1985) Preferential fertilization in *Plumbago*: ultrastructural evidence for gamete-level recognition in an angiosperm. Proc Natl Acad Sci, USA 82:6129-6132

Russell SD (1986) A method for the isolation of sperm cells in *Plumbago zeylanica*. Plant Physiol 81:317-319

Southworth D, Knox RB (1989) Flowering plant sperm cell: isolation from pollen of *Gerbera jamesonii* (Asteraceae). Plant Sci 60:273-277

Theunis CH (1990) Ultrastructural analysis of *Spinacia oleracea* sperm cells isolated from mature pollen grains. Protoplasma 158:176-181

Theunis CH, Van Went JL (1989) Isolation of sperm cells from mature pollen grains of *Spinacia oleracea* L. Sex Plant Reprod 2:97-102

Theunis CH, Pierson ES, Cresti M (1991) Isolation of male and female gametes in higher plants. Sex Plant Reprod 4:145-154

Van Aelst AC, Theunis CH, Van Went JL (1990) Freeze-fracture studies on isolated sperm cells of *Spinacia oleracea* L. Protoplasma 153:204-207

Van Went JL, Kwee H-S (1990) Enzymatic isolation of living embryo sacs of *Petunia*. Sex Plant Reprod 3:257-262

Van Went JL, Willemse MTM (1984) Fertilization. In: Johri BM (ed) Embryology of angiosperms. Springer, Berlin Heidelberg New York Tokyo, pp 273-317

Wagner VT, Kardolus JP, Van Went JL (1989) Isolation of the lily embryo sac. Sex Plant Reprod 2:219-224

Willemse MTM, Van Went JL (1984) The female gametophyte. In: Johri BM (ed) Embryology of angiosperms. Springer, Berlin Heidelberg New York Tokyo, pp 159-196

Zhou C (1987) A study of fertilization events in living embryo sacs isolated from sunflower ovules. Plant Sci 52:147-151

Zhou C, Yang HY (1986) Isolation of embryo sacs by enzymatic maceration and its potential in haploid study. In: Hu SY, Yang HY (eds) Haploids of higher plants in vitro. Springer, Berlin Heidelberg New York, pp 192-203

In Vitro Fertilization in Flowering Plants

E. MATTHYS-ROCHON
Reconnaissance Cellulaire et
Amélioration des Plantes
Université C.Bernard LYON I
43 Bd. du 11 Novembre 1918
69622 Villeurbanne Cedex

INTRODUCTION

Among the most refractory remaining physiological problems concerning sexual reproduction is the mechanism of interaction between the male and female gametes during fertilization. In Angiosperms, the double fertilization process is dependant on numerous events beginning with the deposition of the pollen grain on the stigmatic surface and ending with the double gametic fusion within the embryo sac. During the last decade a renewed interest in this subject has been triggered as a result of the discovery, through cytological studies, of cytoplasmic dimorphism of sperm cells (Russell 1983). In *Plumbago,* Russell (1985) has shown the occurence of a targeted fertilization process: One of the two sperm cells which is plastid-rich preferentially fuses with the egg cell and subsequently forms the embryo, the second (mitochondria rich) associates with the central cell to initiate endosperm formation.

During fertilization, the male gametes are released between the female reproductive cells, with sperm membranes touching both the egg and central cell. Fertilization, therefore appears to be a double cellular fusion process which could involve gamete recognition at the membrane level (Russell 1985).

One has to remember that in flowering plants gametic fusion occurs deep in the ovule. The female cells are enveloped within the gametophyte (embryo sac) and the male cells are continuously in the pollen tube cytoplasmic environment until the fusion events. Consequently, the only way to better understand the gamete recognition and fusion is first to isolate viable gametes and second to use *in vitro* fertilization techniques. This paper will review the main procedures used for the isolation of male and female gametes and present the *in vitro* models developed mainly in Hamburg (lab. Prof.Lörz) and in Lyon (lab. Prof.Dumas).

GAMETE ISOLATION TECHNIQUES

The **male gametes** are isolated either from pollen grains or from the pollen tubes (Theunis et al. 1991). Generally, authors have used an osmotic shock or grinding or both to release the sperms. It is usually necessary to determine a specific protocole for every pollen species especially for pollen which contains a large amount of starch grains (Dupuis et al 1987). The difficulty is to eliminate starch and recover a high yield of intact sperm cells. Results obtained with several species are shown in table 1.

SPECIES	YIELD (a)	VIABILITY(b)	REFERENCES
Dicotyledons			
Plumbago zeylanica	1.7 10^5 cells, 60 to 75%	95%	Russell, 1986
Rhododendron spp.	90 to 270 pairs per style (c)	nd	Shivana *et al.*, 1988
Brassica oleracea	2%	nd	Roeckel *et al.*, 1988
Beta vulgaris	7 10^4 cells	30%	Nielsen and Olesen,1988
Gerbera jamesonii	6 10^4 cells	nd	Southworth and Knox, 1989
Spinacia oleracea	10^5 cells, 5 to 10%	90%	Theunis and Van Went,1989
Monocotyledons			
Zea mays	3 10^6 cells, 20 to 30 %	80 to 90%	Dupuis *et al.*, 1987 Roeckel *et al.*, 1988
Zea mays	1.5 10^6 cells	50%	Cass and Fabi, 1988
Gladiolus gandavensis	65 to 84 pairs per style (c)	nd	Shivanna *et al.*, 1988
Lolium perenne	2%	nd	Van der Maas and Zaal, 1990

(a): Yield was reconsidered from quantitative data of papers, as total number of isolated cells obtained at the end of a single isolation procedure and/or percentage of isolated cells recovered from the initial number of male gametes within pollen grains.
(b): Viability percentage of isolated cells at the end of isolation procedure.
(c): In this study, male gametes have been isolated from pollen tubes growing in style segments by the *in vivo/in vitro* method, and yield has been estimated by the number of isolated pairs of male gametes per style.

TABLE 1: ISOLATION OF MALE GAMETES IN DIFFERENT SPECIES. COMPARISON OF YIELD OBTAINED AND VIABILITY OF ISOLATED CELLS.
From Chaboud et al. 1992

When sperm cells are released in an appropriate medium they are generally elongated and become progressively spherical (Matthys-Rochon et al. 1987 and Matthys-Rochon and Dumas 1988). The FCR test (fluorochromatic reaction, Heslop Harrison and Heslop Harrison 1970) or Evans blue staining are applied to check their viability (table 1). Cytological studies have demonstrated the protoplast nature of the sperm (Dupuis et al. 1987, Wagner et al 1989 a).

As mentionned in the introduction, **the female gametes** are included in the embryo sac which is surrounded by a wall containing cellulose and sometimes callose (Matthys-Rochon et al. 1991). To prepare embryo sacs or female cells an enzymatic digestion and a precise manual dissection are necessary. With this method, it is possible to prepare viable embryo sacs and to release the gametic complex (synergids associated with the egg cell) and the central cells (Theunis et al. 1991). When the time of enzymatic digestion is increased, the female cells are completely individualized in the liquid medium and appear spherical. Table 2 gives the main references relating to the embryo sac or female cell isolation. Cell integrity is usually tested by the fluorochromatic reaction. Cytological studies have shown the protoplast nature of the isolated female cells.

SPECIES	ISOLATION METHOD	VIABILITY	REFERENCES
Dicotyledons			
Nicotiana tabacum	enzymatic digestion, squashing	+ (FCR)	Hu *et al.*, 1985
Antirrhinum majus	enzymatic digestion	+ (FCR)	Zhou and Yang, 1985
Torenia fournieri	enzymatic digestion	+ PhCM	Mol, 1986
Helianthus annuus	enzymatic digestion	+ (FCR)	Zhou, 1987
Plumbago zeylanica	enzymatic digestion, dissection	+ (FCR)	Huang *et al.*, 1989
Monocotyledons			
Lilium longiflorum	enzymatic digestion, suction	+ (FCR)	Wagner *et al.*, 1989b
Zea mays	enzymatic digestion, dissection	+ (FCR, TEM)	Wagner *et al.*, 1988 Wagner *et al.*, 1989c

FCR: fluorochromatic reaction test (according to Heslop-Harrison & Heslop-Harrison, 1970)
TEM: transmission electron microscopy
PhCM: Phase contrast microscope observation

TABLE 2: ISOLATION OF VIABLE FEMALE GAMETOPHYTE *I.E.* EMBRYO SAC IN DIFFERENT SPECIES From Chaboud et al. 1992

IN VITRO TECHNICAL PROCEDURES

In order to study gamete interactions, zygote formation and initiation of embryo development, different *in vitro* models of fertilization has been developed. The first attempt was by Keijzer (1988) in *Torenia Fournieri* . In that species the embryo sac emerges from the integuments and direct access to the female cells is possible. Sperm cells were released from germinated pollen by osmotically bursting. With the aid of microforceps, the **embryo sac** was held and **deposition of male gametes** was ashieved with microforged microcapillaries. Up to now no development has followed this artificial fertilization.

Other experiments have been carried out in maize. Unlike *Torenia* the female gametophyte in maize is deep included in the ovule. *In vitro* experiments therefore, require first the isolation of the embryo sacs or female gametes as described above (Wagner et al. 1989 b).

In Hamburg, E. Kranz has succeeded in ***in vitro* electrofusion of gametes** and production of "zygotes". The latter are able to divide in a nurse culture medium and at the moment a callus with a root is developing (Kranz et al. 1991a, Kranz et al. 1992, and personal communication). Cytological analysis is in progress in Lyon (Joint project: Procope 1991-1992). The observation of one "zygote" using the TEM technique has shown the presence of the male nucleus within the fusion product but no nuclear fusion has yet been observed in other specimens. It has been also noted that the polarity of the *in vitro* "zygote" is less pronounced than that of the natural zygote *in vivo* one. This morphological data may have consequences on further development (Faure et al. 1992 in preparation).

Another way to mimic fertilization is to **microinject nuclei into the female cells.** With that aim, a protocole has been developed: the first attempt consisted of holding the isolated embryo sac with an hemispherical tip pipette and in injecting a fluorescent dye using a microcapillary. However the penetration of the glass tip was unsuccessful. The embryo sac bent and
injection was not possible. As a matter of fact, the wall like structure which limits the embryo sac appeared to be both a resistant and a soft barrier to the capillary. To allow its

penetration it has been necessary to immobilize the embryo sac in an agarose medium. With this protocole we succeeded in accurate injection of lucifer yellow into the egg and central cell (Matthys-Rochon et al. 1991). This preliminary experiment showed that microinjection through the embryo sac wall was possible. Subsequently a protocole for the preparation of intact nuclei from pollen grains has been devised (Matthys-Rochon unpublished results). Injection of male nuclei is now in progress in Lyon.

Finally a very exciting model system of *in vitro* fertilization is to allow "**spontaneous fusion**" without any physical or chemical fusogenic agent. Sperm and egg cells are prepared according respectively to Dupuis et al. 1987 and Wagner et al. 1989 b. The two types of gametes are mixed in an appropriate medium at room temperature. Intergametic contact occurs and the fusion process is followed under an inverted microscope. "Adhesion products" consisting of egg cells on which the sperm cells are "stick" are submitted to cytological analysis. This work is being carried out in Lyon by C.Digonnet.

INTEREST AND PROSPECTS OF *IN VITRO* MODELS IN MAIZE

Artificial and/or spontaneous *in vitro* fertilization models may open different ways of research. **Electrofusion** experiments have shown that every gametic cell of protoplastic nature is able to fuse (Kranz et al. 1991b). So every female cell (synergid, egg or central cell) can be fused with a male gamete. In this fusion process the whole male cytoplasm is transferred to the female cell. It is evident from genetic studies that cytoplasmic organelles are inherited maternally in the majority of plants (Mogensen 1988). Maize sperm cells contain mitochondria. Organelle transfer from the paternal partner is therefore possible.

Microinjection of sperm cells or male nuclei into the embryo sac constitutes an attractive, although difficult, method to mimic double fertilization. The two female cells will receive according to the type of transferred material, either the male cell or the male nucleus. If regeneration of the fertilized embryo sac is possible, it would be possible then to analyse in a different way the fate of the male cytoplasm. Moreover, this model will constitute the only way to follow at the same time the early development of the endosperm and embryo proper. As a consequence, information related to morphological, biochemical, molecular biological changes during the gamete fusion and early embryogenesis, will be available.

The spontaneous *in vitro* fusion model is the most appropriate system to approach the question of sperm dimorphism and determine if specific molecules exist on the gametic membranes and are responsible of gametic recognition. With these ideas in mind, construction of a monoclonal antibody library directed against the cell surface of isolated male gametes is in progress in Lyon by A.Chaboud. However a prerequisite to this manipulation is to collect large quantities of pure gametes which is not perfectly obtained yet. As soon as this difficulty is overcome, monoclonal antibodies will be used both to assess membrane dimorphism of the two male gametes and to attempt gametic fusion inhibition experiments. This will lead to the identification of the putative gamete membrane determinants involved during the recognition process.

In conclusion, success in the isolation of viable and functionnal gametes especially in maize offers the possibility of direct access to the interacting gamete membranes and the development of different *in vitro* intergametic fusion models. This will help to provide a better understanding of the fertilization process in plants. Finally the ability to regenerate a plant from the "zygotes" will permit the development of protocoles for plant transformation *via* the sexual route.

Literature References

Chaboud A., Perez R., Digonnet C., Dumas C. (1992) Gamete recognition in Angiosperms: model and strategy analysis.Society for experimental biology. Seminar series 48. Perspectives in plant recognition. J.A.Callow and J.R. Green (eds). In press.

Dupuis I., Roeckel P., Matthys-Rochon E., Dumas C. (1987). Procedure toisolate viable sperm cells from corn (*Zea mays* L.) pollen grains. Plant Physiol., 85: 876-878

Faure J.E., Mogensen H. L., Kranz E., Digonnet C., Dumas C. (1992) Ultrastructural characterization and three-dimensional reconstruction of isolated maize (*Zea mays* L.) egg cells. In preparation.

Heslop Harrison J. and Heslop Harrison Y. (1970) Evaluation of pollen viability by enzymatically induced fluorescence, intra cellular hydrolysis of fluoresceine diacetate. Stain Technol. 45: 115-120

Keijzer C.J., Reinders M.C., Leferink-ten Klooster H.B.(1988) A micromanipulation method for artificial fertilization in *Torenia*. In: Cresti M., Gori P., Pacini E.(eds) Sexual reproduction in higher plants. Springer, Berlin Heidelberg New york, pp 119-124

Kranz E., Bautor J., Lörz H. (1991a) *In vitro* fertilization of single, isolated gametes of maize mediated by electrofusion. Sex. Plant Reprod. 4: 12-18

Kranz E., Bautor J., Lörz H. (1991b) Electrofusion mediated transmission of cytoplasmic organelles through the *in vitro* fertilization process, fusion of sperm cells with synergids and central cells, and cell reconstitution in maize. Sex. Plant Reprod. 4: 17-21

Kranz E., Lörz H., Digonnet C., Faure J.E. (1992) *In vitro* fusion of gametes and production of zygotes. Int. Rev. Cyt. In press.

Matthys-Rochon, E., Vergne, P., Detchepare, S., Dumas C. (1987) Male germ unit isolation from three tricellular pollen species: *Brassica oleracea, Zea mays and Triticum aestivum.* Plant Physiol. 83: 464-466

Matthys-Rochon E. and C. Dumas (1988) The male germ unit: retrospect and prospects in: H.J.Wilms and C.J. Keijzer (eds) Plant sperm cells as tools for biotechnology Pudoc Wageningen, the Netherlands. pp 51-60

Matthys-Rochon E., Digonnet C., Dumas C. (1991) Microinjection of lucifer yellow into embryo sac cells of *Zea mays*. International symposium on Angiosperm pollen and ovules. Basic and applied aspects. Cômo-Italy

Mogensen H.L. (1988) Exclusion of male mitochondria and plastids during syngamy as a basis for maternal inheritance Proc. Natl. Acad. Sci. USA 85: 2594-2597

Russell S.D. (1983) Fertilization in *Plumbago zeylanica*: gametic fusion and fate of the male cytoplasm. Amer. J. Bot. 70: 416-434

Russell S.D. (1985) Preferential fertilization in *Plumbago*: ultra structural evidence for gamete level recognition in an Angiosperm. Proc. Natl. Acad. Sci. USA 82: 6129-6132

Theunis C.H., Pierson E.S., Cresti M. (1991) Isolation of male and female gametes in higher plants. Sex. Plant Reprod. 4: 145-154

Wagner V.T., Dumas C., Mogensen H.L.(1989a) Morphometric analysis of isolated *Zea mays* sperm. Journal of Cell Science. 93: 179-184

Wagner V.T., Song Y.C., Matthys-Rochon E., Dumas C. (1989b) Observations on the isolated embryo sac of *Zea mays* . Plant Science. 59:127-132

Wide Hybridization in Higher Plants by Applying the Method of Test Tube Pollination of Ovules

M. Zenkteler
Department of General Botany
Adam Mickiewicz University
Poznań, Poland

A b s t r a c t

The method of placental pollination enables to overcome some pre- and postzygotic barriers of incompatibility. By applying this technique it was possible to obtain hybrid embryos and even plants mainly among some species of Solanaceae, Caryophyllaceae and Brassicaceae. The newest findings show that in Brassicaceae fertilization and hybrid embryos developent can occur between species belonging to different genera. No methods are available to enhance hybrid embryo differentiation beyond the globular stage.

I n t r o d u c t i o n

Sexual reproduction constitues an important mechanism for the introduction of variability in population. Hybridization between related species or genera is one approach that is often used to increase variability. In nature only a limited number of species and very few genera undergo natural hybridization. Pre- or postzygotic barriers are limiting the formation of hybrids between distantly related species /Blakeslee, 1945, Brink and Cooper, 1947/. For both self-incompatible and wide crosses there are several levels at which barriers occur: a/ inability of pollen to germinate on the stigma; b/ slow growth or bursting of pollen tubes inside the style; c/ abortion of the embryo inside the embryo sac. Even when fertilization occurs, the hybrid embryos start to degenerate a few days later mainly due to the abnormal endosperm development and/or disturbances occurring in the nucellus and in the integumentary tapetum. Mostly the barriers of hybrid embryo survival occur during the mid to late stages of embryogenesis. In vitro pollination procedures can play an important role in the sexual reproduction processes as removal of the reproductive organs from plant may enable to bypass some pre- and post zygotic barriers of incompatibility. These techniques may be especially useful in raising hybrids that cannot be obtained in nature by conventional methods.

Technical procedures

The majority of species in which successful in vitro fertilization have been achieved are those whose ovaries contain a large number of ovules situated on nutritively rich placentas e.g. Solanaceae, Caryophyllaceae, Papaveraceae. In vitro fertilization can take place in different ways: A/ pollination of stigmas of the pistil; B,C/ a direct pollination of ovules through an opened upper part of ovary or through an opening made in the ovary wall - placental pollination; D/ ovary wall removed and ovules are directly pollinated - -placental pollination; E/ pollination of ovules cultured directly on medium; F/ fertilization of single isolated egg cells (Figs. 1A-F). Explants A,B,C,D, are cultured jointly with a short part of the pedicle. In some cases, depending on the structure of flower, whole petals and only a part of the calyx can be removed. It is very important to leave the calyx as the presence of the accessory organs of the flower has a beneficial effect on the embryo development. Sometimes, the proce-

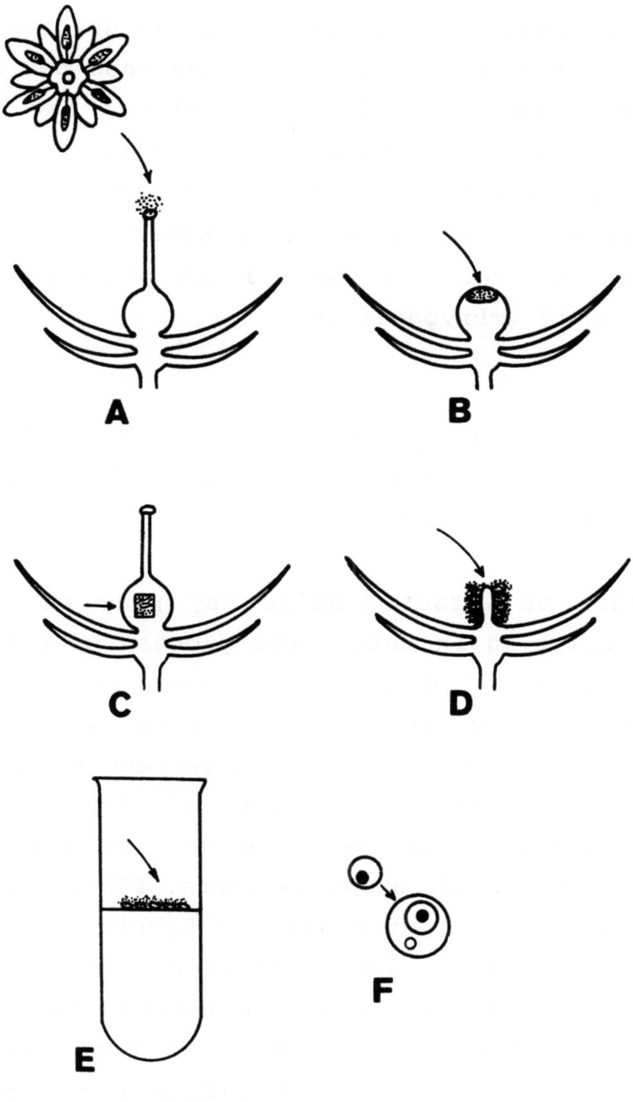

Fig.1.A. - Pollination of stigma of pistil; B-C - A direct pollination of ovules through an opened upper part of ovary /B/ or through an opening made in the ovary wall /C/;D - Ovary wall removed, ovules situated on the placenta ;E - Pollination of ovules cultured directly on medium;F - Fertilization of a single isolated egg cell.

dure which works successfully with one species may not be suitable for another, and in such case one needs to modify the procedure and even develop a new one. All experiments have to be performed in sterile conditions. Little is known about the effects of medium on the process of in vitro fertilization, however, medium plays a very important role in the culture of embryos at an early stage of embryogenesis.

In vitro cross pollination of pistils and ovules-successes and failures

The culture of intact pistils may play an important role in analyzing in detail the process of pollination and fertilization in controlled environment. It also enables to analyze the role of various factors present in the media and in the environmental on self--incompatibility system /Douglas and Conolly, 1989/. However, this procedure seems to be less important for bypassing pre-fertilization barriers. Efforts have been undertaken to obtain interspecific hybrid embryos among some species of Trifolium /Zenkteler, 1991/. Stigmatic cross pollination have been made between the following species: T. rubens x T.pratense in both directions, T.fragiverum x T. repens and T. ambiguum x T. rubens.In all the conditions of crosses only globular embryos were developing which several days after pollination started to degenerate. Cytological analysis of embryo sacs revealed that endosperm was not present.Thus it seems that the applied conditions, mainly the media, were not suitable for the induction of embryos. Successful in vitro hybrid embryos between Trifolium ambiguum used as a pistillate parent and T.repens and T hybridum as the pollen source were obtained by Richards and Rupert /1980/. Those authors noticed that T. ambiguum

Table. Cross pollination of placentas of some species of Brassicaceae

Species	Immature embryos	Viable seeds
Brassica napus x D.tenuifolia	+	
" " x M.arvensis	+	
" " x Sinapis arvensis	+	
" " x S. alba	+	
" " x B.campestris		+
" " x Barbarea vulgaris	+	
" " x E.pallidiflorum	+	
B.oleracea (1) x B. cretica		+
" x Hesperis matronalis	+	
" x Alliaria officinalis	+	
" x E.pallidiflorum	+	
" x E.crepidiflorum	+	
" x E pulchelum	+	
" x E.virgatum	+	
" x Barbarea vulgaris	+	
" x Sinapis arvensis	+	
B.oleracea (2) x D.tenuifolia	+	
" x M.arvensis		+
" x S.alba	+	
Moricandia arvensis x B.napus	+	
" " x B.oleracea (2)	+	
Brassica campestris x B.napus		+
D.tenuifolia x B.oleracea (2)	+	
" x B.napus	+	
" x Moricandia arvensis	+	

D - Diplotaxis
M - Moricandia
E - Erysimum
B.oleracea (1)- Brassica oleracea var.sabellica
B.oleracea (2)- Brassica oleracea var. italica

ovules require various combinations of auxin and cytokinin to enhance embryo differentiation beyond the globular stage.
The technique of a direct pollination of ovules - placental pollination - may be especially useful in circumventing the barriers in prezygotic incompatibility. Overcoming prezygotic barriers does not preclude the existence of postzygotic barriers. Therefore, the combined techniques of in vitro ovule pollination and ovule and embryo culture has considerable potential for the production of hybrid plants. At present, in vitro cross pollination and fertilization of ovules has been accomplished with 56 species, among which only 20 combinations are found which formed viable seeds able to produce hybrid seedlings. The most satisfactory development of hybrid embryos and plants was found only among some species of Nicotiana, Melandrium,Brassica and among two species of Gossypium and Zea /Dhaliwal and King 1978, Stewart 1981, Zenkteler 1990 /.

During the last 3 years much efforts have been undertaken to obtain hybrid plants among several species of Brassicaceae /Zenkteler 1990, 1991/. The excised ovaries of Brassica oleracea var. sabellica and var. italica, B.napus, Diplotaxis tenuifolia and Moricandia arvensis were pollinated with pollen grains of various species (Table 1). Irrespectively of the combinations of crosses pollen germinated on ovules with various intensity. In all of the analysed combinations of crosses some ovules were visibly enlarging (Fig.2A). Embryological analysis of these ovules revealed the presence of embryos at different stages of embryogenesis (Fig.2B). Some embryos developed concomitantly with poor endosperm. The percentage of ovules producing globular hybrid embryos varied depending on the species used as pollinators. Early growth phases of embryos were more or less similar in all the crossed species. The differences in size and structure became marked as the globular embryos reached older stages. The best development of hybrid embryos occurred in crosses of B.napus x Barbarea vulgaris, Sinapis arvensis, S.alba and Diplotaxis tenuifolia.In those crosses the embryos reached nearly the heart stage. Cotyledonary embryos were obtained after pollinating ovules of B.napus with B.campestris, B.oleracea var. sabellica with B.cretica and Sinapis arvensis. Cotyledonary embryos were transferred on a medium and later on they grew into seedlings.Hybrid seedlings of B.oleracea var. sabellica x B.cretica were transferred into the soil where they grow with a high intensity (Fig.2D). In all the remaining intergeneric crosses of Brassicaceae the enlarged ovules with globular embryos degenerated about 12 days after pollination.

It is very difficult to apply the technique of in vitro pollination of single ovules cultured directly on the medium. There are only two papers describing success in culturing individual ovules after self pollination of B.oleracea /Kameya et al.;1966/ and after interspecific cross pollination of Gossypium /Stewart 1981/.We are attempting in vitro fertilization of single ovules of different species of Leguminoseae and Brassicaceae on solidified agar media. Pollen of various species, when dust on the ovules, germinate abundantly (Fig.2C) and in some combinations of crosses pollen tubes inside the embryo sacs are noticed. In two crosses several celled embryos were found. It seems to be possible to apply the technique of in vitro pollination of single ovules, but the procedure needs to be modified. In moment (when writing this article) investigations are carried out in this direction.

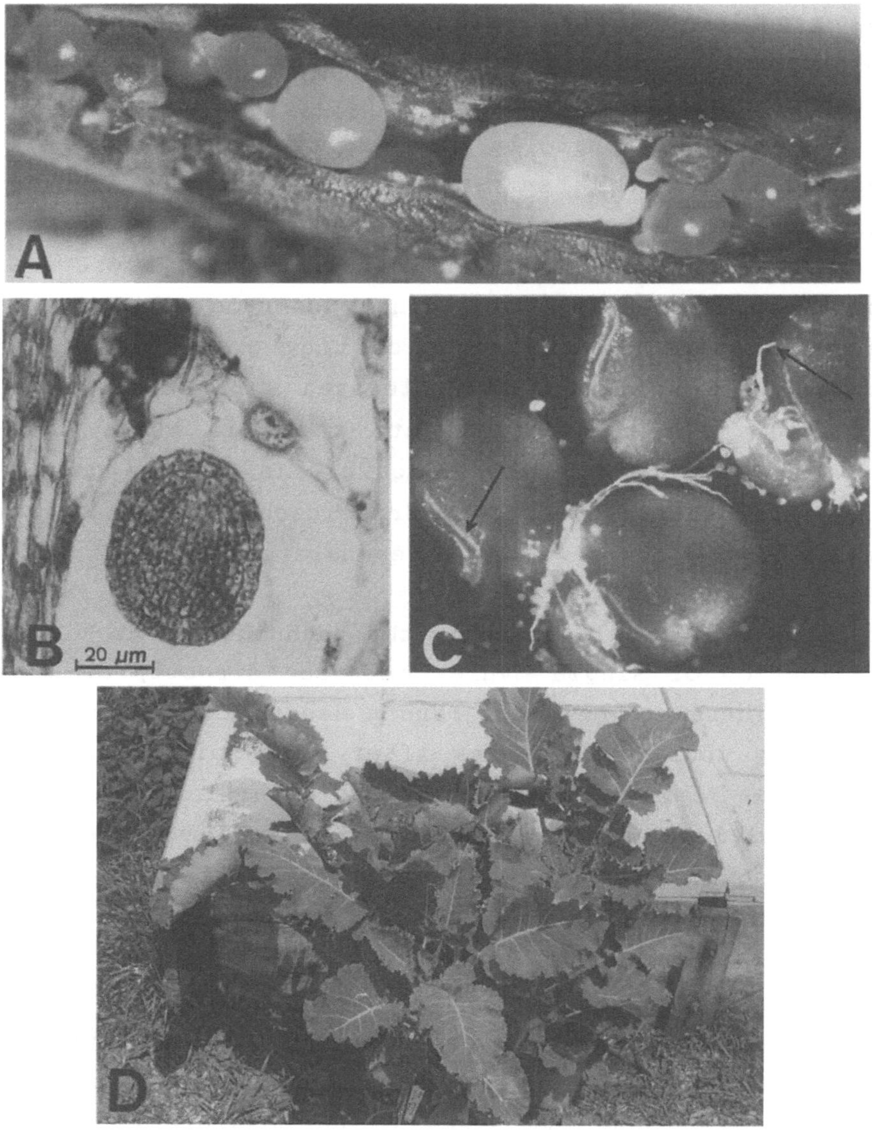

Fig.2. A - *Brassica napus* x *Diplotaxis tenuifolia*, enlarged ovules inside the opened ovary, 6 days after placental pollination, 8x; B - *B.oleracea var.sabellica* x *Sinapis arvensis*, globular embryo and poor endosperm inside the embryo sac, 8 days after placental pollination;C - Germination of pollen grains of *Moricandia arvensis* on ovules of *Brassica napus*, 120x; D - Hybrid plants of *B.oleracea var. sabellica* x *B.cretica* obtained from ovules pollinated in vitro.

Discussion and conclusions

Often new genes of wild species have to be used to improve the domesticated relatives. However, natural genetic and physical barriers prevent the introduction of wild genes by coventional crossing. Much researches is directed toward learning how to overcome these barriers. One of the techniques is the application of test tube pollination and fertilization. This method has been employed successfully to circumvent incompatibility barriers in certain species. The results of investigations carrying out during the last years show that direct interspecific and intergeneric pollination of ovules results in the formation of hybrid embryos and plants among only some species of Brassicaceae, Caryophyllaceae, Solanaceae and among two species of Gossypium and Zea. Those results also show that pollen of various species when dust on the naked ovules germinate with various intensity and several days after pollination globular embryos can be discerned inside embryo sacs. Thus, the direct pollination of ovules enables to overcome some of those barriers which hinder the fusion of gametes of distant species. The main problem is that the postfertilization barriers are usually so strong that the hybrid globular embryos degenerate before reaching the heart shape stage. To date, there are methods to induce further growth of embryos which already reached heart stage but, unfortutunately, no methods are available to culture hybrids at globular stage. In spite of the application of various sophisticated techniques the efforts to induce those small hybrid embryos to grow further are unsuccessful.We have still no knowledge of the nutritive requirements of globular hybrid embryos and the applied artificial culture media are unsatisfactory. It is an urgent need to carry out investigations on in vitro culture of globular hybrid embryos or whole ovules containing embryos in order to provide hybrid plants.

References

Blakeslee A F (1945) Removing some of the barriers to crossability in plants. Proc. Am. Phil. Soc.89: 561-574

Brink R A ,Cooper D C (1947) The endosperm in seed development. Bot.Rev. 13: 423-541

Dhaliwal S, King P (1978) Direct pollination of Zea mays ovules in vitro with Z.mays,Z.mexicana and Sorghum bicolor pollen. Theor. Appl. Genet. 53: 43-46

Douglas G C, Connolly V (1989) Self fertilization and seed set in Trifolium repens L. by in situ and in vitro pollination. Theor. Appl. Genet. 77: 71-75

Kameya T, Hinata K, Mizushima U (1966) Fertilization in vitro of excised ovules treated with calcium chloride in Brassica oleracea L. Proc. Jap. Acad. 42: 165-167

Richards K W, Rupert E A (1980) In vitro fertilization and seed development in Trifolium. In Vitro 16(11): 925-931

Stewart J (1981) In vitro fertilization and embryo rescue. Environ. Exp. Bot. 21: 301-315

Zenkteler M (1990) In vitro fertilization and wide hybridization in higher plants.Crit. Rev. Plant.Sci. 9: 267-279

Zenkteler M (1990 a) In vitro fertilization of ovules of some species of Brassicaceae. Plant Breeding 105:221-228

Zenkteler M (1991) Ovule culture and test tube fertilization. Forum for Applied Biotechnoloy, University of Gent, 25-27 September, in press

Regulation of ABA and Water Stress Responsive Genes in Maize Embryos

Montserrat Pagès, Maria Pla, Anders B. Jensen, Miguel Angel Freire, Cesar Arenas, Eva Poca and Adela Goday.
Departament de Genètica Molecular
Centre d'Investigació i Desenvolupament
CSIC, Jorge Girona Salgado 18-26
08034 Barcelona Spain

ABSTRACT

The maize *rab* genes are expressed in different plant parts in response to abscisic acid and osmotic stress. The pattern of expression of *rab* mRNAs and corresponding proteins was studied during maize embryogenesis and in vegetative tissues of wild type and *viviparous* mutants. ABA induces the synthesis of the mRNAs and accumulation of the proteins in embryo tissues. However, in vegetative tissues subjected to water stress or ABA treatment the levels of mRNA are high but *rab* proteins are absent or poorly accumulated. A chimaeric gene containing the *rab* 17 promoter and the reporter gene CAT is correctly regulated in transgenic tobacco, although lower levels of the enzyme activity have been noticed in vegetative tissues. All these data suggest that more than one mechanism determines the level of expression of the maize *rab* genes.

INTRODUCTION

Almost all plants can withstand extremes of osmotic stress at some stages of their life cycle. This is the case in the cereal seeds where embryos withstand a reduction of its water content of about 80%, while such a severe desiccation kills the cells from other parts of the plant. During growth, plants also endure mild water deficits in periods without significant rainfall. In spite of the considerable degree of research on the desiccation resistance of plants, the factors contributing to the tolerance of the cells to dehydration are largely unresolved.

The plant hormone abscisic acid (ABA) appears to modulate the responses of plants under conditions of water deficit (Davies and Mansfield, 1983). In general there are two basic types of responses to ABA which have been correlated with the regulation of gene expression. First, developmental studies in embryos have shown that ABA accumulates in the stages before desiccation (King 1976, Jones and Brenner 1987) altering the rate of transcription of a specific set of genes (Quatrano 1986). Second, ABA levels increase in water-stressed vegetative tissues (Wright et al. 1969) and alters the expression of genes which can also be induced by exogenous ABA in non-stressed tissues (Mundy and Chua 1988, Gómez et al. 1988, Pla et al. 1989).

RESULTS AND DISCUSSION

EXPRESSION OF MAIZE *rab* GENES AND PROTEINS

A specific set of genes from maize, the *rab* genes, are involved in general plant ABA responses. Several genes showing similar behaviour have been identified in different plant systems (Mundy and Chua 1988, Bartels et al. 1990, Claes et al. 1990). The maize genes *rab* 17 (Vilardell et al 1990) and *rab* 28 (Pla et al 1991) have been identified as ABA-inducible in embryos, and they are also induced by osmotic-stress in vegetative tissues and callus cells (Pla et al. 1989, Vilardell et al. 1990).

The general pattern of expression of *rab* proteins indicate that they are synthesized during normal embryogenesis at the period where desiccation starts, they accumulate in mature and dry embryos, disappearing during the first hours of germination (Goday et al. 1988).

Current evidence indicates that plants subjected to osmotic stress respond with a series of physiological, morphological and biochemical changes which appear to be elicited by the reduction of cell turgor (Davies and Mansfield 1983). Among them, closure of the stomata (Turner 1979), changes in ionic composition (Cheeseman 1988), accumulation of organic osmolites and specific proteins have been described (see figure 1). Little is know about the relation among the various components accumulating in the cell during water-stress, and unfortunately, the function of the *rab* proteins is mainly unknown.

Surprisingly, *rab* proteins are poorly represented in the overall protein pattern of ^{35}S-methionine labelled proteins from ABA-treated or water-stressed vegetative tissues (Goday et al, in preparation) or callus cells (Vilardell et al, 1990). Moreover, by means of immunoblot

```
STRESS   Drought Salt Cold Wounding

STIMULUS Fall in water potential
         Reduced turgor
         Changes in ionic composition

RESPONSE ABA increase
             Synthesis
             Translocation

EFFECTS  Stomatal Closure
         Increase in Organic Osmolytes
                     Polyols
                     Methylamines
                     Amino acids
         Expression of rab genes
                     Function of rab proteins?
```

Figure 1. Diagram of physiological and molecular steps involved in plant responses to osmotic stress.

of total protein extracts from young leaves or callus cells *rab* 17 proteins were found to be absent or poorly accumulated in both types of ABA-treated tissues (see figure 2). All these data seem to indicate that the regulation of the expression of these proteins is more complex than previously suspected and that transcriptional, post-transcripional and post-translational mechanisms are involved in the regulation of *rab* gene expression.

REGULATION OF *rab* GENES IN MAIZE *viviparous* MUTANTS

The maize *viviparous* mutants have great potential for providing information on the regulation of the expression of genes involved in ABA responses (Koornneef 1986, Pla et al 1989). In the maize *viviparous* mutants, the embryo fails to become dormant, but precociously germinate in the mother plant without desiccation. Class I mutants (Robertson 1955) include *vp1* which contains normal levels of ABA in the embryo and seedlings (Neill et al. 1986). The ABA concentration of *vp1* seedlings increases in response to water deficit. However, *vp1* mutant is defective in its responsiveness to this hormone (Robichaud et al. 1980).
Class II mutants, which include, *vp9, vp5, vp7* and *vp2* are all carotenoid deficient and contain subnormal levels of ABA. Seedlings rescued from ABA-deficient *viviparous* kernels contain subnormal levels of ABA and the ABA concentration does not increase in response to water deficit (Neill et al 1986). ABA-deficient *viviparous* embryos respond to exogenously applied ABA as do wild type embryos (Robichaud et al 1980). The effect of ABA on the regulation of *rab* gene expression during embryo development and in the response of maize plants to water stress was studied in wild type and *viviparous* mutants. Synthesis and accumulation of *rab* proteins during

embryogenesis and precocious germination of ABA-deficient *viviparous* mutants closely resembles that found in non-mutant embryo development (see Figure 2). The only difference was the lower amounts of mRNA and protein accumulated in the mutants. The *rab* genes are ABA-inducible in wild type, *vp2* and *vp1* embryos and young leaves. *rab* mRNAs are also induced by water-stress in leaves of wild type and *vp1* plants but not in *vp2* plants (Pla et al 1989, Pla et al 1991). The results obtained on the expression of *rab* mRNAs in the so-called ABA-

rab-17	mRNA	Protein	Phosphorylated forms
Embryo			
Young	-	-	-
+ ABA	++	+	++
Mature	+++	+++	+++
Dry	+++	++++	++++
Seedling			
4 days	-	+	?
+ ABA	++	++	++
Leaves	-	-	-
+ ABA	++	(+)	-
+ w.s.	+++	(+)	-
Callus	-	-	-
+ ABA	++	(+)	-
vp2 Embryo			
Young	-	-	-
+ ABA	+	+	+
Mature	++	++	++
vp2 Leaves	-	-	-
+ w.s.	-	-	-
+ ABA	+	(+)	?
vp1 Embryo			
Young	-	-	-
+ ABA	++	?	?
Mature	(+)	?	?

Figure 2. Expression of *rab*-17 mRNA and protein in different maize tissues. w.s.=water stress,(+) poorly detected.

-insensitive *vp1* has provided new insights concerning the regulatory mechanisms that operate in this mutant. Our results indicate that the product of *vp1* gene is not involved in the complex signal transduction pathway of *rab* gene expression by ABA. These results are in agreement with recent findings concerning the role of *vp1* protein which have structural features consistent with a transcriptional activator (Mc Carty et al 1991). On the other hand the differential regulation of *rab* genes in embryo and vegetative tissues of ABA-deficient mutants indicate a higher sensitivity to ABA in embryo tissues. These results suggest the existence of embryo-specific factors involved in the regulation of *rab* genes (Pla et al 1991). These factors may be part of the developmental programme ensuring the survival of the embryos during dormancy. They can be functionally active before desiccation both in wild type and *viviparous* embryos.

REGULATION OF *rab* PROMOTERS

To elucidate the mechanism by which ABA regulates specific gene expression, we studied the *cis* elements involved in the regulation of *rab* gene expression.
Transient expression studies have shown that the promoter region of *rab* genes contain ABA-responsive DNA motifs which confer ABA responsiveness on reporter genes in cereal protoplasts (Vilardell et al 1991, Pla et al in

MOTIF I	GENE	PLANT
TTGCCGGA CACGTG GCGCGA	Em1a	wheat
GCTCGCGC CACGTG GGCATG	*rab 28*	maize
AGAACCGA GACGTG GCGGGC	*rab 17*	maize
CGACACCG TACGTG GCGCCA	*rab 16*	rice

Figure 3 Homology between ABA-responsive genes upstream regions. Underlined are the sequences conforming the motif I.

preparation). Sequence comparison of these 5' upstream sequences with those from genes expressed during late embryogenesis in other species has identified a conserved motif CACGTG that may be the ABA-responsive element (see figure 4). This motifI has been reported as functionally important in the wheat Em gene and in the rice *rab* 16 gene (Marcotte et al.1989, Mundy et al 1990,).Moreover two leucine zipper proteins ,EMBP-1 and TAF-1, binding to this DNA motif have been recently characterized (Guiltinan et al. 1990, Oeda et al 1991).

The *rab* 5' upstream sequences conferred correct patterns of hormonal and developmental expression upon the CAT reporter gene in stably transformed tobacco. Specifically, a CAT construct containing a large promoter fragment 1.3 kb of *rab* 17, confer the appropiate patterns of expression in transgenic tobacco plants (Vilardell et al 1991). This construct results in high levels of CAT activity in mature seeds, and CAT activity could be increased in ABA-treated immature seeds. The *rab* 17/CAT transgene was also expressed in vegetative tissues under conditions of water-stress. However, the relative percentage of CAT activity is lower in vegetative tissues than in embryos (24% versus 98%). These results again indicate a differential regulation in embryo and vegetative tissues and suggest that different factors may be responsive to the different signals in these two tissues.

Current evidence indicate that regulation of ABA-responsive genes is at the transcriptional level, but control at the post-transcriptional and post-translational level has also been suggested (Quatrano 1986, Goday et al 1988). Furthermore, differences in the level of expression of several ABA responsive genes and proteins have also been reported (Singh et al. 1987, Vilardell et al 1990). All these data suggest that more than one mechanism determines the level of expression of the various ABA-responsive genes.

REFERENCES

Bartels D Schneider K Terstappen G Piatkowski D Salamini F (1990) Molecular cloning of abscisic acid-modulated genes which are induced during desiccation of the resurection plant Craterostigma plantagineum Planta 181 27-34.

Claes B Dekeyser R Villaroel R Van den Bulcke M Bauw G Van Montagu M Caplan A (1990) Characterization of a rice gene showing organ specific expression in response to salt stress and drought Plant Cell 2 19-27

Cheeseman JM (1988) Mechanisms of salinity tolerance in plants Plant Physiol 87 547-550

Davies MJ and Mansfield TA (1983) The role of abscisic acid in drought avoidance. In Addicott FT ed Abscisic Acid Praeger New York 237-268.

Goday A Sánchez-MartÀnez D Gómez J Puigdomènech P and Pagès M (1988) Gene expression in developing Zea mays embryos: Regulation by abscisic acid of a highly phosphorylated 23- to 25-kD group of proteins Plant Physiol 88 564-569.

Gómez J Sánchez-MartÀnez D Stiefel V Rigau J Puigdomènech P and Pagès M (1988) A gene induced by the plant hormone abscisic acid in response to water stress encodes a glycine-rich protein Nature 334 262-264

Guiltinan MJ Marcotte WR and Quatrano RS (1990) A plant leucine zipper protein recognizes an abscisic acid response element. Science 250, 267-270.

Jones RJ and Brenner ML (1987) Distribution of abscisic acid in maize kernel during grain filling Plant Physiol 83 905-909

King RW (1976) Abscisic acid in developing wheat grains and its relationship to grain growth and maturation Planta 132 43-61

Koornneef M (1986) Genetic aspects of abscisic acid In Blonstein AD King PJ (eds) Genetic Approach to Plant Biochemistry pp 35- 54 Springer Verlag Vienna

Marcotte WR Russell SH and Quatrano RS (1989) Abscisic acid responsive sequences from the Em gene of wheat Plant Cell 1 969-976

Mc Carty DR Hattori T Carson ChB Vasil V Lazar M and Vasil VK (1991) The viviparous 1 Developmental gene of maize encodes a novel transcriptional activator Cell 66 895-905

Mundy J and Chua NH (1988) Abscisic acid and water-stress induce the expression of a novel rice gene. EMBO J 7 2279-2286.

Mundy J Yamaguchi-Shinozaki K Chua NH (1990) Nuclear proteins bind conserved elements in the abscisic acid-responsive promoter of a rice RAB gene Proc Natl Acad Sci USA 87 1406-1410.

Neill SJ Horgan R Parry AD (1986) The carotenoid and abscisic acid content of viviparous kernels and

seedlings of Zea mays L Planta 169 87-96.
Oeda K Salinas J Chua NH (1991) A tobacco bZip transcription activator (TAF-1) binds to a G-box-like motif conserved in plant genes EMBO J 10 1793-1802
Pla M Goday A Vilardell J Gomez J Pages M (1989) Differential regulation of ABA-induced 23-25 kDa proteins in embryo and vegetative tissues of the viviparous mutants of maize Plant Mol Biol 13 385-394.
Pla M Gómez J Goday A and Pagès M (1991) Regulation of the Abscisic acid-responsive gene *rab* 28 in maize *viviparous* mutants Mol Gen Genet 230 394-400.
Quatrano RS (1986) Regulation of gene expression by asbcisic acid during angiospem embryo development. Oxford Surveys Plant Mol Cell Biol 3 745-760.
Robertson DS (1955) The genetics of vivipary in maize Genetics 40 745-760.
Robichaud CS Wong J Sussex IM (1980) Control of in vitro growth of viviparous embryo mutants of maize by abscisic acid Dev Genet 1 325-330.
Singh NK LaRosa PCh Handa AK Hasegawa PM and Bressan RA (1987) Hormonal regulation of protein synthesis associted with salt tolerance in plant cells Proc Natl Acad Sci USA 84 739-743.
Turner NC (1979) Drought resistance and adaptation to water deficits in crop plants In Stress physiology in crop plants Musell H and Staples RC eds 343-367 Wileyand Sons Inc
Vilardell J Goday A Freire MA Torrent M Martinez MC Torne JM and Pagès M (1990) Gene sequence developmental expression and protein phosphorylation of *rab*-17 in maize Plant Mol Biol 14 423-432.
Vilardell J Mundy J Stilling B Leroux B Pla M Freyssinet G and Pagès M (1991) Regulation of the maize *rab* 17 promoter in transgenic heterologous systems Plant Mol Biol 17 985-993
Wright ST and Hiron RWP (1969) (-) Abscisic acid the growth inhibitor in detached wheat leaves following a period of wilting Nature 224 719-720.

Gene Expression During Seed Formation and Maturation in *Crucifereae*

M Raynal, L Aspart, P Gaubier, D Depigny, F Grellet , M Delseny*
Laboratoire de Physiologie et Biologie Moléculaire Végétales (URA 565 du CNRS) Université de Perpignan - 66860 Perpignan-Cédex, France

Introduction

Mature seeds contain a significant stock of stored mRNA, the life-span of which is as long as the seed-life (Payne, 1976; Delseny et al., 1977). A basic question in seed biology is the role and function of this stored mRNA. During the last ten years, many plant molecular biologists have addressed this question. As a result, seed development has been extensively studied. Most results concern the easiest genes to deal with, those coding for the storage proteins. However this gives only a partial view of seed development (Dure, 1985; Casey et al., 1986). Trying to answer questions concerning mRNA stored in mature seeds we have been led to analyse gene expression at various developmental stages and to realise that during seed development a number of genes are differentially expressed and sequentially switched on and off.

The aim of this report is to describe our general strategy to identify genes which correspond to mRNAs contributing to seed formation and to those accumulating in mature seeds. We summarize some of our already published results and present an overview of our recent findings. Most of our studies have been made using radish seeds and embryo axes because we found it a convenient model system since these seeds germinate readily in the dark within 12 hours of soaking. Recently we moved to *Arabidopsis* which is a more convenient system to isolate genes and study their regulation.

Results

1 - General strategy

Our general strategy to analyse gene expression during seed formation and maturation consisted in carefully comparing two dimensional gel electrophoresis patterns of proteins prepared in various conditions and at different developmental stages (Aspart et al., 1984; Laroche et al., 1984; Raynal et al., 1987). Coomassie blue stained gels gave a view of the steady state. Analysis of *in vivo* labelled proteins indicate which polypeptides are made at a given stage. Short labelling times usually revealed the primary translation products while longer times or chase experiments reveal precursor product relationships. This type of analysis is indispensable to correlate *in vivo* protein synthesis with the steady state composition. Finally analysis of *in vitro* synthesised proteins indicates which mRNAs are present in the cell and their relative abundance, assuming that all mRNAs are translated *in vitro* with the same efficiency.

A large number of such analyses has led to the conclusion that a significant number of gene sets are differentially expressed at different developmental stages. This conclusion is illustrated by the schema in figure 1 which shows that five large groups of genes have been identified on the basis of this type of analysis.

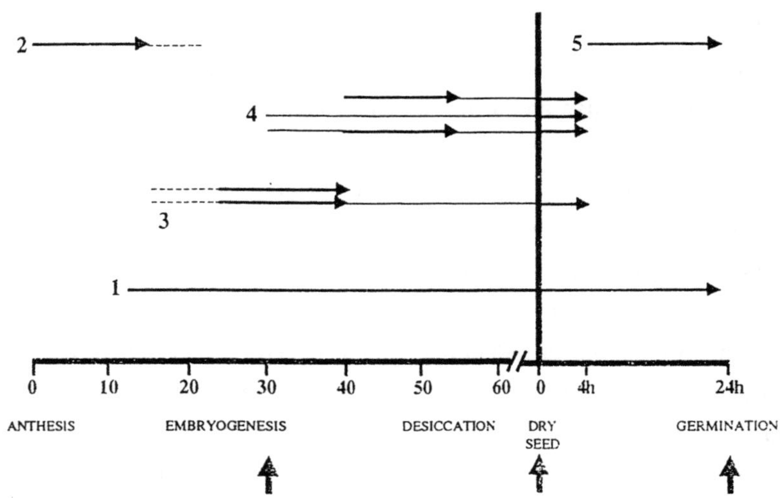

Figure 1.- Differential expression of the main gene families during embryogenesis and germination of the radish seed. Thick arrows indicates the period when synthesis of a protein is detectable *in vivo* and thin arrows the period when only the presence of the mRNA is detected.

Genes which belong to class 1 are constitutive genes expressed throughout seed formation and germination. Class 2 corresponds to genes expressed during early embryogenesis before the accumulation of storage products. Class 3 genes correspond to the accumulation of storage products, mostly proteins and oils in radish. Genes from class 4 are expressed later on and account for the mRNA accumulated in mature and dried seeds. Finally, genes from class 5 are germination specific. This scheme is oversimplified. For instance early embryogenesis between 0 and 20 days has not been studied in detail, but there is certainly some overlap in gene expression between classes 2 and 3. In each class, more careful analysis will certainly reveal subclasses. In some other cases we could detect the mRNA but not the corresponding protein because no antibody was available to ascertain its position in the gels. Therefore, although our data point to extensive transcriptional control of gene expression there is certainly also post-transcriptional and post-translational regulations.

We next decided to focus on genes from groups 3 and 4 and to attempt to isolate a few genes from each class so that their regulation could be studied.

2 - Characterization of genes from class 3 (immature seed).

Since two dimensional gel analysis did not give completely unambiguous data concerning the distinction between some immature seed mRNA products and some stored mRNA ones we decided to investigate first the mRNA for the major storage proteins. These are the 12S cruciferins and the 2S napins. A cDNA library corresponding to immature seed mRNA was prepared (Laroche-Raynal and Delseny, 1986) and analyzed by differential screening.

A first screening was made with single-stranded cDNA homologous to the mRNA used to prepare the library. Clones which light up correspond to mRNAs abundant at this developmental stage. The library was further screened with cDNA made from mature seed mRNA and again with cDNA from 24 hr germinated seedlings. This screening showed that most abundant mRNAs in the immature seed are not present in the mature seed or in the germinated seed. Reciprocally, it also shows that a fairly small amount of RNA abundant in the dry seed corresponds to RNA present in the immature seed. This screening was completed by screening with heterologous napin and cruciferin probes from rapeseed (Crouch et al., 1983; Simon et al., 1985; Laroche Raynal and Delseny, 1986). Napin genes are organized in two closely related subfamilies (Raynal et al., 1991). Cruciferin clones have been characterized by grouping clones in cross-hybridization classes and sequencing some of the members of these classes. The cruciferin genes are organized in two divergent subfamilies which no longer cross-hybridize (Depigny et al., 1991). The expression of these various gene families was analyzed using gene specific probes. It was observed that one napin subfamily was expressed slightly earlier than the other and that the latter was expressed at about the same time as the two types of cruciferin genes. Recent random sequencing work in *Arabidopsis* revealed that one of the four napin genes and one of the three cruciferin genes are more abundantly expressed than the others. In any case these studies clearly establish that the mRNAs for major storage proteins are no longer present at a significant level in dry seeds. This was a strong indication that stored mRNA has another function than coding for storage proteins.

The sequencing work was extended to other cDNAs and at this stage it was possible to characterize an immature seed-specific, extensin-like clone as well as several as yet unidentified genes.

3 - Characterization of genes from class 4 (mature seeds).

In order to isolate cDNA clones representative of the abundant mRNAs present in mature seeds, a new library was prepared from mature seed polyadenylated mRNA and screened in a similar way. The first screening was with homologous single stranded cDNA. Among these positive clones we kept only those which did not hybridize with either an immature seed cDNA or a germinated seed one. In this way 15 cDNAs were selected and further characterized by northern blot hybridization, cross-hybridization and sequencing. We have already reported the characterization of two of them which correspond to proteins and genes already described in other species and sometimes in other tissues.

Clone 8B6 (Raynal et al., 1989) for instance is 65% homologous to the so call Em- protein gene from wheat (Litts et al., 1987) and to the D19 gene of cotton (Baker et al., 1988). We recently isolated two different homologous genes from *Arabidopsis* as well as several *Arabidopsis* cDNA clones. These data indicate that the two genes are functional. Use of gene specific probes demonstrate that they are differentially regulated, one being expressed earlier and at a lower level than the other. Studies on their promoters using transgenic plants are now in progress. Clone 6F5,also named LEA 2 (Raynal et al., 1990), has blocks of homology with a group of proteins which are induced by ABA in cotton embryos and in water-stressed tissues from maize (Vilardell et al., 1990), rice (Yamaguchi et al., 1989) or barley (Close et al., 1989). A similar gene is also induced in salt-stressed tomato (Godoy et al., 1990). In these species this gene is named RAB 17 (Responsive to ABA) or dehydrin. Fig. 2 shows these homologies. Particularly striking is the conservation of the block of 8 serine residues and the adjacent residues as well as a repeat of this sequence at the carboxyl end of the sequence.

```
1   ME-H--QGQH-GHV-TSRVDEYGNPVG-------TGAGH-AQMGTAGM-------------------
2   MENY--QGQH-GYG-ADRVDVYGNPVGAGQYGGGATAPG-GGHGAMGM-------------------
3   ME-Y--QGQHGGH-ASSRADEHGNPAVT---TGN--APT-----GMG--------------------
4   ME-Y--QGQH-GH-ATDKVEEYGQPVAG--HGGFTGGPT-GTHG-----------------------
5   ME-YGQQGQ-RGHGRTGHVDQYGNPVGGVEHGTGGMRHGTGTTGGMG--------------------
6   MAQY-GN-QD-Q---MRKTDEYGNHVQETGVYQGTGTGGMMGGTGTGG-------------------
7   MADL-KD-E-RGNP-IHLTDAYGNPVQSLDEFGNPM-HITGVASSAPQYKDSVTGNIAEYPTEAPPAGV
8   MA-HFQN-QY-SAPEVTQTDAYGNPTRRTDEYGNP-IPTQET-------------------------
                 *          * ****

--------GTHGTAGTGRQFQPMREEHKTGGVLQRSG--SSSSSSSEDDGMGGRRKKGIKEKIKEKLPGGN-K
--------GGHAGAGAGGQFQPAREDRKTGGILHRSG-SSSSSSSSEDDGMGGRRKKGIKEKIKEKLPGGN-K
--------------AGHIQ-EPAREDKKTDGVLRRSG-SSSSSSSSEDDGMGGRRKKGIKEKIKEKLPGGN-K
------------AAGVGGAQLQATRDGHKTDGVLRRSG--SSSSSSSEDDGVGGRRKKGMKEKIKEKLPGGAHK
-----QLGEHGGAGMGGGQFQPAREEHKTGGILHRSG--SSSSSSSEDDGMGGRRKKGIKEKIKEKLPGG-HK
--------GGTGGETGTQGMGTGTHHHEGQQQLRRSD----SSSSSEDDGEGGRRKKGLKEKIMEKMPG----
AAGTGAAATTAAGVTTSETTTGQEHHGSLGEHLRRSG--SSSSSSSEDDGQGGRRKKRIKDKIKDKLGGGKHK
---------------GRGILGTGGHHHGGHHGLHRTGSSSSSSSSSEDEG-TGKKKKGLKERLKEKIPG-N-K
                                   * ***  *********** ******* ******* **   *

GEQQHAMGGTGTGTGTGTGTGGAYGQQGH-GTGMT-TGTTGAHGTTTTD-TGEKKGIMDKIKEKLPG--QH--
GNNQQQQQMMG-------NTGGAYGQQGH--AGMT-GAGTGVHGAEYGNA-GEKKGFMDKIKEKLPG--QH--
GNNQQQQQEH----TTTTGGAYGPQGH-DTKIA-TGARGGTAATTADAGGEKKGIVDKIKEKLPG--QH--
DAAGQQQQ----------------------TAMA-GEYAGTHGT---EATGEKKGVMDKIKEKLPGG-QH--
-DDQHA----------TATTGGAYGQQGHTGSAYGQQGHTGGAYATGTEGTGEKKGIMDKIKEKLPG--QH--
--------------------------------QHE----GEYGQTTGEEKKGMMDKIKDKIPGM--H--
DEQ------------TPTTATTTGPTTTTTTTGAADQHH--------------EKKGILEKIKEKLPGHHNHHP
---EHQSQ----------ATSTTTPGQG---PTY-HQHHREERSDGQGEAP----WSPQPLISCLWSAISY--
                                          ****    *** ****     *
```

Fig. 2.- Comparison of the predicted amino-acid sequences of the rice (1-3), barley (4), maize (5), tomato (6), radish (7) and cotton (8) RAB 17-like proteins. Dashes indicate gaps introduced in the protein sequence to maximize homology. Asterisks indicate amino acid identity in seven out of eight aligned sequences.

Again several homologous genes have been isolated from *Arabidopsis* and are being characterized in various laboratories. This protein has been shown to be phosphorylated in maize (Vilardell et al., 1990) and this prompted us to isolate protein-kinase cDNA from a dry seed library. So far two distinct kinase cDNA have been characterized in our group (P. GALLOIS, unpublished results).

We are now characterizing additional cDNA clones from the radish dry seed library and fishing out the corresponding genes from *Arabidopsis* cDNA and genomic library.

4 - Pattern of expression of various genes.

In order to confirm the scheme presented in Fig. I, RNA was prepared from the various developmental stages. This RNA was fractionated according to its size by denaturing agarose gel electrophoresis and blot transferred onto nylon membranes. The blots were then repeatedly successively hybridized with the various probes. Part of these results are shown in Fig. 3.

The first clone, pBA 3, corresponds to a napin cDNA (Raynal et al., 1991) and it is clear that napin mRNA specifically accumulates at two stages and that expression of these genes is restricted to the period of accumulation of storage products. Identical patterns were obtained with the cruciferin family probes (not shown). The second clone, pAB7, has also been isolated from the immature seed library. Its expression pattern is slightly broader than the napin ones. In addition, the same gene (or a closely related one) is apparently switched on again during germination because an RNA product is clearly detected after 48 hrs of germination. This clone was sequenced but showed no homology with any sequence in the data banks.

The next set of cDNA clones, except the last one, were all isolated from the mature seed library. All corresponding mRNAs have their maximum steady state level in either the mature dry seed or in a late maturation stage. These results confirm the validity of the differential screening. An interesting point to note is that the two identified mRNAs, Em like (8B6) and RAB 17-like (6F5), are not coordinately expressed, the RAB 17-like mRNA reaching its maximum steady state level before complete desiccation of the seed when the storage protein mRNAs are still present. These results also confirm our previous observation that most stored mRNA disappears during the imbibition period (Delseny et al., 1977).

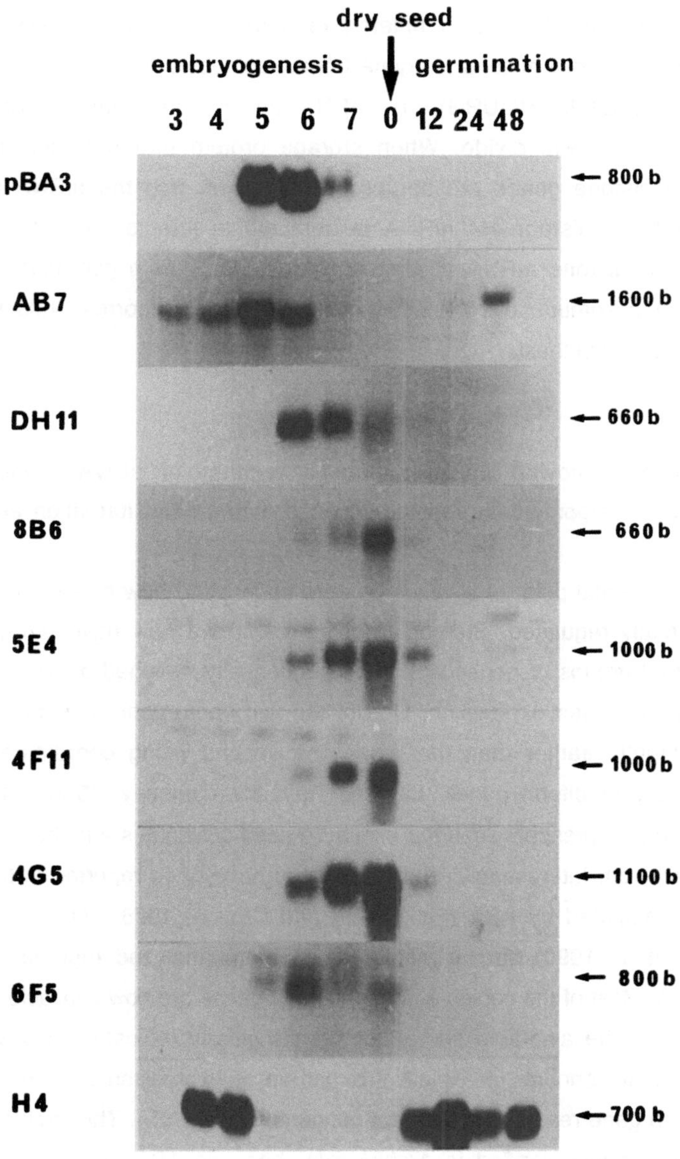

Figure 3.- Northern blot analysis of the expression of different genes during embryogenesis and germination of radish seeds. 10 µg of total RNA prepared from immature seeds at stage 3, 4, 5, 6 and 7, dry seeds (o), seeds germinated during 12, 24 and 48 h were electrophoresed under denaturing conditions on parallel gels and blotted to nylon membranes. Each membrane was hybridized with a different clone: clones BA3, AB7, DH11, 8B6, 5E4, 4F11, 4G5and 6F5 were isolated from radish cDNA libraries and H4 is a *Arabidopsis thaliana* histone H4 genomic clone.

The last clone (H4) shows a completely different pattern of expression. Indeed, it corresponds to a histone H4 gene from *Arabidopsis* (given to us by C. GIGOT, IBMP, STRASBOURG). Its mRNA accumulates during early embryogenesis when cells divide. When storage protein mRNA begins to accumulate, the histone genes are abruptly switched off and the mRNA is rapidly degraded. No histone H4 mRNA is detected in later stages of the developing seeds. Histone mRNA accumulation resumes during germination, with an apparent maximum at 24 hrs, a time which roughly corresponds to the maximum of DNA synthesis.

Discussion

This work has allowed us to isolate a number of cDNA probes corresponding to developmentally regulated genes during seed formation and germination.

From a fundamental point of view we have to understand how these genes are developmentally regulated. Starting from the napin we now have probes corresponding to 4 groups of genes which are sequentially switched on and off with some overlap in their expression. One of the two napin gene subfamilies is expressed slightly earlier than the other, the second being coordinately expressed with the cruciferin genes (Laroche Raynal et Delseny, 1986). The RAB 17-like gene is expressed when the immature seed dessicates and the Em-like gene still later on. Interestingly, all these genes have been reported in the literature to be regulated by ABA (Finkelstein and Crouch, 1986; Litts et al., 1987; Vilardell et al. 1990) but they have never been analyzed together in the same species. Most of the copies of these various genes are now available in *Arabidopsis* or will be available soon. Our results clearly indicate that ABA alone is probably not enough to explain this pattern of expression even if we assume that each gene reacts to a different concentration of ABA. This problem can now be most easily clarified in *Arabidopsis* because ABA deficient and insensitive mutants are characterized in this species.

Another important point to elucidate is the function of proteins encoded by mature seed mRNA. A first hint should be obtained from tissue and subcellular localization experiments. Several lines of evidence suggest that these proteins have something to do with controlling water exchange and preserving membrane integrity during dessication. The best indications for such a role are that these genes can be induced in immature seeds by artificial dehydration and soaking (Aspart et al., 1984), that some but not all of them can be induced by water stress in other tissues, (Close et al., 1989; Yamaguchi-Shinozaki et al., 1989; Villardell et al., 1990) and that several of these proteins are highly hydrophilic (Baker et al., 1988; Raynal et al., 1989). Other evidence indicate that some of these genes might have regulation functions.

From a practical point of view, this set of probes will allow us to assay for a correlation between the desiccation tolerance phase and accumulation of these mRNAs and their translation products in a much more precise manner. This set of probes should also allow us to follow and assay for the integritry of some mRNAs which are normally translated during imbibition. Besides many possible reasons for loss of seed viability, one can imagine that specific sets of mRNA have been altered in seeds with reduced viability and germinability. This possibility can now be explored with appropriate tools. Finally these probes will be very useful in mapping these genes on RFLP maps and evaluating their contribution in QTL such as germination vigour or drought tolerance.

Present work in our group is carried out with the aim of further characterizing the isolated genes, their regulation and function as well as identifying new genes involved in seed formation.

Acknowledgements. This work was supported by a grant from the Ministry of Agriculture (Biologie des Semences) and by the Centre National de la Recherche Scientifique (URA 565). The authors thank Marie-Thérèse Lacoste for typing the manuscript.

REFERENCES

- Aspart L, Meyer Y, Laroche M, Penon P (1984) Developmental regulation of the synthesis of proteins encoded by stored mRNA in radish embryos. *Plant Physiol* 76: 664-673.
- Baker JC, Steele C, Dure L III (1988) Sequence and characterization of 6 Lea proteins and their genes from cotton. *Plant Mol Biol* 11: 277-291.
- Casey R, Domoney C, Ellis N (1986) Legume storage proteins and their genes. *Oxford Surveys of Plant Molecular and Cell Biology* 3: 1-95.
- Close TJ, Kortt AA, Chandler PM (1989) A cDNA-based comparison of dehydration-induced proteins (dehydrins) in barley and corn. *Plant Mol Biol* 13: 95-108.
- Crouch ML, Tenbarge KM, Simon AE, Ferl R (1983) cDNA clones for *Brassica napus* seed storage proteins: evidence from nucleotide sequence analysis that both subunits of napin are cleaved from a precursor polypeptide. *J Mol Appl.Genet* 2: 273-283.

- Delseny M, Aspart L, and Guitton Y (1977) Disappearance of stored polyadenylic acid and mRNA during early germination of radish (*Raphanus sativus* L.) embryo axes. *Planta* 135: 125-128.
- Depigny D,Raynal M, Grellet F, Delseny F (1991) Structure, expression and evolution of storage protein genes in radish and related crucifers. *Proc. of VIIIth International Rapeseed Meeting* Saskatoon (in press).
- Dure L III (1985) Embryogenesis and gene expression during seed formation. *Oxford Surveys of Plant Molecular and Cell Biology* 2: 179-197.
- Finkelstein RR, Crouch M (1986) Rapeseed embryo development in culture on high osmoticum is similar to that in seed. *Plant Physiol* 81: 907-912.
- Godoy JA, Pardo JM, Pintor-Toro JA (1990) A tomato cDNA inducible by salt stress and abscisic acid: nucleotide sequence and expression pattern. *Plant Mol Biol* 15: 695-705.
- Laroche-Raynal M, Aspart L, Delseny M, Penon P (1984) Characterization of radish mRNA at three developmental stages. *Plant Sci Lett* 35: 139-146.
- Laroche-Raynal M, Delseny M (1986) Identification and characterization of the mRNA for major storage proteins from radish. *Eur J Biochem* 157: 321-327.
- Litts JC, Colwell GW, Chakerian RL, Quatrano RS (1987) The nucleotide sequence of a cDNA clone encoding the wheat Em protein. *Nucl Acids Res* 15: 3607-3618.
- Payne PI (1976) The long-lived messenger ribonucleic acid of flowering plant seeds. *Biol Rev* 51: 321-363.
- Raynal M, Aspart L, This P, Delseny M (1987) Biosynthesis of cruciferin polypeptides in immature radish seeds. *Plant Physiol Biochem* 25: 439-444.
- Raynal M, Depigny D, Cooke R, Delseny M (1989) Characterization of a radish nuclear gene expressed during late seed maturation. *Plant Physiol* 91: 829-836.
- Raynal M, Gaubier P, Grellet F, Delseny M (1990) Nucleotide sequence of a radish cDNA clone coding for a late embryogenesis abundant (LEA) protein. *Nucl Acids Res* 18: 6132
- Raynal M, Depigny D, Grellet F and Delseny M (1991) Characterization and evolution of napin-encoding genes in radish and related crucifers. *Gene* 99: 77-86.
- Simon AE , Tenbarge KM, Scofield SR, Finkelstein RR, Crouch ML (1985) Nucleotide sequences of a cDNA clone of *Brassica napus* 12S storage protein shows homology with legumin from *Pisum sativum* . *Plant Mol Biol* 5: 191-201.
- Vilardell J , Goday A, Freire MA, Torrent M, Martinez C, Torne H, Pages M (1990) Gene sequence, developmental expression and protein phosphorylation of RAB 17 in maize. *Plant Mol Biol* 14: 423-432.
- Yamaguchi-Shinozaki K, Mundy J, Chua NH (1989) Four highly linked rab genes are differentially expressed in rice, *Plant Mol Biol* 14: 29-39.

In Vitro Culture of Zygotic Embryos: Its Use in Soya and Sunflower Improvement

H. Serieys
INRA Station de génétique et d'amélioration
des plantes de Montpellier.
Domaine de Melgueil.
34130 Mauguio. France

INTRODUCTION

A primary objective in soybean and sunflower breeding programs is to increase the genetic variability available for use in germplasm collections and then to introgress desirable traits into commercial varieties. Wild perennial species of the subgenus *Glycine/glycine* may be of potential use since they possess tolerance to physiological traits such as drought and salt tolerance, day neutrality (Broué & al., 1982) or resistance to soybean rust, yellow mosaic virus, powdery mildew (Newell & Hymowitz, 1982). Similarly, the wild species of *Helianthus* genus show great interest for new sources of disease resistance, cytoplasmic male sterility (Serieys, 1987) or tolerance to stress environments (Serieys, 1980; Chandler & Beard, 1983). Hybridization of the cultivated types with the wild perennial species appeared difficult, because the embryos aborted at an early stage, either in soybean (Collins & al., 1984; Gyu & Jin, 1990) or within *Helianthus* genus (Chandler & Beard, 1983). The main application of immature embryo culture was related to "embryo rescue". This method, used in the interspecific hybrids when early embryo abortion happened, is well documented both in soybean (Newell & Hymowitz, 1982; Gyu & Jin, 1990) and sunflower (Chandler & Beard, 1983; Espinasse & al., 1985; Krauter & al., 1991).

Zygotic embryo culture was also employed to shorten the cycle duration, to break down the strong dormancies occurring in interspecific embryos or to generate reactive material for production of somatic embryos. An example of prevalent application of zygotic embryo culture by private firms was shortening cycle length of sunflower in breeding programs.

The major factors acting on soybean or sunflower zygotic embryo development and the recent contributions of immature embryo culture in the genetic improvement of these two important crops will be presented.

A) FACTORS AFFECTING ZYGOTIC EMBRYO DEVELOPMENT

Numerous references indicate that efficiency of the embryo culture was highly dependant on both genetic nature and age of the embryo at plating and nature of the growing medium.

1) Developmental stage of the embryo

The success of interspecific embryo culture is strongly influenced by the age and size of the embryo at the time of its culture. In soybean and related interspecific hybrids, successful developments were reported either by direct culture of 10 to 33 days old embryos, (Chan & Lin, 1967; Newell & Hymowitz, 1982; Gyu & Jin, 1990; Cutter & Bingham, 1975) or after embryo transfer on a "nurse endosperm" (Broué & al., 1982). The embryo survival at globular/early heart stage varies from 0 to 10 %, to reach more than 80 % at early cotyledon stage (Tilton & Russel, 1984). In sunflower, the interspecific embryos were generally rescued from the mother plant between 3 to 8 days after anthesis (Chandler & Beard, 1983; Alissa & al., 1986; Krauter & al., 1991). In vitro rescuing of interspecific embryos between sunflower (*Helianthus annuus L.*) and other *Helianthus* species shows that survival plant was 12 % at globular stage (3-4 days), 42 % at young heart-shaped stage (4-5 days), 46 % at differentiated stage (Krauter & al.,

1991) and close to 100 % for embryos larger than 2 mm (Espinasse & al., 1985). Embryo rescue was limited by the inability to grow small sunflower embryos, less developed than the vascular stage. So, an alternative method: ovule culture, was used when abortion occurred early in the development of the embryo, when its removal may be unpractical or when appropriate media formulations are lacking. Then, ovule culture relays the development to the point where embryo rescue techniques can be successfully implemented. Application of this method gave rise to original interspecific hybrids either in soybean, by growing 11-33 days old ovaries (Newell & Hymowitz, 1982; Broué & al., 1982) or in sunflower, by culturing 2-4 days old ovules (Espinasse & al., 1991).

2) Culture media composition

The growth response of immature embryos appears linked to the ability of the medium to provide the embryo requirements at the various developmental stages. Few mineral media were utilized in soybean (Table 1) or sunflower (Table 2) immature embryo culture. Commonly used basal media were defined by Murashigue & Skoog =MS (1962), Gamborg =B5 (1968), White (1963), Monnier (1976), Williams (1978), Huff & Dybing =HD (1980) and Nistch & Nistch (1969). The Gamborg's B5 salt medium, complemented in vitamins, sugars and growth regulators was the most used for smaller embryos in soybean (Newell & Himowitz, 1982; Tilton & Russel, 1984) or in sunflower (Chandler & Beard, 1983; Espinasse & al., 1985; Krauter & al., 1991), while Murashigue and Skoog or Nistch and Nistch media were generally employed for older immature embryos (Plotnikov, 1983; Alissa & al., 1986; Aspiroz & al., 1988).

Sucrose was the most used carbohydrate source in both soybean and sunflower. But, sucrose substitution with glucose, in the early steps of embryo maturation, promoted radicle development in soybean (Roumet 1991, personal comm.). In sunflower embryos, a critical developmental stage was defined around 1.5-2 mm length; before this stage, embryos could only grow on a medium with high osmolarity. So, raising the sucrose concentration from 30 to 150 g/l favoured maturation and decreased the incidence of premature germination of young sunflower embryos (Chandler & al., 1983;

TABLE 1: SUMMARY OF RECENT EMBRYO CULTURE MEDIA USED FOR THE REGENERATION OF SOYBEAN (*Glycine* sp.) PLANTS.

MEDIA AND METHODS	REFERENCES
1°) INTERSPECIFIC EMBRYO RESCUE	
*** 15-20 days old embryos**	
White's basal salts medium + (3 mg/l glycin, 0.15 mg/l thiamin, 10 mg/l nicotin. ac., 0.2 mg/l adenin, soya cotyledon extracts, 20 g/l sucrose).	Chan & al. (1967)
EMM[1]: Gamborg's B5 medium modified EGM[2]: Gamborg's B5 medium modif.+ 5 mg/l GA_3.	Cutter & al. (1975)
*** 15-25 days old embryos**	
EMM: GAMBORG's B5 medium + (100 mg/l inositol, 10 mg/l thiamin, 1 mg/l pyridoxin, 1 mg/l nicotinic acid, 0.1 mg/l IBA, 20 g/l sucrose). EGM: previous medium + 0.25 mg/l 2-iP.	Chung & al. (1990)
*** 0.2 to 0.5 mm lengh embryos**	
Gamborg's B5 nutrient salts + Williams's vitamins + 30 g/l sucrose.	Coble & al. (1990)
2°) EMBRYO CULTURE (Cultivated type)	
*** 2-21 days old embryos**	
Gamborg's B5 medium + 0.1 mg/l IBA.	Tilton & al. (1984)
3°) OVULE/EMBRYO RESCUE	
*** 10-21 days old embryos**	
EMM: Gamborg's B5 medium + (0.17 g/l KH_2PO_4, 1.65 g/l NH_4NO_3, 2 mg/l glycin, 1 mg/l BAP, 65.1 mg/l Fe, 0.64 mg/l kinetin, sucrose 30-100 g/l). EGM: previous medium + 10 g/l sucrose, without growth regulators.	Newell & al. (1982)
*** 11-33 days old embryos**	
Williams's medium.	Broué & al. (1982)

([1]) EMM: Embryo maturation medium, ([2]) EGM: embryo germination medium.

TABLE 2: SUMMARY OF MEDIA USED FOR SUNFLOWER EMBRYO CULTURE.

MEDIA AND METHODS	REFERENCES
1°) INTERSPECIFIC EMBRYO RESCUE.	
* 3 to 7 days old embryos	
EMM[1]: Gamborg's B5 salts + 1 mg/l (nicot.acid & pyrid.), 10 mg/l thiamin, 4 g/l inosit., amino-acids, 0.05 mg/l NAA, 120 g/l sucrose. EGM[2]: Gamborg's B5 salts + 10 g/l sucrose.	Chandler & al. (1983)
EMM: MS medium with 1/2 nutrient salts + 30 g/l sucrose, 1 mg/l (thiamin, pyridoxin, nicot. acid, glycin), 0.01 mg/l biotin, 0.01-0.1 mg/l ANA & BAP.	Alissa & al. (1986)
* Embryos < 1.5 mm	
EMM: Gamborg's B5 mofified medium. MS modified medium. ERM: MS medium.	Krauter & al. (1991)
* Embryos > 1.5 mm	
EMM: MS modified medium. ERM: MS medium	"
2°) IMMATURE EMBYRO CULTURE (*H. annuus* sp)	
* 2 to 14 days old embryos	
EMM: Huff & Dybing medium Gamborg's (B5) modified. EGM: Huff & Dybing medium + 10 g/l sucrose, without inositol.	Espinasse & al.(1985)
* 3 to 5 days old embryos	
EMM: Monnier medium + (462 mg/l KCl, 220 mg/l $CaCl_2$, 120 g/l sucrose). EGM: MS modif. Alissa (1986), hormons free.	Denat & al. (1991)
3°) OVULE CULTURE	
* 2-4 days old embryos	
MS medium Gamborg's B5 medium +/- (vitamins and amino acids). Nistch and Nistch medium + (0.1 mg/l ANA, 60-120 g/l sucrose).	Espinasse & al.(1991)

([1]) EMM: Embryo maturing medium, ([2]) EGM: Embryo germinating medium

TABLE 2 (continued): SUMMARY OF MEDIA USED FOR SUNFLOWER EMBRYO CULTURE.

MEDIA AND METHODS	REFERENCES
4°) SHORTENING DURATION CYCLE	
*** 10 days old embryos**	
Nistch and Nistch's salts medium + (1-1.5 mg/l IAA, 2-3 mg/l NAA & kinetin, 1-2 g/l activated coal, 10-20 g/l sucrose).	Plotnikov (1983)
*** 10-18 days old embryos**	
MS modified by Alissa (1986) + (0.1 g/l inositol, B group vitamins, 20 g/l sucrose) without phytohormons.	Aspiroz & al. (1988)
5°) BREAKING INTERSPECIFIC EMBRYOS DORMANCY	
*** 45-50 days, mature embryos**	
EGM: White's medium + (Fe EDTA (MS), 250 mg/l casein hyrolysate, 250 mg/l yeast extract, 0.05 mg/l NAA & kinetin, 0.01 mg/l GA3, 15 g/l glucose, 15 g/l sucrose). ERM: same medium without growth regulators.	Bohorova (1985)

([1]) EMM: Embryo maturing medium, ([2]) EGM: Embryo germinating medium

Denat & al. 1991; Espinasse & al., 1991). On the opposite, lowering concentration (10-30 g/l) enhances germination and rooting of older and larger embryos. In the same way, decreasing sucrose concentration from 100 to 10 g/l (Newell & Hymowitz, 1982) improved soybean ovule germination. Similarly, Coble & Schapaugh (1990) growing immature embryos (0.2-0.5 mm length) on a 30 g/l sucrose medium, enhanced plant survival from perennial soybean accessions. Sunflower embryos smaller than 2 mm required high osmolarity associated with low nitrogen concentration (C/N=10.6), in contrast to embryos of soybean, cotton and barley for which increased concentration in NH_4 successfully replaced high osmolarity. The high sucrose content (9%) associated with low nitrogen concentration explained the superiority of Gamborg's B5 medium for the survival of young sunflower embryos (Espinasse &

al., 1985).

Growth regulators such as IBA (Chung & Kim, 1990; Tilton & Russel, 1984), kinetin and BAP (Newell & Hymowitz, 1982), or GA_3 (Cutter & Bingham, 1975) were frequently used to improve soybean embryo development and rooting. Therefore, growth regulators were generally deleted from germinating media of very young sunflower embryos (embryo rescue), due to callogen effects (Chandler & Beard, 1983; Denat & al., 1991). Inversely, when older (7 to 12 days) embryos were cultured to shorten cycle length or break down dormancy; IAA, NAA and BAP phytohormons were found to increase significantly vigour of germinating embryos (Plotnikov, 1983; Bohorova & al., 1985; Alissa & al., 1986).

3) <u>Genetic effects</u>

In vitro embryo development was reported to be highly dependant on the genetic origin (inter/intraspecific variations) of the cultured material both in soybean (Newell & Hymowitz, 1982; Broué & al., 1982) and sunflower (Chandler & Beard, 1983; Espinasse & al., 1985).

B) APPLICATIONS OF ZYGOTIC EMBRYO CULTURE

1) <u>Interspecific hybrid embryo rescue</u>

In soybean significant results were got in interspecific hybridization through embryo or ovule rescue. Successful crossing *Glycine max X G. tomentella* were obtained either by growing ovules from 10-21 days old pods (Newell & Hymowitz, 1982) or from 15-25 days old embryos (Chung & Kim, 1990; Coble & Schapaugh, 1990). Likewise, *(Glycine tomentella X G. canescens) X G.max* hybridization was realized, else by ovule culture, or transfer 11-33 days old embryos on a nurse endosperm (Broué & al., 1982). Other crosses were also secured through embryo culture techniques: *Glycine argyrea X G. canescens) X G. max* (Brown & al., 1985), *Glycine max X G. clandestina* (Singh & al., 1987).

In sunflower, the most significant results in embryo rescue were got from crosses between the annual and perennial sections, where early embryo abortion usually occurred around the 5th days after anthesis. The following "hard to secure" hybrids were only obtained through latter methods: *H. angustifolius X H. annuus, H. anomalus X H. annuus, H. cusickii X H. annuus, H. deserticola X H. annuus, H. gracilentus X H. annuus* (Chandler & Beard, 1983), *H. pumilus X H. annuus* (Serieys, 1988-1990)

Ovule culture, as suggested by Espinasse & al. (1991), provides additional improvement for sunflower interspecific hybrid rescue. Anyway, in vitro ovule/embryo rescue methods proved their efficiency to overcome interspecific post-zygotic barriers where embryo degenerates prior to full maturity due to the inability of the endosperm to carry out its normal role in supplying nutriments.

2) Shortening cycle duration

Shortened vegetative cycle duration also helps in breeding programs, were only a few plants are needed per generation, such as rapid fixation of genotypes by single seed descend method (SSD), transferring a fertile inbred line into a sterile cytoplasm or introducing a desirable gene into an inbred line through recurrent crossing. A particularity of soybean/sunflower vegetative cycle lies in the maturation phase of the seed which takes 50 to 60 % of the total cycle duration. So, embryo culture method represents a way to by-pass the long maturation phase in these crops. Plotnikov (1983), growing plants from 10 days old immature embryos, on a modified Nistch and Nistch medium, reduced by half the duration of the sunflower cycle (within 55 to 75 days). So, from any fertility maintainer lines, it was possible to create isogenic cytoplasmic male sterile lines in about one year. Similarly, Alissa & al. (1986) growing 7 days old embryos on a modified Murashigue and Skoog medium, obtained 97 % viable plants and 4 generations in less than one year. Later, Aspiroz & al. (1988) growing 10-18 days old embryos also obtained 4 cycles in 320 days. According to earliness of genotypes, it took between 78 and 101 days to realize the complete cycle (Fig 1). Some morphological modifications were observed in plants issued from immature embryo culture. Main effect concerns the reduction of the dimension of the vegetative part: plant height, head and leaf size (Aspiroz & al., 1988). Such modifications were also reported by Tilton & Russel (1984) on soybean. But, not any durable genetic effect was registered in subsequent normally cultivated progenies (Serieys unpublished data).

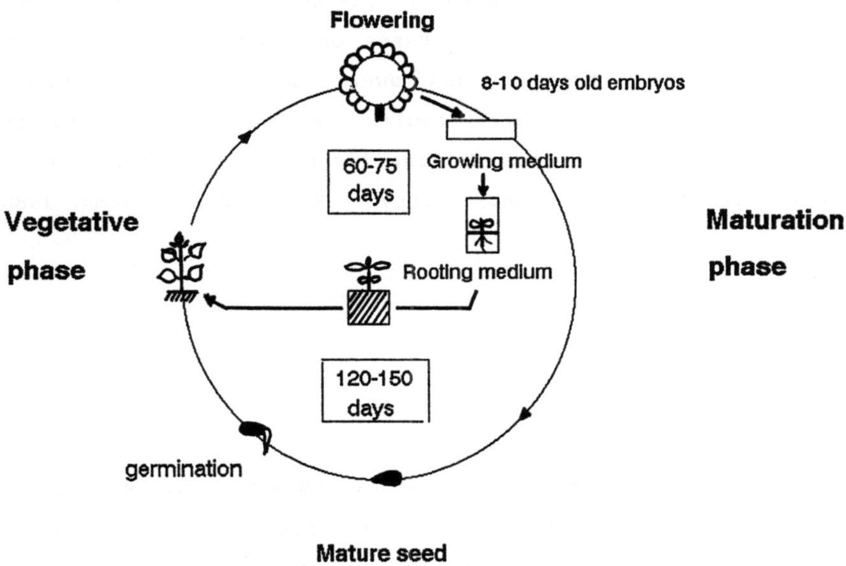

FIG 1: Embryo culture and shortening sunflower cycle duration.

In soybean, adaptation of embryo culture methods to shorten cycle duration is now in progress. According to Roumet (personal comm.), it is now possible to get between 60 to 100 % germinated embryos, within 12 days of culture in a modified liquid Gamborg'B5 medium, by plating embryos from 12 to 17 days old pods. Therefore, shoot development remains slow and incomplete. So, additional research is still necessary to speed up the generation turn over.

3) <u>Breaking down seed dormancy</u>

Another problem found in conventional interspecific hybridization is related to hybrid seed dormancy in *Helianthus* genus (Alissa & al., 1986). Since the mature seeds (45-50 days old) from interspecific hybrids *H. annuus X H. scaberimus* and *H. annuus X H. hirsutus* fail to germinate, Bohorova & al. (1985) growing mature embryos on a modified White's medium, obtained between 33 % and 100 % seed germination.

4) Support for somatic embryogenesis

In order to develop somatic regeneration, immature zygotic embryos were often grown on an induction medium to promote the development of embryoids, embryo-like structures or shoots, directly on the surface of the zygotic sunflower explants (Freyssinet & Freyssinet, 1988). Similar results were found when immature embryos of *Glycine max* and *G. gracilis* accessions where grown to produce high levels of somatic embryos (Komatsuda & Ohyama, 1988).

CONCLUSIONS

Zygotic embryo culture was successfully used (via ovule culture or embryo rescue) to improve the efficiency of interspecific hybridization programs by overcoming post-zygotic barriers, such as early embryo abortion or strong seed dormancies in sunflower or soybean. Other important applications concern the shortening of the cycle duration, this methodology in now currently used in breeding programs specially for rapid genotype fixation (SSD) or in creation of isogenic female analogues needful for cytoplasmic hybrid production of sunflower. Recurrent selection methods using male sterility are now developing in soybean breeding programs but, the evaluation of the genetic value of derived lines requires more rapid fixation techniques, which are not still available. Finally, the acquired knowledge from young zygotic embryo culture may usefully benefit to other somatic, androgenetic or gynogenetic regeneration methods.

REFERENCES

Alissa A, Jonard R, Serieys H, Vincourt P (1986) La culture d'embryons isolés in vitro dans un programme d'amélioration du tournesol. CR Acad Sci Paris, t 303, série III, 5:161-164.

Aspiroz HS, Vincourt P, Serieys H, Gallais A (1988) La culture in vitro des embryons immatures dans l'accélération du cycle de selection des lignées de tournesol et ses effets morphovégétatifs. Helia 10:35-38.

Bohorova N, Atanasov A, Georgieva-Todorova (1985) In vitro organogenesis, androgenesis and embryoculture in the genus *Helianthus L.* Pflanzenzuchtg 95:35-44.

Broué D, Douglass J, Grace JP, Marshall DR (1982) Interspecific hybridization of soybeans and perennial *Glycine* species indigenous to Australia via embryo culture. Euphytica 31:715-724.

Brown AHD, Grant JE, Burdon JJ, Grace JP, Pullen R (1985) Collection and utilisation of wild perennial *Glycine*. In World Soybean Research Conference (Shibles Red) West press, Colorado 345-352.

Chan K, Lin FM (1967) Studies on the method of embryo culture in the cultivated and the wild forms of Soybean. J taiwan agric res 16:8-14.

Chandler JM, Beard BH (1983) Embryo culture of *Helianthus* hybrids. Crop Sci 23:1004-1007.

Chung GH, Kim JH (1990) Production of interspecific hybrids between *Glycine max* and *G. tomentella* through embryo culture. Euphytica 48(2):97-101.

Coble CJ, Schapaugh WT (1990) Nutrient culture medium components affecting plant recovery from immature embryos of three *Glycine* genotypes and an interspecific hybrid grown in vitro. Euphytica 50(2):127-133.

Collins GB, Taylor NL, Deverna JW (1984) In vitro approaches to interspecific hybridization and chromosome manipulation in crop plants. Gene manipulation in plant improvment, genetics symposium 16 /1984/ Columbia Mi, USA 323-383.

Cutter GL, Bingham ET (1975) Soybean embryo culture studies. Soybean genet Newslett 2:52-53.

Denat JF, Coumans M, Serieys H, Jonard R (1991) Définition d'un milieu de culture permettant le développement d'embryons immatures de tournesol de trois jours. Application à l'obtention d'hybrides interspécifiques. CR Acad Sci Paris, t 313, série III,145-151.

Espinasse A, Lay C, Dybing CD (1985) Factors controlling in vitro development of sunflower embryos. Agronomie 5 (9):825-832.

Espinasse A, Volin J, Dybing CD, Lay C (1991) Embryo rescue through in ovulo culture in *Helianthus*. Crop Sci 31 (1):102-108.

Freyssinet M, Freyssinet G (1988) Fertile plant regeneration from sunflower (*Helianthus annuus* L) immature embryos. Plant science 56 (2):177- 181.

Gamborg OL, Miller A, Ojima K (1968) Nutrient requirements of suspension cultures of soybean root cells. Exp Cell Res 50:151-158.

Gyu Hwa Chung, Jin Hyeong Kim (1990) Production of interspecific hybrids between *Glycine max* and *G. tomentella* through embryo culture. Euphytica 48:97 -101.

Huff A, Dybing CD 1980) Factors affecting shedding of flowers in soybean (*Glycine max* (L) Merill. J Exp Bot, 31 (122):751-762.

Komatsuda T, Ohyama K (1988) Genotypes of high competence for somatic embryogenesis and plant regeneration in soybean (*Glycine max*). Theoretical and applied genetics 75 (5):695-700.

Krauter R, Steinmetz A, Friedt W (1991) Efficient interspecific hybridization in the genus *Helianthus* via "embryo rescue" and characterisation of the hybrids. Theor Appl Genet 82: 521-525.

Monnier M (1976) Culture in vitro de l'embryon immature de *Capsella Bursa-pastoris* Moench. Rev Cyt Biol Véget 39:1-120.

Murashigue T, Skoog F (1962) A revised medium for rapid growth and bioassays with tobacco tissue cultures. Physiol plant 15:473-497.

Newell CA, Hymowitz T (1982) Successful wide hybridization between the soybean and a wild perennial relative *G. tomentella hayata*. Crop Sci 22:1062-1065.

Nistch JP, Nistch C (1969) Haploid plants from pollen grains. Science (Washington DC) 163:85-87.

Plotnikov VA (1983) Use of method of culturing young embryos for accelerated development of sunflower cytoplasmic male serility analogues. Tsitologiya i Genetika, 17 (6):40-43.

Serieys H (1980) Utilisation des espèces sauvages d'*Helianthus* pour l'amélioration du tournesol cultivé. Proc IX th Internat Sunflower Conference Torremolinos/Spain (1): 107-121.

Serieys H (1987) Characterisation of some new cytoplasmic androsterilitiy sources from *Helianthus* genus. Helia 10: 9-13.

Serieys H In (1988-1990) FAO Progress report. Subnetwork Genetics and breeding. Evaluation of morphological physiological, biochemical and taxonomic aspects of wild *Helianthus*.

Singh RJ, Kollipara KP, Hymowitz T (1987) Inter subgeneric hybridization of soybeans with wild perennial species, *Glycine clandestina* Wendl. Theor Appl genet 74:391-396.

Tilton VR, Russel SH (1984) In vitro culture of immature soybean embryos. J Plant Physiol Vol 115:191-200.

White PR (1963) The cultivation of animal and plant cells. New York: Ronald Press.

Williams E (1978) A hybrid between *Trifolium repens* and *Trifolium ambiguum* obtained by the aid of embryo culture. New Z J Bot 16:499-509.

Histology of Somatic Embryogenesis

N. Michaux-Ferrière and J. Schwendiman
BIOTROP-CIRAD
B.P. 5035
34032 Montpellier Cedex
France

Asexual multiplication is a unique process by which plants give rise to new individuals identical to the parent plant. This potential can be directly extended by microcutting, an *in vitro* vegetative propagation technique. Cell totipotency is also specific to the plant kingdom and expressed in *in vitro* cultures by the ability to regenerate embryos from somatic cells. Somatic embryogenesis was demonstrated in the 1950s in carrots (Steward *et al.*, 1958); subsequently it has been obtained easily in certain plants but with much more difficulty in others (Ammirato, 1983a). The many advantages of this method, which combines enhanced regeneration (theoretically true to type) with rejuvenation, have incited numerous recent studies aimed at optimizing experimental conditions for somatic embryogenesis in monocots (Vasil, 1987), dicots (Raghavan, 1986 ; Rangaswamy, 1986 ; Ammirato, 1989).

The difficulty of obtaining somatic embryos in certain species, particularly ligneous plants (Bajaj, 1989), or in genotypes having other interesting characteristics (yield, pest and disease resistance), has prompted studies to enlarge the understanding of mechanisms involved in this process. The impact of a number of factors on the embryogenic potential of cultured explants has been recently studied: water parameters (Imamura & Harada, 1980; Etienne *et al.*, 1991a), hormonal balance (Carman, 1989; El Hadrami *et al.*, 1991) mineral balance (Etienne *et al.*, 1991a), addition of polyamines (El Hadrami *et al.*, 1989) or antioxidant agents (Housti *et al.*, 1991) to the culture medium, and carbon sources (Meijer & Brown, 1987; Etienne *et al.*, 1991b). Moreover, the importance of atmospheric gas content (O_2, CO_2, C_2H_4) in culture vessels (Carman, 1988; Vain *et al.*, 1989; Auboiron *et al.*, 1990), the timing of subculturing (Michaux-Ferrière & Carron, 1989), the type of medium (solid, liquid or semi-liquid) and the quality of the solid culture medium (gelrite, agar, gelose, cellulose blocks) have also been investigated.

There has been an increase in the systematic use of histocytology to complement *in vitro* culture studies (Profumo *et al.*, 1986; Guiderdoni & Demarly, 1988; Michaux-Ferrière & Carron, 1989; Dhed'a *et al.*, 1991). By this approach, cytological phenomena that occur during explant culturing and the formation of calli and embryos capable of germinating can be investigated. Histocytological analysis shows whether the obtained structures are abnormal or immature, or actual formed embryos. By comparison to a reference model (zygotic ontogenesis or a well controlled system able to regularly produce somatic embryos), this analysis can also be used to pinpoint blockages in cultures expressing no embryogenic potential, or dead end routes taken by calli during their development.

I. Embryogenic cells

Generally, in studies carried out in our laboratory (banana, oil palm, rice, cocoa, coffee, cotton, rubber tree and papaya) and other reported studies (Lu & Vasil, 1985; Williams & Maheswaran, 1986), somatic embryos were formed by the development of cells which are termed "embryogenic" due to their individual characteristics. They resemble meristematic cells; they are not very bulky, have high RNP, a dense cytoplasm, a single and quite large nucleolus (indicating an intense RNA metabolism) and small starch granules (Fig. 1).

Cells expressing this set of characteristics are usually localized on the periphery of calli which have formed from explants. These cells gather into clusters which can vary in size and number. Embryogenic cells have occasionally been observed within cultured tissues without any prior callus formation (Châtelet *et al.*, 1992).

Regardless of the origin of the cells (meristematic or parenchymatous) embryogenic characteristics are acquired by a dedifferentiation process. However, this induction process is not yet fully understood. Embryogenesis is dependent upon the type and physiological status of the cultured explant, as well as the experimental conditions. If auxin stress is confirmed to be essential in many species (Carman, 1990), then embryogenic cells could also be induced in the presence of a single cytokinin, as reported in coffee (Yasuda *et al.*, 1985). Embryogenesis is induced at a specific stage of callus development. This stage can vary for explants of the same quality and genotype, depending especially on the culture conditions and the hormonal balance. Nevertheless, suitable culture conditions and the correct cytophysiological status of the callus for embryogenesis can occur at different points during callus development (Michaux-Ferrière *et al.*, 1989).

Within the same embryogenic cluster, cellular development is not synchronised. There is a time period during which embryogenic cells correspond with the classical definition. During this time period (the "window of embryogenesis"), the cells are capable of dividing or of evolving into embryos, depending upon the culture conditions. Subsequently, the cells which did not form proembryos begin the process of differentiation: vacuolisation, decrease in nuclear and nucleolar volume and the accumulation or complete disappearance of starch granules.

Experience has shown that embryogenic cells usually only develop into embryos if the culture conditions have been modified. The modified medium on which these cells are able to express their embryogenic potential is characterized by a reduced auxin-cytokinin balance (Ammirato, 1983b; El Hadrami *et al.*, 1991) or even a complete absence of auxin. In order to obtain an optimal response, the medium must be changed when the callus contains the greatest number of typically embryogenic cells.

II. Ontogeny of somatic embryos

1. Direct or indirect embryogenesis

Direct embryogenesis (Fig. 2) (Pence *et al.*, 1979) occurs when embryos are generated by the development of embryogenic cells initiated on the explant without preliminary callus formation. This type of embryogenesis often occurs when the explants are immature zygotic embryos (Maheswaran & Williams, 1986). As relatively short amount of time is required for embryo formation by this process, the risk of obtaining variants is minimized.

When embryogenic cells are obtained from a callus on which cells are frequently multiplied, differentiated and dedifferentiated, the process is termed indirect embryogenesis. There are many reports in the literature that embryos can be generated at different times during callus development, either after only a few weeks or up to several months of culture. However, after a long period of culture in the presence of auxins, there is, in theory, the risk of inducing somaclonal variations.

There have not yet been any specific studies carried out to determine a link between somaclonal variations and the type of embryogenesis. Moreover, a correlation has not been established between the percentage of variants obtained and the length of culture in medium containing synthetic auxins.

2. Uni- or multicellular origin of somatic embryos

Embryogenic cells develop into embryos when cultured in experimental conditions which allow them to express their potential. Two ontogenic conditions should be noted:

- Unicellular origin

Embryogenic cells are isolated from one another by an important modification of their cell walls, in particular gelification of the middle lamella. The individual cells which are surrounded by a polysaccharide mucilage, undergo polarized divisions, thus forming globular proembryos (Fig. 3). The preferential localization of starch results from early polarization of these proembryos. Epidermization occurs and they acquire a bilateral symmetry. The successive steps of somatic embryo ontogenesis include: growth of one or more cotyledons, provascular strands, root and shoot apex initiation and progressive accumulation of amylaceous and lipoprotein reserves (Michaux-Ferrière *et al.*, 1987; Michaux-Ferrière & Carron, 1989) (Fig. 4).

- Multicellular origin

In this case, somatic embryos are formed from cell clusters instead of single cells. Diverse pathways can be observed: in *Quercus* (El Maataoui *et al.*, 1990), coffee (Pierson *et al.*, 1983) and many other plants (Haccius, 1978; Maheswaran & Williams, 1985 and 1986), somatic embryos are formed from epidermal and sub-epidermal cell clusters. This type of ontogenesis from immature zygotic embryos or somatic embryo structures is an example of direct multicellular embryogenesis and is commonly termed budding or adventive embryogenesis. In *Coffea arabica*, early globular proembryos (in 30 day old calli) were formed from embryogenic cells clumped in small spherical masses, isolated from the rest of the callus by a highly differentiated layer of cells (unpublished results). In oil palm and rubber tree, very precise histological analysis of compact potentially embryogenic calli revealed an original ontogenesis of multicellular-derived embryos (Schwendiman *et al.*, 1990; Michaux-Ferrière *et al.*, 1992). In nodular parenchymatous calli, clusters of cells (closely resembling meristematic cells) which were within or on the periphery of the callus demonstrated synchronised development (Fig. 5). These small cells have a very high nucleoplasmic ratio, a particularly dense cytoplasm, and an especially voluminous, stainable nucleolus. They differ from embryogenic cells by their lack of reserve materials, by their very thin cell walls and their frequent cell divisions. They resemble embryonic cells in the initial stages of zygotic ontogenesis (Rondet, 1965; Jones & Rost, 1989a). Such structures can be considered to be globular proembryos (Fig. 6). They develop an epidermis, the provascular strands differentiate, reserve materials accumulate and polarization occurs. These globular structures produce cotyledons and become actual somatic embryos by the formation of root and shoot apices.

- Uni- and multicellular embryo initiation conditions

There is insufficient precise data available in the literature to establish why an embryogenic structure will develop from a uni- or multi-cellular origin. In rubber tree, using the same genotype (PB 260), the origin of embryos varied according to the experimental conditions. Somatic embryos of unicellular origin were first obtained (Michaux-Ferrière & Carron, 1989). The use of a new experimental technique, with a culture medium that differed from the initial medium (new mineral and hormonal balance, presence of $AgNO_3$ and spermidine), resulted in the formation of multicellular-derived embryos (Michaux-Ferrière *et al.*, 1992). Although the latter culture conditions gave rise to embryos of multicellular origin with clone PB 260 (and also with PB

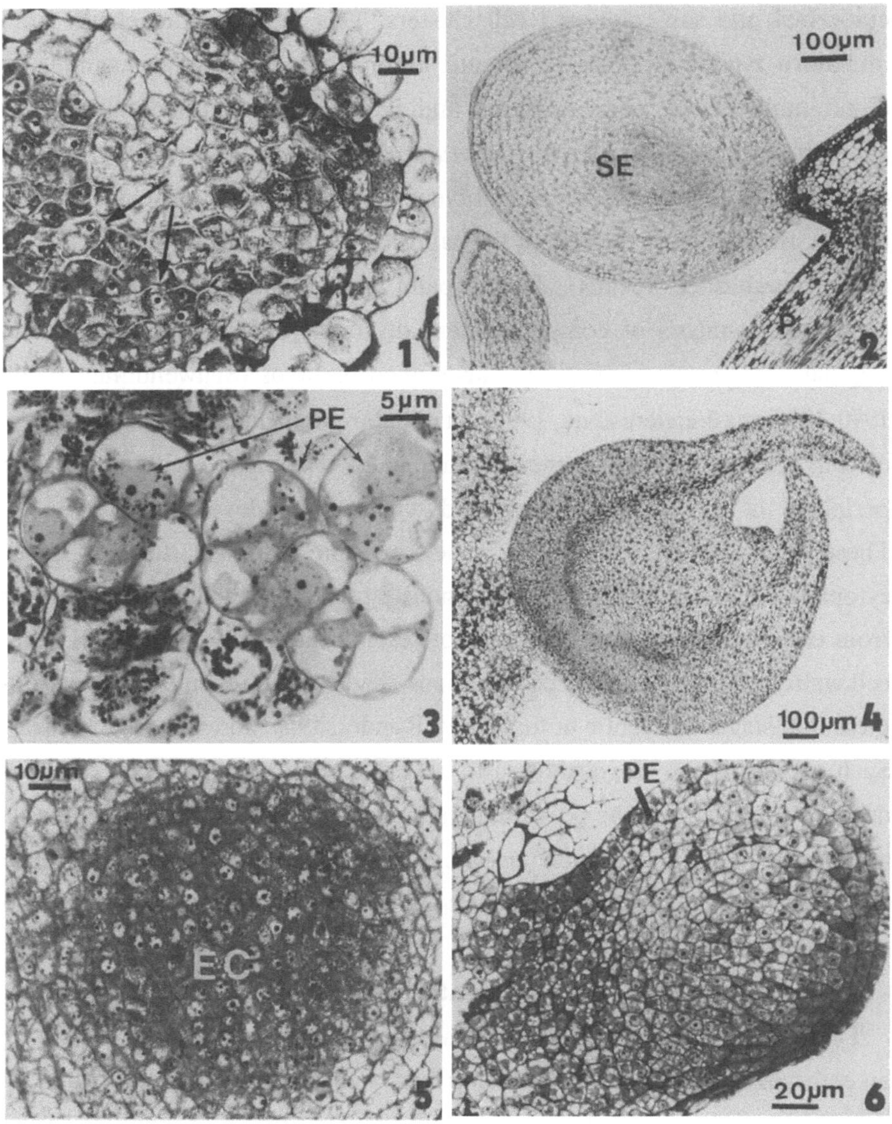

Figure legends

Fig.1 : A cluster of embryogenic cells in a callus of *Coffea arabica*.
Fig. 2 : Cacao somatic embryo (SE) formed on a petiole (P) of a young leaf of a germinating embryo.
Fig. 3 : *Coffea arabica*. Proembryos (PE) of unicellular origin.
Fig. 4 : *Hevea brasiliensis*. Bipolar somatic embryo.
Fig. 5 : Group of embryonic cells (EC) in a callus of *Elaeis guineensis*.
Fig. 6 : *Hevea brasiliensis*. Proembryo (PE) of pluricellular origin.

235), embryos obtained using the same conditions but with the two other clones (RRIM 600 and PR 107) were of unicellular origin. Thus the response can vary according to the genotype used. In the examples when the two modes of expression of somatic embryogenesis were initiated almost simultaneously (oil palm, rubber tree) only the multicellular origin was expressed. Under certain conditions, in coffee, the two routes can be expressed successively; in such cases multicellular induction occurs first.

3. Comparison of somatic and zygotic ontogenesis

The development of a zygotic embryo and that of a somatic embryo of unicellular origin are comparable, with the progression through the ontogenetic stages being the same.

In dicots, the globular stage is followed by a transient heart stage corresponding to the acquisition of bilateral symmetry: epidermization, and then the development of the procambial strands and root meristem occur successively. The shoot apex is formed later during the cotyledon stage. Nevertheless, an intermediary stage (termed oblong) has been observed in some plants between the globular and heart stages in somatic embryos (Schiavone & Cooke, 1985). This would correspond to premature individualization of the root apex and procambium and is associated with axial elongation of the embryo induced by polarized auxin transport (Schiavone & Cooke, 1987). Monnier (1988) suggested that this stage, which is sometimes observed during *in vitro* ontogenesis of zygotic embryos, is induced by the culture conditions.

In monocots, the shoot apex develops rapidly after emergence of the scutellum or cotyledon. The provascular strands and root apex are formed thereafter (Schwendiman *et al.*, 1988 ; Jones & Rost, 1989a and b).

Multicellular-derived embryos obtained by budding show the same ontogenetic stages as zygotic embryos, according to Williams & Maheswaran (1986). However, in rubber tree, globular proembryos formed from meristematic cell clusters seemed to develop rapidly from the globular to the cotyledon stage; there was no clearly defined heart stage.

Development of the metabolism of reserve materials during ontogenesis was found to be identical in somatic and zygotic embryogenesis. This development resulted in a precisely timed progressive acquisition (qualitative and quantitative) of starch and/or lipoprotein reserves. Recent studies, in our laboratory, of coffee and rubber tree have demonstrated the rapidity of somatic embryo formation (from embryogenic cells or meristematic clusters) relative to the time required to develop a zygote after fertilization.

III. Conformity of somatic embryos

Without histological analysis, somatic embryo formation can only be confirmed by germination. However, the low percentage of embryos capable of evolving into plantlets, as determined with a large sample of plants, cannot be explained simply by an inadequacy of the induction medium.

Somatic embryos, when compared to zygotic embryos of the same species and whose ontogenesis and mature structure are taken as the reference, often develop abnormally or are immature. Such embryos are thus unable to germinate (Michaux-Ferrière *et al.*, 1992). The most frequently observed abnormalities, were double or triple vascular systems, caused by poor polarization of auxin transport (Chée & Cantliffe, 1989); amylaceous and/or protein reserves that were excessively high, absent or very reduced; or the non-development of shoot (rubber tree) or root (palm) meristems.

A study of zygotic ontogenesis in rubber tree showed that the apical meristem was late in forming, much later than the root meristem and just before the accumulation of lipoprotein reserves. The absence of shoot apices in somatic embryos or the detection of very low protein reserves should therefore not be considered abnormal, but rather as being due to embryo immaturity probably caused by rapid somatic embryo formation. We decided to add an additional phase to the initial culture medium to enable maturation of the formed structures (Michaux-Ferrière *et al.*, 1992). This resulted in an increase in the percentage of somatic embryos capable of germinating.

Relatively important morphogenic abnormalities were observed regardless of the pathway of somatic embryo formation. At the present time, there is insufficient precise data to allow an understanding of whether or not a multicellular origin will induce more abnormalities than a unicellular origin.

IV. Conclusion

The control of somatic embryogenesis for species or genotypes that are poorly suited for this technique would require an overall physiological, cellular and molecular understanding of the processes involved. Precise histocytological and histochemical information on embryogenic calli and embryos during culturing would be essential for determining the times when calli or embryogenic structures have deviated from the normal pathway.

Results from our laboratory on many monocot and dicot plants have allowed us to identify the basic cellular somatic embryogenesis process. The expression conditions varied in accordance with the explants, experimental conditions, culture media and genotypes. It is important to note that the development of somatic embryos, although often not an exact copy of zygotic embryo development, always retains the general ontogenic characteristics of the plant group concerned: chronology of meristem development, type of reserve materials and formation of one or more cotyledons.

A great number of embryogenic cells are initiated during the somatic embryogenesis process. However, not many of these are able to develop into proembryos at the right time in given culture conditions. Only a few of the globular structures present continue their development. Consequently, the number of bipolar embryos formed is relatively low in comparison to the number of globular proembryos or embryogenic cells which were initiated. Moreover, considering all of the problems occurring during the conversion of embryos into plantlets, the present yield of the somatic embryogenesis technique, even for plants which are considered to react well, is clearly much lower than the real potential. Improvement of culture conditions for somatic structures according to their stages of development could probably enhance the efficiency of this multiplication technique.

Current studies on embryogenesis in liquid medium in many plants should solve these problems and considerably increase the number of plantlets regenerated. The aim is to obtain a self-maintained cell suspension which can, when required, be directed towards somatic embryogenesis. Controlling such a system should eliminate certain drawbacks of embryogenesis in solid medium (asynchronism, the need to regularly use new explants) while promoting protoplast formation and thus access to genetic transformations. Once again, histocytology should be used constantly as a guide for *in vitro* cultures, to indicate whether cultures are developing well or deviating; and to suggest further experiments, and improvements, in culture conditions.

References

Ammirato PV (1983a) Embryogenesis. *In* Handbook of plant cell culture, vol. 1, D.A. Evans, W.R. Sharp, P.V. Ammirato and Y. Yamada (eds), pp. 82-123. New York, Macmillan Publishing Co.

Ammirato PV (1983b) The regulation of somatic embryo development in plant cell cultures : suspension culture techniques and hormone requirements. Biol. Technology 1 : 68-74.

Ammirato PV (1989) Recent progress in somatic embryogenesis. Newsletter IAPTC 57 : 2-16.

Auboiron E, Carron MP, Michaux-Ferrière N (1990) Influence of atmospheric gases particularly ethylene on somatic embryogenesis of *Hevea brasiliensis.* Plant Cell Tissue and Organ Culture 21 : 31-37.

Bajaj YPS (1989) Biotechnology in agriculture and forestry 5. Trees II. Springer-Verlag Berlin, 622 p.

Carman JG (1988) Improved somatic embryogenesis in wheat by partial simulation of the *in-ovulo* oxygen growth regulator and desication environments. Planta 175 : 417-424.

Carman JG (1989) The *in-ovulo* environment and its relevance to cloning wheat via somatic embryogenesis. *In vitro* cell Dev. Biol. 25 : 1155-1162.

Carman JG (1990) Embryogenic cells in plant tissue culture. Occurrence and behavior. *In vitro* cell Dev. Biol. 26 : 746-753.

Châtelet P, Michaux-Ferrière N, Dublin P (1992) Potentialités embryogènes du nucelle et du tégument interne de graines immatures de cacaoyer (*Theobroma cacao* L.). C.R. Acad. Sci. Paris. Sous presse.

Chée R, Cantliffe DJ (1989) Inhibition of somatic embryogenesis in response to 2, 3, 5-triiodobenzoic acid and 2,4-dichlorophenoxyacetic acid in *Ipomoea batatas* (L.) Lam. cultured *in vitro*. J. Plant Physiol. 135 : 398-403.

Dhed'a D, Dumortier F, Panis B, Vuylsteke D, De Langhe E (1991) Plant regeneration in cell suspension cultures of the cooking banana cv "bluggoe" (*Musa* spp. ABB group). Fruits 46 : 125-135.

El Hadrami I, Michaux-Ferrière N, Carron MP, D'Auzac J (1989) Polyamines, a possible limiting factor in somatic embryogenesis of *Hevea brasiliensis*. C.R. Acad. Sci. Paris 308 : 205-211.

El Hadrami I, Carron MP, D'Auzac J (1991) Influence of exogenous hormones on somatic embryogenesis in *Hevea brasiliensis*. Ann. Bot. 67 : 511-515.

El Maataoui M, Espagnac H, Michaux-Ferrière N (1990) Histology of callogenesis and somatic embryogenesis induced in stem fragments of cork oak (*Quercus suber*) cultured *in vitro*. Ann. Bot. 66 : 183-190.

Etienne H, Berger A, Carron MP (1991a) Water status of callus from *Hevea brasiliensis* during induction of somatic embryogenesis. Phys. Plantarum 82 : 213-218.

Etienne H, Montoro P, Carron MP (1991b) The effect of water parameters on the development of *Hevea brasiliensis* calli in *in vitro* culture. Ann. Sci. Forestières 48 : 253-265.

Guiderdoni E, Demarly Y (1988) Histology of somatic embryogenesis in cultured leaf segments of sugarcane plantlets. Plant Cell Tissue and Organ Culture 14 : 71-88.

Haccius B (1978) Question of unicellular origin of non-zygotic embryos in callus cultures. Phytomorphology 28 : 74-81.

Housti F, Coupé M, D'Auzac J (1991) Facteurs enzymatiques du brunissement *in vitro* et capacité embryogène des cals d'*Hevea brasiliensis*. C.R. Acad. Sc. Paris 313 : 461-466.

Imamura J, Harada H (1980) Effects of abscisic acid and water stress on the embryo and plantlet formation in anther culture of *Nicotiana tabacum* cv Sansun. Z. Pflanzenphysiol. 100 : 285-289.

Jones TJ, Rost TL (1989a) Histochemistry and ultrastructure of rice (*Oryza sativa*) zygotic embryogenesis. Amer. J. Bot. 76 : 504-520.

Jones TJ, Rost TL (1989b) The developmental anatomy and ultrastructure of somatic embryos from rice (*Oryza sativa*) scutellum epithelial cells. Bot. Gaz. 150 : 41-49.

Lu CY, Vasil IK (1985) Histology of somatic embryogenesis in *Panicum maximum (Guinea Grass)*. Amer. J. Bot. 72 : 1908-1913.

Maheswaran G, Williams EG (1985) Origin and development of somatic embryoids formed directly on immature embryos of *Trifolium repens in vitro*. Ann. Bot. 56 : 619-630.

Maheswaran G, Williams EG (1986) Primary and secondary direct somatic embryogenesis from immature zygotic embryos of *Brassica campestris*. J. Plant Physiol. 124 : 455-463.

Meijer EGM, Brown DCW (1987) Role of exogenous reduced nitrogen and sucrose in rapid high-frequency somatic embryogenesis in *Medicago sativa*. Plant Cell Tissue and Organ Culture 10 : 11-19.

Michaux-Ferrière N, Dublin P, Schwendiman J (1987) Etude histologique de l'embryogenèse somatique à partir d'explants foliaires de *Coffea arabica* L. Café, Cacao, Thé 31 : 103-110.

Michaux-Ferrière N, Carron MP (1989) Histology of early somatic embryogenesis in *Hevea brasiliensis* : the importance of the timing of subculturing. Plant Cell Tissue and Organ Culture 19 : 243-256.

Michaux-Ferrière N, Bieysse D, Alvard D, Dublin P (1989) Etude histologique de l'embryogenèse somatique chez *Coffea arabica* induite par culture sur milieux uniques de fragments foliaires de génotypes différents. Café, Cacao, Thé 33 : 207-217.

Michaux-Ferrière N, Grout H, Carron MP (1992) Origin and ontogenesis of somatic embryos in *Hevea brasiliensis* (Euphorbiacée). Am. J. Bot. : 79, sous presse.

Monnier (1988) Embryogenèse zygotique et somatique. *In* J.P. Zryd (Ed.) Cultures de cellules. Tissus et organes végétaux. Fondements théoriques et utilisation pratiques. Presses polytechniques romandes. Lausanne. 119-134.

Pence VC, Hasegawa PM, Janick J (1979) Asexual embryogenesis in *Theobroma cacao* L.. J. Am. Soc. Hortic. Sci. 104 : 145-148.

Pierson ES, Van Lammeren AM, Schel JHN, Staritsky G (1983) *In vitro* development of embryoids from punched leaf discs of *Coffea canephora*. Protoplasma 115 : 208-216.

Profumo P, Gastaldo P, Dameri RM, Caffaro L (1986) Histological study of calli and embryoids from leaf explants of *Aesculus hippocastanum* L.. J. Plant Physiol. 126 : 97-103.

Raghavan V (1986) Embryogenesis in Angiosperms. A developmental and experimental study. Cambridge University Press. Developmental and Cell Biology series 17 : 115-151.

Rangaswamy NS (1986) Somatic embryogenesis in angiosperm cell tissue and organ culture. Proc. Indian Acad. Sci. (Plant Sci.) 96 : 247-271.

Rondet P (1965) Mise en place des méristèmes chez les angiospermes au cours de l'embryogenèse. Bull. Soc. Fr. Phys. Vég. 11 : 25-26.

Schiavone FM, Cooke TJ (1985) A geometric analysis of somatic embryo formation in carrot cell cultures. Can. J. Bot. 63 : 1573-1578.

Schiavone FM, Cooke TJ (1987) Unusual patterns of somatic embryogenesis in the domesticated carrot : developmental effects of exogenous auxines and auxine transport inhibitors. Cell differentiation 21 : 53-62.

Schwendiman J, Pannetier C, Michaux-Ferrière N (1988) Histology of somatic embryogenesis from leaf explants of the oil palm *Elaeis guineensis*. Ann. Bot. 62 : 43-52.

Schwendiman J, Pannetier C, Michaux-Ferrière N (1990) Histology of embryogenic formations during *in vitro* culture of oil palm *Elaeis guineensis* Jacq. Oléagineux 45 : 409-418.

Steward FC, Mapes MO, Mears K (1958) Growth and organized development of cultured cells. II. Organization in cultures grown from freely suspended cells. Am. J. Bot. 45 : 705-708.

Vain P, Yean H, Flament P (1989) Enhancement of production and regeneration of embryogenic type II callus in *Zea mays* L. by $AgNO_3$. Plant Cell Tissue and Organ Culture 18 : 143-151.

Vasil IK (1987) Developing cell and tissue culture systems for the improvement of cereal and grass crops. J. Plant Physiol. 128 : 193-218.

Williams EG, Maheswaran G (1986) Somatic embryogenic factors influencing coordinated behaviour of cells as an embryogenic group. Ann. Bot. 57 : 443-462.

Yasuda T, Fujii Y, Yamaguchi Y (1985) Embryogenic callus induction from *Coffea arabica* leaf explants by benzyladenine. Plant Cell Physiol. 26 : 595-597.

Cell Wall Glycoprotein Encoding Genes in Somatic and Zygotic Embryogenesis

S.C. De Vries, H. Booij, J. Cordewener, F. Van Engelen, M. Hartog, T. Hendriks, A. De Jong, E. Meijer, P. Sterk, A. van Kammen.
Department of Molecular Biology
Agricultural University Wageningen
Dreijenlaan 3
6708 HA Wageningen
The Netherlands

Introduction

In carrot, somatic embryos can develop after simple culture manipulations from single embryogenic cells (Komamine et al. 1990) or from clusters of embryogenic cells designated proembryogenic masses (Halperin 1966). Most embryogenic carrot cultures are maintained over many subcultures with 2,4-D either as the sole growth regulator, or in the presence of both 2,4-D and cytokinin. Therefore, cells that have embryogenic potential may either be continuously formed from non-embryogenic cells, or constitute an independent self-propagating subpopulation. In carrot cultures, depletion of the population of embryogenic cells eventually leads to loss of the embryogenic potential. In Figure 1 a schematic representation of the acquisition and expression of embryogenic potential in carrot cultures is presented. In other culture systems such as alfalfa, non-embryogenic cells can be maintained in media containing NAA, while a short exposure to a high concentration of 2,4-D is sufficient for these cells to acquire embryogenic potential (Bögre et al. 1990). It is unlikely that 2,4-D is unique in its rôle in the establishment of embryogenic potential. For instance, Smith and Krikorian (1990) have described a culture system in which cells derived from carrot zygotic embryos maintain their embryogenic potential indefinitely in the absence of any growth regulator, solely by subtle adjustments of the medium pH. In carrot, it is not

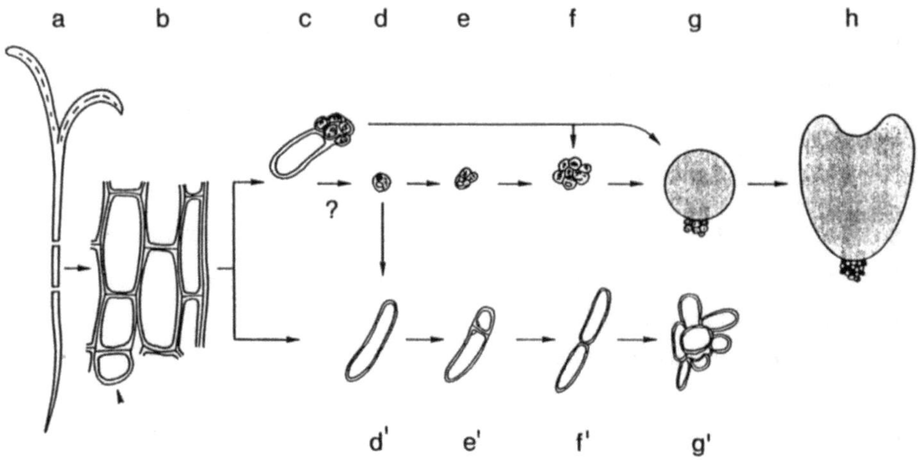

Fig. 1. Formation of carrot somatic embryos from suspension cells (Backs-Hüsemann et al. 1970, De Vries et al. 1988, Komamine et al. 1990, Dudits et al. 1991). Excised pieces of seedling roots (a,b) are immersed in liquid medium with auxin. Cell divisions become visible in the explant tissue (arrowhead in b) and result in the release of non-embryogenic, vacuolated single cells (c,d'). Some of these cells divide asymmetrically to produce clusters of small embryogenic cells (c). From these clusters small isodiametric cells may be released (d), that are capable of development into embryogenic clusters (e,f). Somatic embryos can develop from a surface cell of embryogenic cell clusters after culture dilution in medium without growth regulators (g,h). Cell division of vacuolated cells without restriction of expansion is shown in d'-g'. This pathway predominates in cultures that have lost their embryogenic capacity but also occurs in embryogenic cultures (reprinted from Van Engelen and De Vries 1992 with permission).

simply the removal of 2,4-D that triggers embryo development, but rather the change in cell density that appears to allow the formation of globular embryos, even in the presence of 2,4-D.

Advances in the molecular analysis of genes expressed in plant somatic and zygotic embryos have been reviewed by Dudits et al. (1991) with an emphasis on the control of cell division in the early phases of embryo development, and by Sterk and De Vries 1992). In this short overview we will present some recent findings that indicate that the expression of genes encoding carrot secreted glycoproteins reflect the developmental state of cells in suspension cultures. In addition, two secreted glycoproteins have been identified that have a promoting effect on arrested carrot somatic embryos.

Secreted medium proteins reflect the developmental states of suspension cells.

Previously we reported that the initiation of somatic embryogenesis is accompanied by changes in the abundance of several secreted glycoproteins (De Vries et al. 1988a). The typical embryogenic *in vitro* plant cell culture consists of heterogenous cell populations, and an added complication to studying the significance of secreted components of the conditioned medium is the precise identification of the cells from which they originate (Van Engelen et al. 1991). To resolve this question, we screened a cDNA expression library with antisera raised against all proteins secreted by an embryogenic suspension culture. Individual proteins corresponding to the cDNA clones obtained were then identified by Western blotting using specific antisera raised against the *E. coli*-produced lacZ-cDNA fusion proteins. This strategy has allowed identification of cDNA clones corresponding to individual secreted proteins. One of these, designated EP1, was found to be synthesized only by elongated, non-embryogenic cells in both embryogenic and non-embryogenic suspension cultures. Immunolocalization showed that the EP1 protein is present in extracellular matrix material that also reacts with antibodies specific for pectin (Van Engelen et al. 1991). Sequencing of the corresponding cDNA did not reveal homology to other plant genes, nor did biochemical assays reveal any of the enzyme activities commonly found in plant cell conditioned media (Van Engelen et al. 1992b). In contrast, a second protein designated EP2 was only found in media conditioned by embryogenic cells, and has been identified on the basis of sequence homology, antigenicity and biochemical activity as a lipid transfer protein. Based on the epidermis specific expression pattern of the encoding gene in embryogenic cell clusters, somatic and zygotic embryos, shoot apical meristems, flowers and seeds, we have postulated that the EP2 protein reflects the presence of an insoluble water-repellent cuticle-like coating of the embryo protoderm. The fact that this protein and the expression of the encoding gene is tightly associated with somatic as well as zygotic embryo development suggests an important function for such a coating (Sterk et al. 1991).

Together with the observation on the presence of specific carbohydrate moieties on the surface of some cells in embryogenic tissue cultures of carrot (Pennell et al. 1992), these findings demonstrate that the surfaces of suspension cells are not at all uniform and may very well reflect important developmental changes. However, most suspension cells, regardless of their developmental state, all express plasma membrane AGP epitopes that are developmentally regulated in the intact plant tissue (Pennell et al. 1991, Knox et al. 1991).

In *Citrus* cultures, glycoproteins have been identified in the conditioned medium that disappear from the culture medium upon initiation of embryogenesis (Gavish et al. 1991). Satoh et al. (1986) described the analysis of two secreted glycoproteins, GP57 and GP65, from carrot cultures. GP57 was secreted by non-embryogenic as well as embryogenic cells in an auxin-dependent fashion; in contrast GP65 was only secreted by embryogenic cells in the absence of 2,4-D.

The evidence available to date therefore suggests a more general picture that encompasses the occurrence of various secreted proteins and cell wall carbohydrates, either associated with embryogenic cells and early somatic embryo formation, or with different types of non-embryogenic cells. In most cases the biochemical function of these secreted proteins has not been elucidated, while sofar only in the case of the carrot EP1 and EP2 genes it has been established that the expression pattern is indeed identical in somatic as well as zygotic embryogenesis.

Evidence for the embryo-rescue activity of the conditioned medium and identification of the responsible components.

Several observations have pointed to components of the conditioned growth medium that can positively influence somatic embryo development.

As early as 1966, it was noted by Halperin that "conditioning of the medium by means of cell death, leakage or active secretion may release unknown substances essential to embryo development". Later it was observed that dilution of embryo cultures to very low cell densities resulted in reduced numbers of embryos and retarded development while the addition of conditioned medium from an embryo culture at normal cell density could partially reverse these effects (Hari 1980).

The speed with which a population of carrot cells newly released from hypocotyl tissue, acquires the ability to produce somatic embryos was increased considerably by the addition of unfractionated cell-free medium conditioned by an established embryogenic cell line (De Vries et al. 1988b).

The fungal antibiotic tunicamycin inhibits somatic embryogenesis at an early, preglobular stage of development, but does not inhibit unorganized proliferation of carrot cells. Rescue of embryo development was observed when concentrated conditioned medium from an untreated culture was added together with the tunicamycin. The tunicamycin-complementing agent was heat-labile and protease-sensitive, suggesting that secreted glycoproteins were the causative factor (De Vries et al. 1988a). Recently the purification of the secreted glycoprotein able to rescue embryogenesis in tunicamycin-inhibited embryo cultures was reported. It was demonstrated that, although tunicamycin affects the correct glycosylation of many secreted proteins, only one of these appeared to be active in the assay. This protein was identified as a, most likely single, peroxidase isoenzyme with a pI of 7.6. A similar rescue was possible with a preparation of cationic horse radish peroxidase isoenzymes. Inactivation of the peroxidase activity by removal of the heme group resulted in loss of rescue activity. Thus the enzymatic properties of the peroxidase appear to be indispensable for its embryo-rescue activity (Cordewener et al. 1991).

Analysis of a temperature-sensitive variant cell line, ts11, has provided additional evidence for a promoting rôle of extracellular proteins. Ts11 somatic embryos develop normally at 24°C. At 32°C the embryos do not progress beyond the globular stage and abnormal development ensues, unless the medium is supplemented with medium conditioned by wild-type embryos. Secreted proteins appear to be involved, since the effect was shown to be protease-sensitive. The proteins secreted by ts11 at 32°C were found to contain less fucose when compared to the wild-type, so the original mutation in this line may affect protein glycosylation (LoSchiavo et al. 1990). Like tunicamycin-arrested embryos, blocked ts11 embryos also respond to the addition of a single secreted 32 kD glycoprotein. This protein has been identified on the basis of its amino acid sequence and biochemical activity as an unusual form of plant acid endochitinases (De Jong et al. 1992).

In addition to these secreted glycoproteins that have an embryo-promoting effect on arrested somatic embryos, there also appear to be secreted proteins that do not have a measurable effect at all on either arrested or normal embryogenesis (Van Engelen et al. 1991) or appear to be inhibitory to embryo development. In Citrus cultures, a 53-57 kDa family of secreted glycoproteins has been found that effectively inhibit somatic embryo formation (R. Fluhr, personal communication). The biochemical identity of these proteins is sofar not known.

These results demonstrate that with the aid of inhibition/complementation experiments individual proteins, present in the conditioned medium, can be identified that are essential for somatic embryogenesis in carrot. With the identification of inhibitory secreted proteins, it could well be that correct somatic embryo formation from suspension cells depends at least in part on a balance between the activities of several secreted proteins.

Acknowledgements.

I like to thank many colleagues for stimulating discussions and providing unpublished information. Original research from our group is supported by the European Commission Biotechnology Programs BAP and BRIDGE, the Netherlands Organization for Scientific Research (NWO) and the Stichting Innovatiefonds Plantenveredeling.

References.

Backs-Hüsemann D, Reinert J (1970) Embryobildung durch isolierte Einzelzellen aus Gewebekulturen von Daucus carota. Protoplasma 70: 49-60

Bögre L, Stefanov I, Abrahám M, Somogyi I, Dudits D (1990) Differences in responses to 2,4-dichlorophenoxy acetic acid (2,4-D) treatment between embryogenic and non-embryogenic lines of alfalfa. In: HJJ Nijkamp, LHW van der Plas, J van Aartrijk (eds). Progress in Plant Cellular and Molecular Biology pp 427-436. Kluwer Academic Publishers, Dordrecht, the Netherlands

Cordewener J, Booij H, Van der Zandt H, Van Engelen F, Van Kammen A, De Vries SC (1991) Tunicamycin-inhibited carrot somatic embryogenesis can be restored by secreted cationic peroxidase isoenzymes. Planta 184: 478-486

De Jong AJ, Cordewener JHC, Lo Schiavo F, Terzi M, Vandekerckhove J, van Kammen A, De Vries SC (1992) A secreted chitinase can rescue arrested somatic embryos of a temperature sensitive carrot cell variant (submitted)

De Vries SC, Booij H, Janssens R, Vogels R, Saris L, LoSchiavo F, Terzi M, van Kammen A (1988a) Carrot somatic embryogenesis depends on the phytohormone-controlled expression of correctly glycosylated extracellular proteins. Genes Develop 2: 462-476

De Vries SC, Booij H, Meyerink P, Huisman G, Wilde HD, Thomas TL, van Kammen A (1988b) Acquisition of embryogenic potential in carrot cell suspension cultures. Planta 176: 205-211

Dudits D, Bögre L, Györgyey J (1991) Molecular cellular approaches to the analysis of plant embryo development from somatic cells in vitro. J Cell Science 99: 475-484

Fry SC (1990) Rôles of the primary cell wall in morphogenesis. In: HJJ Nijkamp, LHW van der Plas, J van Aartrijk (eds) Progress in Plant Cellular and Molecular Biology pp 504-513. Kluwer Academic Publishers, Dordrecht, the Netherlands

Gavish H, Vardi A, Fluhr R (1991) Extracellular proteins and early embryo development in *Citrus* nucellar cell cultures. Physiol Plant 82: 606-616

Halperin W (1966) Alternative morphogenetic events in cell suspensions. Am J Bot 53: 443-453

Hari V (1980) Effect of cell density changes and conditioned media on carrot cell embryogenesis. Z Pflanzenphysiol 96: 227-231

Knox JP, Linstead PJ, Peart J, Cooper C, Roberts K (1991) Developmentally regulated epitopes of cell surface arabinogalactan proteins and their relation to root tissue pattern formation. Plant J 1: 317-326

Komamine A, Matsumoto M, Tsukahara M, Fujiwara A, Kawahara R, Ito M, Smith J, Nomura K, Fujimura T (1990) Mechanisms of somatic embryogenesis in cell cultures - Physiology, biochemistry and molecular biology. In: HJJ Nijkamp, LHW van der Plas, J van Aartrijk (eds) Progress in Plant Cellular and

Molecular Biology pp 307-313. Kluwer Academic Publishers, Dordrecht, the Netherlands
LoSchiavo F, Giuliano G, Pitto L, De Vries SC, Bollini R, Genga AM, Nuti-Ronchi V, Cozzani F, Terzi M (1990) A carrot cell variant temperature-sensitive for somatic embryogenesis reveals a defect in the glycosylation of extracellular proteins. Mol Gen Genet 223: 385-393
Pennell RI, Booij H, Van Kammen A, De Vries SC, Scofield GN, Janniche L, Roberts K (1992) A cell-specific cell wall carbohydrate epitope in the precursor cells of carrot somatic embryos (submitted)
Pennell RI, Janniche L, Peart JM, Scofield GN, Kjellbom P, Roberts K (1991) Developmental regulation of a plasma membrane arabinogalactan protein epitope in oilseed rape flowers. The Plant Cell (in press)
Satoh S, Kamada H, Harada H, Fujii T (1986) Auxin-controlled glycoprotein release into the medium of embryogenic carrot cells. Plant Physiol 81: 931-933
Smith DL, Krikorian AD (1990) pH control of carrot somatic embryogenesis. In: HJJ Nijkamp, LHW van der Plas, J van Aartrijk (eds) Progress in Plant Cellular and Molecular Biology pp 449-453. Kluwer Academic Publishers, Dordrecht, the Netherlands
Sterk P, De Vries SC (1992) Molecular markers for plant embryos In: SynSeeds: Application of synthetic seeds to crop improvement. K. Redenbaugh (ed) CRC Press (in press)
Sterk P, Booij H, Schellekens GA, van Kammen A, De Vries SC (1991) Expression pattern of the carrot EP2 gene encoding an extracellular lipid transfer protein. The Plant Cell 3: 907-921
Van Engelen FA, De Vries SC (1992) Extracellular proteins in plant embryogenesis. Trends in Genet (in press)
Van Engelen FA, Hartog M, Sturm A, Taylor B, Thomas TL, van Kammen A, De Vries SC (1992b) Epidermis-specific expression of the carrot EP1 gene encoding a secreted protein of 52/54 kDa (in preparation)
Van Engelen FA, Sterk P, Booij H, Cordewener JHG, Rook W, van Kammen A, De Vries SC (1991) Heterogeneity and cell type-specific localization of a cell wall glycoprotein from carrot suspension cells. Plant Physiol 96: 705-712

Progress in Research on Apomixis and Its Transfer to Major Grain Crops

Yves H. Savidan
APONET-International Network for Apomixis Research
ORSTOM/CIMMYT
Apdo. Postal 6-641
06600 México D.F.
Mexico

Introduction

Seeds are generally considered a product of sexual reproduction, and yet seeds are produced asexually by hundreds of species of higher plants. Apomixis--a term that originally described all types of asexual reproduction, including vegetative propagation--is now used to denote asexual reproduction by means of seeds. This type of reproduction has long been regarded merely as a botanical curiosity and, in his review, Stebbins (1950) stated that this process was controlled by complex interactions between recessive genes, leaving little hope for breeders ever making practical use of apomixis. The transfer of such a characteristic from a wild species to its cultivated relative was unthinkable.

Forty years later, the utilization of apomixis is seen as one of the biggest challenges for plant breeding in the 1990s. In the six months preceeding the XIII EUCARPIA Congress, international workshops on the potential use of apomixis in plant breeding have taken place in China, USA, and France; three books on the subject have been announced; and the transfer of apomixis to wheat, maize, rice, sorghum and pearl millet is under study.

These circumstances are the result of progress in the last ten to fifteen years toward understanding apomixis, elucidating the genetic factors that control it, and developing tools for screening large numbers of progeny for their modes of reproduction at each hybrid generation.

Apomixis research is now a very dynamic part of studies on reproductive biology and its application in plant breeding. This paper reviews the status of apomixis knowledge and research.

What is apomixis?

Gustafsson (1947) gave a simple and still accurate classification of the different types of apomixis (Fig. 1). Apomixis is the result of both a failure of reduction (diplospory when the embryo sac results from the megaspore mother cell or apospory when it results from a nucellar cell) and a failure of fertilization (parthenogenesis). The products are perfect copies of the mother plant.

Gametophytic apomixis		**adventitious embryony**
MMC	Nucellar cells	sporophyte
↓	↓	↓
↓ **diplospory**	↓ **apospory**	↓
↓	↓	↓
unreduced ES	unreduced ES	↓
↓	↓	↓
↓ **parthenogenesis**	↓ **parthenogenesis**	↓
↓	↓	↓
embryo	embryo	embryo

Fig.1 Apomixis is the result of both a failure of reduction (diplospory when the embryo sac [ES] results from the megaspore mother cell [MMC], or apospory when it results from a nucellar cell) and a failure of fertilization (parthenogenesis). The products are perfect copies of the mother plant.

Adventitious embryony is well known in Citrus. Although there is no real sporophyte- gametophyte-sporophyte alternance for the production of a new individual, the somatic embryo needs the presence of a meiotic embryo sac to develop. Gametophytic apomixis is more frequent and much more important for the breeder because it is widespread in many families, especially the Composeae, Rosaceae and Gramineae. Brown and Emery (1958) even described gametophytic apomixis as a characteristic of the Panicoideae, the subfamily which includes all tropical and subtropical grasses and grains.

Maize, sorghum, and pearl millet have many relatives that reproduce apomictically. In the Poaceae, the genus Poa includes several apomictic species, and Elymus rectisetus is an apomictic relative of wheat. Beets also have two apomictic relatives, while most of the diversity of the rubber producing Parthenium (guayule) is fixed in apomictic biotypes.

Because no satisfactory source of male sterility has been disseminated, apomixis has potential for lowering the cost of producing hybrid seed of rice and wheat, and can also help fix the characteristics of other crops. The main advantage of apomixis, however, will probably be for farmers who do not benefit from high yielding varieties because none has been developed for their production conditions or because they lack reliable access to seeds, inputs, and/or credit.

Farmers in Africa, for example, typically grow crops using their own seed. The varieties are largely heterogeneous, although some common characters make them recognizable to farmers. On the one hand, such heterogeneity results in low production, but on the other it is a safeguard against evolving stresses in the local environment. In "hostile" environments, improved varieties or landraces with major alterations in their genetic makeup may quickly succumb to stresses, whereas a locally adapted variety with only minute genetic changes will conserve its "rusticity." This suggests that apomixis could feasibly be conferred to local varieties without reducing their adaptation. Furthermore, inasmuch as farmers select seed for planting mainly on the basis of grain yield (i.e., the non-reproductive organs do not enter into their selection criteria), farmer selection in an apomictic variety would result in the variety becoming homogenous for grain production but would maintain the heterogeneity for non-reproductive plant parts. Thus, contrary to the common conception of apomixis as reducing genetic diversity, in the circumstances described here it would actually fix the diversity for plant type in local varieties and ensure their continued ability to withstand local stresses. Another way to keep this heterogeneity would be the transfer of a facultative apomixis, i.e. one in which only part of the progeny are maternal and the rest come from sexual reproduction and recombination.

These discussions point up the potential usefulness of apomixis to farmers in developing countries. A more fundamental question, however, is whether or not the trait can actually be transfered to crops of interest. The sections that follow address this issue, discussing the genetic bases of apomixis, describing screening techniques, and citing research on its transfer to basic food crops.

The genetic control of apomixis

As a mode of reproduction, apomixis could be regarded as a complex physiological process, with each step under the control of specific genes. But the difference between apomixis and normal sexual reproduction could possibly be traced to a single step, further deviations being a consequence of this first and unique event.

In studies on apospory in plants as distinct as a small Renunculaceae (Nogler, 1984a) and a huge tropical forage (Savidan, 1982), it was found that aposporous embryo sacs developed precociously (as compared to meiotic embryo sacs) and from several nucellar cells at the same time. This results in a sort of "competition" between embryo sacs of different origins, and the meiotic embryo sacs are seldom favored in position or size to compete with their unreduced aposporous counterparts. The difference in timing of development also means that when the pollen tube (whose development is synchronized with that of meiotic embryo sacs) reaches the micropile, the aposporous embryo sac which is close to it has been mature for quite some time. Ultrastructural observations made in several sexual species showed that at the time of egg fertilization, this cell is surrounded by an incomplete cell wall whose synthesis has begun prior to pollen tube growth through the style. If this process is linked with embryo sac maturity and not with pollen tube growth, at the time the pollen reaches the aposporous embryo sacs, the egg cell may be surrounded by a complete cell wall, avoiding any possibility of fertilization. This hypothesis, which presents parthenogenesis as a consequence (i.e., a secundary effect) of apospory, is currently being studied in The Netherlands and Argentina, using different materials.

The genetic control of apospory has been especially well documented in Ranunculus auricomus (Nogler, 1984b) and Panicum maximum (Savidan, 1981, 1982). Both analyses are based on a long series of crosses, backcrosses, testcrosses, sibcrosses, selfing, and led to a similar conclusion: all data fit perfectly with a monogenic inheritance, with the gene of apomixis being dominant. The strength of these analyses is illustrated in the Figure 2, which summarizes the data obtained using Panicum maximum. Many new crosses were made after this analysis was

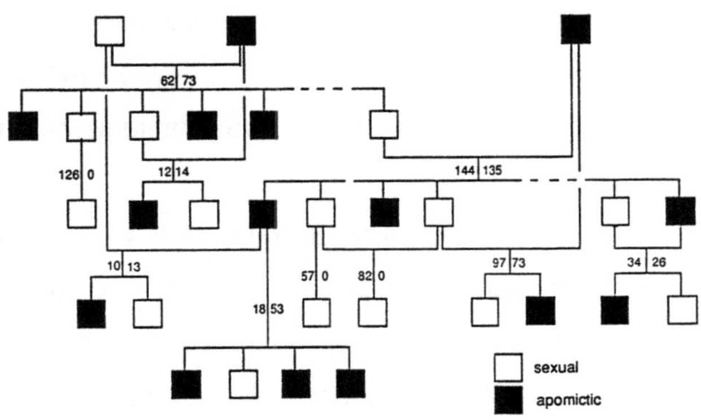

Figure 2 - Genealogy studies for the analysis of apomixis genetic control in Panicum maximum. Apomicts are shown in black squares, while sexual plants are shown in white squares. Numbers above horizontal lines refer to the observed segregations (Savidan, 1982).

concluded and thousands of new hybrids were produced for breeding purposes, all of which has confirmed the one-to-one segregations observed in the analysis for sexual x apomictic combinations. The sexual tetraploids used, coming from colchicine treatment of wild diploids, have the genotype **aaaa**. All the apomicts tested in these crosses have the genotype **Aaaa**. A simple population genetics model (Savidan and Pernès, 1982) explains why wild populations show a predominance of **Aaaa** genotypes among the apomicts and why sexuality is found only at the diploid level.

Other genetic analyses (Harlan et al., 1964; Taliaferro and Bashaw, 1966; Hanna et al., 1973; and others) have postulated different models of inheritance, mostly of the two-gene type. These models, though, are more strongly supported by the fact that apomixis appears to be a two-step process (i.e., failure of reduction plus failure of fertilization) than by any conclusive data. Most of these studies utilized inappropriate germplasm and/or tetraploid sexual "off" types found in apomictic populations. The studies we made in Panicum (Savidan, 1982) clearly show that all levels of sexuality can be found among facultative apomicts, up to a situation in which the plant is functionally sexual but still has an 'apomictic' genotype. The point is that no natural sexual will produce an apomictic progeny when selfed, whereas 'sexual' tetraploid off-types will. Obviously, genetic analyses based on such off-types are consequently more difficult to interpret.

Two other analyses using diploids as the source of sexuality gave one-to-one segregations in the F1 progeny. One study (Voigt and Bashaw, 1972) involved a diplosporous species, Eragrostis curvula, but failed to obtain conclusive evidence. The second (doValle et al., 1991), which utilizes the genus Brachiaria, is underway and looks very promising.

The first possible observation of the effect of the "apomixis" gene was made by Crane and Carman (1987) on Elymus rectisetus. Whereas in sexual plants the megaspore mother cells are isolated from the surrounding nucellar cells by a thick callous wall, megaspore mother cells of diposporous Elymus do not show any callous isolation. This could allow proteins to pass back and forth between reproductive and somatic cells. In aposporous materials--where somatic cells begin development prior to meiosis and later divide to form the embryo sac, a structure normally produced only by the reproductive cell--the implication is that something has passed from the reproductive cell to the neighbouring nucellar cells. Furthermore, the fact that the reproductive cells in diplosporous Elymus are unable to follow up with normal meiosis afterwards suggests that at least some of the proteins which regulate this division are also missing.

New studies are needed at the ultrastructural level to determine whether the presence of a callose wall is the only difference between meiotic and ameiotic plants, or if other changes can be observed in the cytoplasms of megaspore mother cells and surrounding nucellar cells. But from the first observations made in Elymus and the simple inheritance of apomixis as evident in Ranunculus and Panicum, it seems that apomicts could be considered as little more than a meiotic mutant. Some meiotic mutants, as reported in reviews by Golubovskaya and Khristolyubova (1985) and Curtis (1985), replace meiosis by mitosis. The main difference with the case of the apomictic gene may simply be that these mei--mutants usually affect both female and male meioses. It would be interesting to see whether callose deposits around the megaspore mother cells is also affected in some mei-mutants. Several programs on apomictic species are now investigating this issue.

Screening for apomixis

There is a tremendous need for simple screening tools to separate apomictic from sexual progenies in genetic analyses, transfer attempts, or the manipulation of apomixis in breeding programs. Older studies used the morphological and sometimes cytological heterogeneity of the progeny to recognize sexuality. Progeny-tests based on the morphological uniformity or heterogeneity were especially popular, because they did not require any equipment or technical skill. They led, however, to many errors, mainly because apomixis is more often facultative than was previously thought. Screening tools developed in recent years are more rapid and reliable.

Clearing techniques have been developed since 1971 to allow the observation of ovule development from early meiosis to embryo development, without having to make inclusions in the paraffine or time--consuming cuttings and cytological observations. After being cleared, the structures are observed *in situ* without cut or squash, utilizing only the light reflection from the cell walls. The first of these techniques, presented by Herr (1971), was used in the study of the genetic control of apomixis in Panicum (Savidan, 1975, 1982). Other techniques were later proposed (Crane, 1978; Young et al., 1979; Herr, 1982), and the choice depends both on the material and, especially, the nature and thickness of the tissues surrounding the ovule. The aposporous apomixis found in tropical grasses is especially easy to detect, given that the aposporous embryo sacs have a 4-nucleate structure while the meiotic embryo sacs are of the classical Polygonum type (antipodals usually dividing more). Diplospory differs from meiosis at two stages of observation: prophase is replaced by a long resting stage in which the nucleus appears to be stretched in an elongated megaspore; in place of the tetrad of reduced megaspores, at the end of meiosis, a dyad of unreduced megaspores appears.

Precocious progeny tests can be made using biochemical markers. The simplest involves a couple of isozyme systems and must be applied on young seedlings, ten to fifteen from each hybrid, and the results compared with those for the mother plant. Restriction fragment length polymorphisms (RFLPs) and random amplified polymorphic DNAs (RADPs) are also being used in some cases, and their popularity is certain to grow in coming years. These markers, currently used to analyze diversity at the DNA level and construct linkage maps, may soon be employed to study genome segments associate with apomixis.

Finally, a new technique has been introduced by a group studying apomixis in Chinese chive, Allium tuberosum (Kojima and Kawaguchi, 1989). In this method, unpollinated ovules are cultured *in vitro*. If the plant is sexual, only rare haploids may grow; if the plant is apomictic, the culture medium functions as the endosperm and most ovules produce an embryo that later develops into a young seedling.

Attempts to transfer apomixis in grain crops

An attempt to transfer apomixis to maize was made in early 1960s by a group of scientists from the USSR. Unfortunately, their results reflect their lack of appropriate germplasm, screening tools, and environmental conditions. Even so, they succeeded in producing F1 hybrids from crosses between maize and its apomictic relative Tripsacum dactyloides, and backcrossed these progeny to get 38-chromosome apomictic plants having a full set of maize chromosomes and the haploid set (x=18) from Tripsacum.

I am participating in a collaborative project, begun in 1989, between the French Institute of Scientific Research for Cooperative Development (ORSTOM) and the International Maize and Wheat Improvement Center (CIMMYT). We have established a collection of over 1,000 accessions of various Tripsacum species and aim to produce the first apomictic maize within six to eight years, if only conventional methods are used (i.e. backcrossing).

The transfer of apomixis to pearl millet is being attempted by researchers of the US Department of Agriculture at Tifton (Dujardin and Hanna, 1989), and has reached the fifth backcross, producing apomictic hybrid derivatives that have already recovered the chromosome number of the pearl millet recurrent parent and part of its fertility. There are still a few questions regarding this program: 1) the plants are probably substitution lines and not pure recombined pearl millet (too many deleterious characters are still present), 2) we know nothing about the possibilities of recombination between the chromosomes of the apomictic species and the chromosome of the crop, 3) in pearl millet, there are no translocations like the B-A translocation in maize to favor the required recombinations (Beckett, 1991).

An attempt is underway at the Utah State University to transfer apomixis to wheat from an Australian hexaploid grass, Elymus rectisetus (Carman and Wang, 1991). Crosses between the two and the first backcross have been successful. The second backcross is now being attempted.

Finally, several Chinese labs are looking for apomixis in rice (Yuan et al., 1990) and a project to transfer apomixis to sorghum from related species of the Andropogoneae is taking place in India (Bharathi et al., 1991) and USA.

References

Bharathi M Murty UR Visarada KBRS Annapu·· ıa A (1991) Possibility of transferring obligate apomixis from Cenchrus ciliaris L. to Sorghum bicolor (L.) Moench. Apomixis Newsletter 3:13-14.

Beckett JB (1991) Cytogenetic, genetic and plant breeding applications of B-A translocations in maize. in: Chromosome ingineering in plants, PK Gupta T Tsuchiya eds, Elsevier Amsterdam (in press)

Brown WV Emery WHP (1958) Apomixis in the Gramineae: Panicoideae. Amer.J.Bot.45: 253-263

Carman JG Wang RRC (1991) Apomixis and wide hybridization in Elymus. Apomixis Newsletter 3:24-25

Crane CF (1978) Apomixis and crossing incompatibilities in some Zephyranthaceae. PhD Thesis, Univ.of Texas, Austin

Crane CF Carman JG (1987) Mechanisms of apomixis in Elymus rectisetus from Eastern Australia and New Zealand. Amer.J.Bot.74:477-496

Curtis C (1985) Meiotic mutants of maize. PhD diss Univ of Missouri, Columbia.

Dujardin M Hanna WW 1989 Developing apomictic pearl millet - characterization of a BC_3 plant. J.Genet.Plant Breed.43:145

Golubovskaya IN Khristoluybova NB (1985) The cytogenetic evidence of the gene control of meiosis, maize meiosis, and mei-genes. in: Plant Genetics, M Freeling ed., Alan R Liss New York, pp723-738

Gustafsson A (1946-7) Apomixis in higher plants. Lunds Univ.Arsskr.N.F.II 42:1-67 (1946), 43:71-179 (1947) & 43:183-370 (1947)

Hanna WW Powell JB Millot JC Burton GW (1973) Cytology of obligate sexual plants in Panicum maximum Jacq and their use in controlled hybrids. Crop Sci.13:695-697

Harlan JR Brooks MH Borgaonkar D deWet JMJ (1964) Nature and inheritance of apomixis in Bothriochloa and Dichanthium. Bot.Gaz.125:41-46

Herr JM (1971) A new clearing-squash technique for the study of ovule development in angiosperms. Am.J.Bot.58:785-790

Herr JM (1982) An analysis of methods for permanently mounting ovules cleared in four--and-a-half type clearing fluids. Stain Tech.57:161-169

Kojima A Kawaguchi T 1989 Apomictic nature of Chinese chive (Allium tuberosum Rottl.) detected in unpollinated ovule culture. Jap.J Breed.39:449-456

Nogler GA (1984a) Gametophytic apomixis. in: Embryology of angiosperms, BM Johri ed., Springer-Verlag Berlin, pp475-518

Nogler GA (1984b) Genetics of apospory in apomictic Ranunculus auricomus V. conclusion. Bot.Helvetica 94:411-422

Savidan YH (1975) Hérédité de l'apomixie. Contribution à l'étude de l'hérédité de l'apomixie sur Panicum maximum Jacq. (analyse des sacs embryonnaires). Cah.ORSTOM, ser.Biol.10:91-95

Savidan YH (1981) Genetics and utilization of apomixis for the improvement of guineagrass (Panicum maximum Jacq.). Proc. XIV Int.Grassl.Congr., Lexington KY, pp182-184

Savidan YH (1982) Nature et hérédité de l'apomixie chez Panicum maximum Jacq. Travaux et Documents ORSTOM, No153, 159p

Savidan YH Pernès J (1982) Diploid-tetraploid-dihaploid cycles and the evolution of Panicum maximum Jacq. Evolution 36:596-600

Stebbins GL (1950) Apomixis in relation to variation and evolution. in: Variation and evolution in plants, Columbia Univ.Press, pp380-419

Taliaferro CM Bashaw EC (1966) Inheritance and control of obligate apomixis in breeding buffelgrass, Pennisetum ciliare. Crop Sci.6:473-476

Valle CBdo Leguizamon G Guedes NR (1991) Interspecific hybrids of Brachiaria. Apomixis Newsletter 3:10-11

Voigt PW Bashaw EC (1972) Apomixis and sexuality in Eragrostis curvula. Crop Sci.12: 843-847

Young BA Sherwood RT Bashaw EC (1979) Cleared-pistil and thick-sectioning techniques for detecting aposporous apomixis in grasses. Can.J.Bot.57:1668-1672

Yuan L.P. Li Y.C. Deng H. 1990 Studies on rice twin seedlings (3rd Report). Ann.Meeting Coop.Program on Rice Biotechnology, Nov.1990, 2p

Cryopreservation of Embryos

F. Engelmann

Laboratoire de Ressources Génétiques et Amélioration des Plantes Tropicales

ORSTOM (Institut Français de Recherche Scientifique pour le Développement en Coopération)

911 Av. Agropolis, BP 5045

34032 Montpellier Cédex 01

France

Introduction

For the last decade, plant cell cultures have become subjects of intensive research with emphasis on important biotechnological applications such as large scale production of clonal plants, genetic engineering, production and isolation of mutants, selection of variants, screening for resistance to various stresses and creation of new plant species through somatic cell hybridization. In this frame, somatic embryogenesis is widely used, due to its numerous advantages. However, this development leads to the production of an increasing number of strains. This creates problems for the management of laboratories. Moreover, many strains are faced with somaclonal variation and progressive loss of organogenic potentialities. Finally, genetically engineered lines are of extreme value, due to the amount of work necessary for their development. Therefore, safe and efficient techniques are needed for the long term conservation of this new germplasm.

On the other hand, the utilization of a restricted number of highly selected lines poses the problem of the conservation of the wild genotypes which often possess interesting agronomic characteristics. For this purpose, zygotic embryos represent a choice material, since they allow to conserve a good representation of the genetic variability of a given species with a limited number of samples.

Today, only cryopreservation (liquid nitrogen -196°C) can ensure the long term storage of germplasm. Resistance to freezing in liquid nitrogen has been shown for more than 70 plant species (Kartha, 1985 ; Dereuddre and Engelmann, 1987 ; Engelmann, in press) using various materials, such as cell suspensions, protoplasts, calluses, meristems and embryos.

This article presents a brief overview of the work carried out on cryopreservation of somatic and zygotic embryos and tries to identify the critical points for the successful freeze-preservation of such materials. Technical aspects, including the new cryopreservation techniques are dealt with by Dereuddre in the same volume.

Freezing characteristics of somatic and zygotic embryos

From a cryopreservation point of view, somatic and zygotic embryos have common characteristics. They are large and complex structures, formed of differentiated tissues, i.e. many different cellular types, depending on their developmental stage. However, somatic and zygotic embryos represent two different categories of material : somatic embryos are structures which have been neoformed *in vitro* and are thus highly hydrated. Somatic embryo cultures often comprize heterogenous structures in size and developmental stage, since in many embryogenic systems, synchronous development of embryos is difficult to achieve. They are generally of relatively smaller size than zygotic embryos (several hundred um to several mm).

On the contrary, zygotic embryos generally have lower water levels. They develop in the seed and are cryopreserved before being introduced *in vitro*. They are morphologically well defined structures (cotyledon(s), shoot and root pole, embryonic axis). Homogenous populations can be obtained by screening according to their developmental stage. Depending on their maturity, their size can be much larger than that of somatic embryos (up to more than 1 cm in the case of coconut).

For cryopreservation, the challenge is different with these two types of material. In most of the cases, when using somatic embryos, only the proliferation capacities (i.e. a sufficient amount of living cells to restart proliferation) and not necessarily the structural integrity of the embryos has to be preserved. On the contrary, with zygotic embryos (and somatic embryos cryopreserved at late developmental stages), the whole structure has to be preserved in order to regenerate directly a plant from each individual embryo.

Successive steps of a cryopreservation process

A standard cryopreservation technique comprizes successive steps (choice of starting material, pregrowth, freezing, thawing and post-treatment) for which optimal conditions have to be defined for each new material.

Choice of starting material

For cryopreservation, meristematic cells are the most likely to withstand freezing, due to their characteristics : they are of small size, have a very dense cytoplasm, few vacuoles (i.e. a low water level), their nucleo-cytoplasmic balance is very high. Therefore, in both cases, embryos comprizing the maximal amount of cells with such characteristics will be used preferentially. Young somatic embryos (globular stage) will be selected. With such structures, thermal gradients between external and internal cell layers are reduced and cryoprotective substances can act more efficiently and homogeneously. It may be sometimes necessary to set up a special medium in order to produce embryos of the right type. This is the case with oil palm : only a two-month culture on a medium enriched with sucrose will ensure the production of finger like, shiny white clumps of embryos which are the only ones likely to withstand freezing (Engelmann and Dereuddre, 1988a).

For the same reasons, immature zygotic embryos may appear as optimal candidates for cryopreservation due to their reduced size and their lower differentiation. However, one drawback for their utilization is that if *in vitro* culture of mature embryos can be easily achieved, it is not the case with immature ones, for which very complicated media formulations generally have to be set up. This is illustrated with coconut immature embryos, which can withstand freezing with high survival rates (Assy-Bah and Engelmann, in press) but for which *in vitro* culture conditions leading to the production of whole plants are not yet well defined (Engelmann and Assy-Bah, 1991). One will thus have to look for embryos at a stage where they are simultaneously most likely to survive freezing and easy to cultivate *in vitro*. The size and structural complexity of mature embryos can be reduced by cutting off the cotyledons and using embryonic axes instead of whole embryos.

Pregrowth

Before freezing in liquid nitrogen, the embryos have to be submitted to a pregrowth phase during which they will be partially dehydrated. Indeed, plant tissues contain high amounts of freezable water which will be converted to ice during freezing and will be harmful for cell survival. They are therefore extremely sensitive to freezing injury. Protection against freezing injury has to be imposed artificially through various strategies, using cryoprotectants or other means.

Three types of pretreatments can be used with embryos : standard pretreatment using cryoprotective substances, desiccation an a combination of both treatments. The first method is mainly used with somatic embryos and the other two with zygotic embryos.

Use of cryoprotective substances

With this type of pretreatment, the embryos are cultured for various durations (several minutes to hours) in presence of cryoprotective substances. The following parameters have to be determined : the nature and concentration of cryoprotectants and the duration of the pretreatment. The cryoprotective substances used vary greatly one from the other by their molecular weight and their structure. The most commonly employed are dimethylsulfoxide (DMSO), sucrose, sorbitol, mannitol, polyethyleneglycol. The pretreatment may be carried out on solid medium only, in liquid medium only or successively with solid and liquid medium.

The cryoprotective substances may be used separately as for *Citrus sinensis* (Marin and Duran-Vila, 1988) or in combination, like sucrose + DMSO for coffee (Bertrand-Desbrunais *et al.*, 1988) or cassava (Sudarmonowati and Henshaw, 1990). The same technique is used with zygotic embryos of several species : solid medium containing high sucrose levels with *Capsella bursa pastoris* (Monnier and Leddet, 1980), or a mixture of glucose and glycerol with immature embryos of coconut (Assy-Bah and Engelmann, in press), liquid medium (propanediol and sucrose) in the case of *Juglans* and *Prunus* embryonic axes (de Boucaud *et al.*, 1991 ; de Boucaud and Brison, 1991).

Sucrose is very often employed in cryoprotective treatments and appears to play a central role in cryopreservation. It has an important osmotic effect but its absorption and its accumulation seems to be necessary for survival. Indeed, when pretreated on 0.75 M sucrose, the water level of *Capsella* embryos dropped within 24 hours down to a plateau. However, resistance to freezing in liquid nitrogen was obtained only after a minimal culture duration of 6 days (Monnier and Leddet, 1980). The same observation has been reported in the case of oil palm somatic embryos (Engelmann, 1986).

Desiccation

This technique is preferentially used with zygotic embryos or embryonic axes of various species. The explants are placed in the air current of a laminar flow cabinet in open Petri dishes and their dehydration regularly followed by weighing. However, these dehydration conditions cannot be very precisely controlled, due to the possible variations of the ambient temperature or humidity and they are not necessarily reproducible. Other methods such as the use of silicagel, as recommended by Uragami *et al.* (1990) for *Asparagus* stem segment desiccation may allow more reproducible conditions. The dehydration duration varies according to the size of the embryos and their initial water content. Embryos must have a minimal size. Immature embryos of coconut were too small (5 to 15 mg) to be dehydrated using this technique and therefore had to be pretreated using the conventional way (Assy-Bah and Engelmann, in press). The optimal desiccation duration is generally comprized between 2 and 4 hours, depending on the species. The water content is thus drastically reduced from an initial average level of 50-60% down to 10-16%. Prolonged desiccation durations rapidly lead to a dramatic loss in the viability of unfrozen controls caused by dehydration injury.

Cryoprotective substances combined with desiccation

It is also possible to combine both techniques described above. Mycock *et al.* (1989) could freeze pea embryonic axes after a pretreatment with glycerol and DMSO followed by desiccation. Assy-Bah and Engelmann (in preparation) submitted mature embryos of coconut to a 4-hour desiccation under the laminar flow followed by a culture on a medium containing high levels of glucose and glycerol. In this case, the water content of the embryos could be decreased to an extremely low level (5 to 8%).

Freezing

Two different protocols can be used : a two-step freezing, comprizing a controlled slow freezing followed by the immersion of the samples in liquid nitrogen, or a one-step rapid freezing performed by plunging directly the samples in liquid nitrogen.

The two-step process is generally employed with somatic embryos. During slow freezing, crystallization takes place firstly in the external medium and water flows out of the cells to the external ice, thus continuing the dehydration of the samples. The two variables which allow to control the intensity of this dehydration are the cooling rate and the prefreezing temperature , i.e. the temperature at which the controlled freezing stops. These parameters may have to be very precisely determined. In the case of *Citrus sinensis*, embryos have to be frozen at 0.5°C.min^{-1} down to -42°C (Marin and Duran-Vila, 1988). On the contrary, oil palm embryos can withstand a wide range of freezing rates (from 0.1 to 200°C.min^{-1}, Engelmann and Dereuddre, 1988b). If the freezing conditions are not optimal, the survival rate of the material is lowered. The type of recovery can also be modified (direct regrowth versus adventive embryogenesis), as observed by Florin (1989) with carrot somatic embryos.

When rapid freezing is performed, the remaining internal water cristallizes in the form of microcrystals which are of a size unharmful to the cellular structures. However, the samples have to be sufficiently dehydrated before freezing. Rapid freezing is commonly employed with zygotic embryos since their water content can be drastically reduced without damages during the pretreatment. It could be interesting to carry out thermal analysis experiments in order to determine if crystallization still takes place or if the remaining water vitrifies. Indeed, the conditions of this freezing procedure may be comparable to that of a vitrification process, that is low water content (obtained here by extensive dehydration instead of very concentrated cryoprotective solutions) and rapid freezing. This may explain for the high survival rates observed during cryopreservation of zygotic embryos.

Storage

No alteration to the material is known to take place during storage, provided that it remains at the temperature of liquid nitrogen. As concerns the effects of storage duration, no modification was observed in the recovery rate of oil palm somatic embryos after 15 months in liquid nitrogen (Engelmann, 1991). A similar result was obtained recently after 52 months in storage (Engelmann, unpublished).

Thawing

Thawing is generally rapid, using a water-bath thermostated at 40°C. This is made in order to avoid recrystallization and ice crystal growth which may occur during warming and produce intracellular damages. This is particularly important with material produced *in vitro* (somatic embryos) which still contains high levels of water. Carrot (Withers, 1979) and *Citrus sinensis* (Marin and Duran-Vila, 1988) are exceptions to this procedure. Oil palm somatic embryos are capable of reproliferation after slow or rapid thawing (Engelmann, unpublished).

Post-treatment

After the freeze-thaw cycle, culture conditions different from the standard procedure may have to be defined in order to stimulate the regrowth of the embryos. Indeed, regrowth of frozen somatic embryos is often delayed when compared with unfrozen controls. With *C. arabica* and *C. canephora* somatic embryos, a progressive reduction in the sucrose concentration of the medium and a culture in the dark allowed for a 20% improvement of the initial procedure (Bertrand-Desbrunais, 1991). With oil palm somatic embryos, auxin has to be added transitorily to the medium in order to stimulate reproliferation (Engelmann *et al.*, 1985). With zygotic embryos, initial regrowth is generally rapid. However, differences with unfrozen controls sometimes appear during the further development of the embryos. Chin *et al.* (1988) observed that the haustorium of cryopreserved *Howea* and *Veitchia* embryos did not develop.

The same observation has been made recently with coconut mature embryos (Assy-Bah and Engelmann, in preparation).

Conclusion

Cryopreservation has now been successfully applied to embryos of a large number of species from tropical and temperate origins (see Table 1). The techniques employed range from classical ones including cryoprotection and controlled freezing in the case of somatic embryos to simpler ones using desiccation and rapid freezing in the case of zygotic embryos. However, cryopreservation is still far from being routinely used for long-term storage of plant germplasm. Additional research is still needed in order to simplify, standardize and develop cryopreservation techniques. Another field of research concerns the development of techniques alternative to cryopreservation, such as storage at positive temperature of desiccated embryos. The current increasing amount of research dealing with germplasm conservation should lead in the near future to making available various efficient techniques for the safe long-term conservation of plant germplasm.

Table 1 : list of somatic and zygotic embryos successfully frozen in liquid nitrogen

Somatic embryos	
Citrus sinensis	Marin and Duran-Vila, 1988
Coffea arabica	Bertrand-Desbrunais *et al.*, 1988
Coffea canephora	Bertrand-Desbrunais, 1991
Daucus carota	Withers, 1979
Elaeis guineensis	Engelmann *et al.*, 1985
Manihot esculenta	Sudarmonowati and Henshaw, 1990
Zygotic embryos	
Aesculus hypocastanea	Pence and Dresser, 1988
Araucaria excelsa	Pritchard and Prendergast, 1986
Brassica napus	Withers, 1982
Capsella bursa pastoris	Monnier and Leddet, 1980
Carva	Pence and Dresser, 1988
Castanea	Pence and Dresser, 1988
Cocos nucifera	Chin *et al.*, 1989
Elaeis guineensis	Grout *et al.*, 1983
Fagus	Pence and Dresser, 1988
Hevea brasiliensis	Normah *et al.*, 1986
Hordeum vulgare	Withers, 1982
Howea fosteriana	Chin *et al.*, 1988
Juglans	Pence and Dresser, 1988
Manihot esculenta	Marin *et al.*, 1990
Musa	More *et al.*, 1991
Phaseolus vulgaris	Zavala and Sussex, 1986
Pisum	Mycock *et al.*, 1989
Prunus persica	de Boucaud and Brison, 1991
Quercus	Pence and Dresser, 1988
Triticum aestivum	Bajaj, 1983
Triticale	Bajaj, 1983
Veitchia merrillii	Chin *et al.*, 1988
Zea mays	Delvallée *et al.*, 1989

References

Assy-Bah B, Engelmann F (in press) Cryopreservation of immature embryos of coconut (*Cocos nucifera* L). Cryo-Lett 13

Bajaj YPS (1983) Cryopreservation of germplasm of cereals - Progress and prospects. Proc 6th Intl Wheat Genet Symp Kyoto Japan : 565-574

Bertrand-Desbrunais A (1991) La conservation des ressources génétiques des caféiers. Thèse de Doctorat Université Paris 6 : 259 p

Bertrand-Desbrunais A, Fabre J, Engelmann F, Dereuddre J, Charrier A (1988) Reprise de l'embryogenèse adventive d'embryons somatiques de caféier (*Coffea arabica*) après leur congélation dans l'azote liquide. C R Acad Sci Paris 307 Sér III : 795-801

de Boucaud MT, Brison M (1991) Cryopreservation of embryonic axes of *Prunus persica* L Batsch. Proc Cryo 91, KU Leuven, 7-12 july 1991 : 102.

de Boucaud MT, Brison M, Ledoux C, Germain E, Lutz A (1991) Cryopreservation of embryonic axes of recalcitrant seed : *Juglans regia* L cv Franquette. Cryo-Lett 12 : 163-166

Chin HF, Krishnapillay B, Alang ZC (1988) Cryopreservation of *Veitchia* and *Howea* palm embryos : non-development of the haustorium. Cryo-Lett 9 : 372-379

Chin HF, Krishnapillay B, Hor YL (1989) A note on the cryopreservation of embryos from young coconuts (*Cocos nucifera* var Mawa). Pertanika 12 : 183-186

Delvallée I, Guillaud J, Beckert M, Dumas C (1989) Cryopreservation of immature maize embryos after freeze-hardening in the ear and *in vitro*. Plant Sci 60 : 129-136

Dereuddre J, Engelmann F (1987) The use of cryopreservation for setting up banks of plant germplasm. Proc Coll Franco-Britannique IAPTC Angers 8-9 oct 1987 : 48-78

Engelmann F (1986) Cryoconservation des embryons somatiques de palmier à huile (*Elaeis guineensis* Jacq). Mise au point des conditions de survie et de reprise. Thèse de Doctorat Université Paris 6 : 228 p

Engelmann F (1991) Le développement actuel de la cryoconservation des embryons somatiques de palmier à huile. Proc XVIIIè Cong Intl du Froid, Montréal, 10-17 août 1991 : 305

Engelmann F (in press) *In vitro* conservation of tropical plant germplasm - a review. Euphytica : in press

Engelmann F, Assy-Bah B (1991) Maintenance of coconut genetic resources - *In vitro* techniques for medium and long term conservation. Proc Intl Workshop on Coconut Genetic Resources Cipanas Indonesia 8-11 oct 1991 : in press

Engelmann F, Dereuddre J (1988a) Effets du milieu de culture sur la production d'embryoïdes destinés à la cryoconservation chez le palmier à huile (*Elaeis guineensis* Jacq). C R Acad Sci Paris 306 Sér III : 515-520

Engelmann F, Dereuddre J. (1988b) Cryopreservation of oil palm somatic embryos : importance of the freezing process. Cryo-Lett 7 : 220-235

Engelmann F, Duval Y, Dereuddre J (1985) Survie et prolifération d'embryons somatiques de Palmier à huile (*Elaeis guineensis* Jacq) après congélation dans l'azote liquide. C R Acad Sci Paris 301 Sér III : 111-116

Florin B (1989) Etude de différentes voies de conservation d'embryons, de tissus et de cellules de végétaux cultivés *in vitro* : applications de l'hypoxie et de la cryoconservation. Thèse de Doctorat, Université de Tours : 259 p

Grout BWW, Shelton K, Pritchard HW (1983) Orthodox behaviour of oil palm seed and cryopreservation of the excised embryo for genetic conservation. Ann Bot 52 : 381-384

Kartha KK (ed) (1985) Cryopreservation of Plant Cells and Organs. CRC Press Boca Raton Florida

Marin ML, Duran-Vila N (1988) Survival of somatic embryos and recovery of plants of sweet orange (*Citrus sinensis* L Osb) after immersion in liquid nitrogen. Plant Cell Tissue Organ Cult 14 : 51-57

Marin ML, Mafla G, Roca WM, Withers LA (1990) Cryopreservation of cassava zygotic embryos and whole seeds in liquid nitrogen. Cryo-Lett 11 : 257-264

Monnier M, Leddet C (1980) Action du saccharose sur la résistance au gel des embryons immatures de capselle. Bull Soc Bot Fr 127 : 71-77

More A, Abdelnour A, Villalobos V (1991) Cryopreservation of *Musa* zygotic embryos. Proc 4th IPBNet Conf Biotechnology for Tropical Crop Improvement in Latin America San Jose Costa Rica 14-18 january 1991 : 81

Mycock DJ, Berjak P, Blakeway FC, Watt P (1989) Development of a model system for cryostorage : applicability to recalcitrant material. Proc Intl Conf The Impact of Biotechnology in Agriculture Amiens France 10-12 july 1989 : CP1

Normah MN, Chin HF, Hor YL (1986) Desiccation and cryopreservation of embryonic axes of *Hevea brasiliensis* Muell - Arg. Pertanika 9 : 299-303

Pence VC, Dresser BL (1988) Embryo cryostorage as a technique for germplasm preservation of several large-seeded tree species. Abst Beltsville Symp in Agric Res XIII Biotic Diversity and Germplasm Preservation - Global Imperatives 9-11 may 1988 : 24

Pritchard HW, Prendergast FG (1986) Effects of desiccation on the *in vitro* viability of embryos of the recalcitrant seed species *Araucaria hunsteinii* K Schum. J Plant Physiol 37 : 1388-1397

Sudarmonowati E, Henshaw GG (1990) Cryopreservation of cassava somatic embryos. Abstr VIIth Intl Cong Plant Tissue and Cell Culture, Amsterdam : 378

Uragami A, Sakai A, Nagai M (1990) Cryopreservation of dried axillary buds from plantlets of *Asparagus officinalis* L grown *in vitro*. Plant Cell Rep 9 : 328-331

Withers LA (1979) Freeze preservation of somatic embryos and clonal plantlets of carrot (*Daucus carota*). Plant Physiol 63 : 460-467

Withers LA (1982) The development of cryopreservation techniques for plant cell, tissue and organ culture. Proc 5th Intl Congr Plant Cell Tissue Culture Fujiwara A Ed Tokyo : 793-794

Zavala ME, Sussex IM (1986) Survival of developing wheat embryos and bean axes following cryoprotection and freezing in liquid nitrogen. J Plant Physiol 122 : 193-197

Cryopreservation of In Vitro Cultures of Plant Cells and Organs by Vitrification and Dehydration

J. Dereuddre
Laboratoire de Cryobiologie Végétale,
Université P. et M. Curie, 12, rue Cuvier,
75230 Paris, France.

INTRODUCTION

The development of cryopreservation has been considerable during the past decade. However, while cryopreservation of most cell suspension cultures can be used routinely without major problems, difficulties still remain when freezing organized and/or macroscopic structures from *in vitro* cultures (meristems and somatic embryos).

Conventional procedures initially involve equilibration of plant cells and organs in medium containing cryoprotectants (sugars, polyols, DMSO, glycerol, proline, ethylene glycol). Intracellular crystallization can be avoided during the slow-cooling step by the dehydration of the cells: loss of water is induced by the formation of ice in the cryoprotective medium and the resultant difference in water potentials between the intra- and extracellular media. To avoid recrystallization, rapid thawing is generally required.

With this procedure, more than 70 species have been cryopreserved in liquid nitrogen in the form of cell suspensions, yet only a few have been cryopreserved in the form of meristems and somatic or haploid embryos. For cell suspensions, regrowth occurs if a certain percentage (generally about 50%) of cells survives after rewarming. In the case of organized structures like shoot-tips, direct regrowth requires survival of the apical dome as a whole, or a major part of it. With embryos, plantlet recovery by direct regrowth implies survival of the two apical meristems and of hypocotyl tissues.

Two new procedures for cryopreservation have been recently proposed which may replace freeze-induced cell dehydration. Removal of freezable water from the cells can be achieved by exposure to concentrated solutions medium (vitrification procedure) or by evaporation in sterile air flow (dehydration procedure).

VITRIFICATION OF SOLUTIONS

Vitrification is the formation of a glasslike, noncrystalline solid at temperatures below the freezing point. Crystallization of a solution involves two phenomena: the formation of ice nuclei (i.e. a group of molecules of water forming an ice-like cluster), and their subsequent growth to give ice crystals. The first event is favoured by low temperatures, the second by highest temperatures. So, to obtain glass transitions, concentrated solutions of sugars and polyols are generally employed.

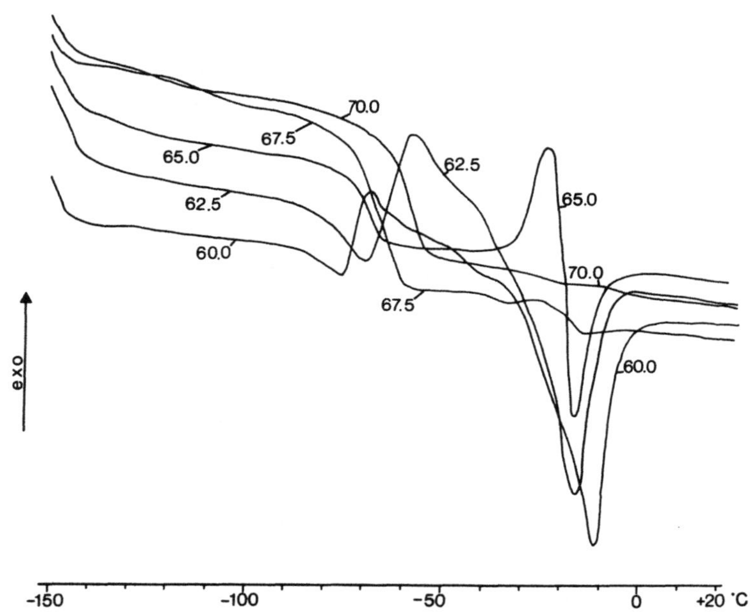

Fig.1. Warming thermograms of sucrose solutions containing (% total weight): 60 %, 62.5%, 65%, 67.5% and 70% sucrose. Three events are apparent: glass transition, a crystallization (devitrification) peak and an endothermic peak of melting. Glass transition without recrystallization was only obtained with 70% sucrose (Dereuddre and Kaminski, 1991).

The water content of the vitrification solution is often considered as the unfreezable water content of the medium. this is close to 0.56g of water per g of sucrose (Franks, 1985). However, this value is not high enough to ensure devitrification without ice formation during thawing. During cooling, vitrification can easily be obtained with sucrose concentrations ranging from 60 to 70% (% total weight). On rewarming, glass transition was only obtained with solutions containing around 30% water (0.43g of water per g of sucrose, i.e. an osmolality of about 7). For solutions of higher water contents, from 40% to 32.5% (% weight), warming thermograms generally show three events (fig.1): a glass transition with an inflexion point corresponding to the transition of amorphous solid water to hypercooled liquid water, an exothermic peak corresponding to the crystallization of the hypercooled water (devitrification peak), and the endothermic melting peak of ice. The size of the two peaks decreases as the concentration of sucrose increases.

Hence, in vitrification and dehydration procedures, the problem of the resistance of plant cells and organs to liquid nitrogen shifts to the one of their tolerance of dehydration.

VITRIFICATION PROCEDURES

The vitrification procedure consists in placing cells and organs in extremely concentrated solutions of permeating and/or non-permeating cryoprotectants. This technique was first applied to protoplasts and cell suspensions. It has been also been extended to shoot-tips.

The vitrification solution (of approximately 19 osmolals) used for *Brassica campestris* protoplasts (Langis and Steponkus, 1990) and cells (Langis et al., 1989), contains ethylene glycol (40% weight), sorbitol (15% weight) and bovine serum albumin (6% weight). Vitrification was achieved after transfer of cells into polypropylene straws by direct immersion in liquid nitrogen and rapid thawing.

Cryopreservation of *Asparagus* embryogenic cell suspensions (Uragami et al., 1989) has been achieved with a solution containing 85% of the following solution: 22% glycerol, 15%

ethylene glycol, 15% propylene glycol and 7% DMSO in medium enriched with 0.5M sorbitol.

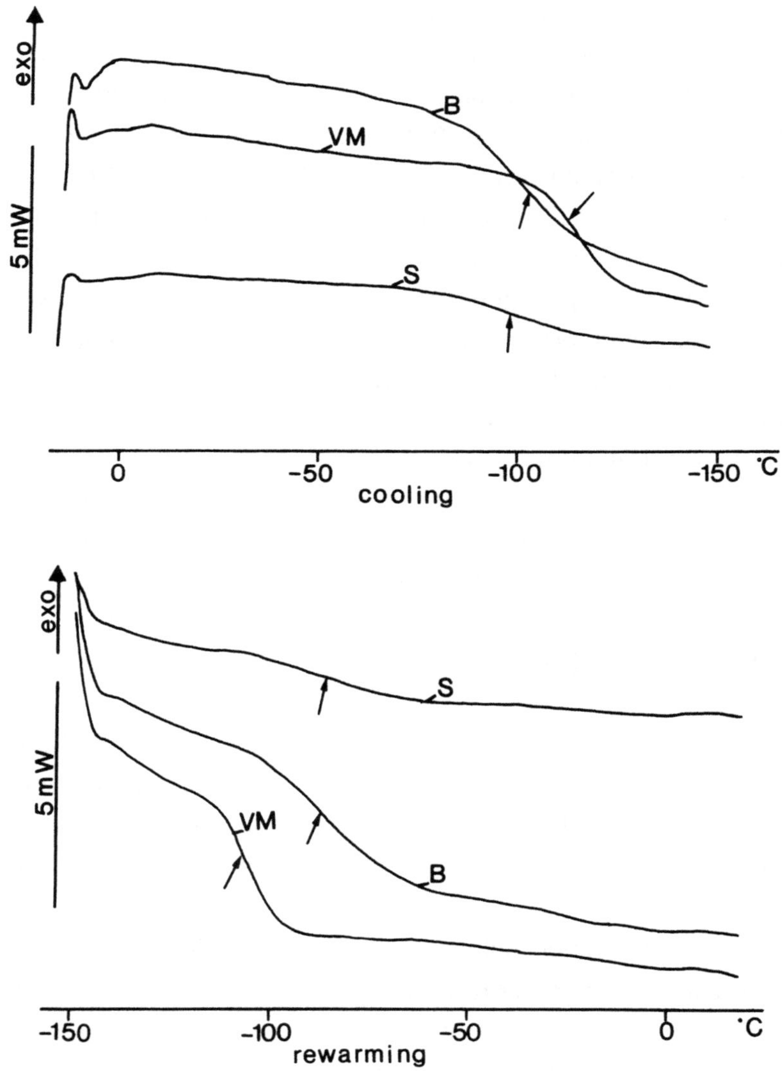

Fig. 2: Thermograms of cooling from +20°C to -150°C and rewarming from -150°C to +20°C of vitrification medium (VM), beads treated with loading medium (B) and shoot-tips extracted from the beads (S). The cooling and rewarming rates were 10°C/min. Vertical bars represent 5 mW, arrows indicate the middle temperature of glass transitions (Tannoury et al., 1991).

For nucellar orange cells, a vitrification solution containing 30% glycerol, 15% ethylene glycol and 15% DMSO in medium containing 0.4M sucrose was proposed (Sakai et al., 1990 and 1991). This medium has also been used for white clover shoot-tips (Yamada et al., 1991).

For shoot-tips of mint, Towill (1990) used a vitrification medium containing 35% ethylene glycol, 1M dimethylsulfoxide and 10% polyethylene glycol 8000. Ultrarapid cooling (4800°C/min) and rewarming (9000°C/min) was required to obtain 57% survival and direct regrowth of shoot-tips.

For shoot-tips of carnation, the vitrification procedure has been applied after encapsulation (Tannoury et al., 1991). After overnight preculture in culture medium enriched with sucrose (0.75M), trapped shoot-tips were transferred to vitrification medium containing 6 g sucrose/ 4 g water/ 6 g ethylene glycol (medium 6-4-6). Survival appeared to be independent of the cooling procedure, i.e. direct cooling in liquid nitrogen or two-step cooling (0.5°C/min to 20°C/min from 0 to -150°C), and of rewarming rate. This independence has been related to glass transition of vitrification medium, encapsulated material and shoot-tips (fig. 2).

DEHYDRATION PROCEDURE

Dehydration of organs (shoot-tips or somatic embryos) involves evaporation at room temperature. It may be performed after encapsulation in alginate beads (artificial seeds).

For encapsulation, somatic embryos or shoot-tips are suspended in culture medium supplemented with 3% Na-alginate. The mixture is then dispensed from a pipette with sterile tip into apex culture medium containing 100 mM calcium chloride.

Beads of about 4 mm in diameter containing 1 to 3 somatic embryos or shoot-tips are precultured for several days in media supplemented with sucrose (0.3M to 1.5M). Coated organs are dried under sterile air flow at ambient temperature and humidity for 0 to 6 hrs. After dehydration, artificial seeds are transferred into cryovials without liquid medium for cooling. Two cooling procedures can be used: rapid cooling, by direct immersion of vials in liquid nitrogen, and two-step cooling. In two-step cooling samples are progressively cooled (0.5°C/min) from 20°C to -40°C, -80°C or -150°C, before immersion in liquid nitrogen. In both cases, beads are slowly rewarmed in air, at room temperature.

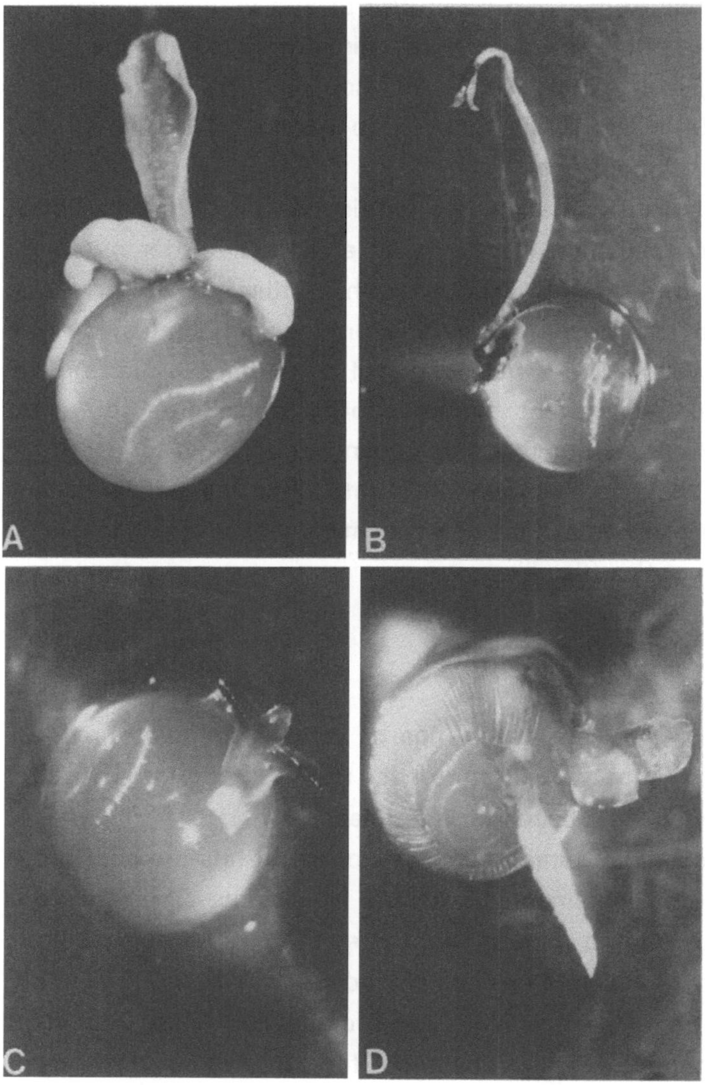

Fig.3. Plantlet recovery from shoot-tips of pear (A), *Solanum* (B), carnation (C) and somatic embryo of carrot (D) after encapsulation, dehydration in air, cooling in liquid nitrogen, slow rewarming and subculture in Petri dishes (photographs from C. Scottez, J. Fabre, M. Tannoury and J. Dereuddre).

The encapsulation-dehydration procedure has been applied to cryopreservation of somatic embryos of carrot (Dereuddre et al., 1991a) and shoot-tips of *in vitro* plantlets of pear (Dereuddre et al., 1990b), *Solanum* (Fabre et Dereuddre, 1990), grape (Plessis et al., 1991), and carnation (unpublished results).

Resistance of shoot-tips and somatic embryos to dehydration and subsequent cooling in liquid nitrogen is generally obtained after specific pretreatments like cold-hardening of *in vitro* donor plantlets and preculture with sucrose. For carrot somatic embryos and carnation shoot-tips, best results were obtained after overnight preculture with 0.3M and 0.75 sucrose, respectively. For pear shoot-tips (Dereuddre et al., 1990b), resistance was strongly enhanced by two-month cold-hardening of *in vitro* plantlets. For plants which cannot be cold-hardened by *in vitro* culture, increased duration of preculture has been proposed. For *Solanum phureja*, a tuber-bearing species (Fabre and Dereuddre, 1990; Fabre, 1991), plantlet recovery increased from 0% to 15%, 26% and 40% after respectively 1, 2, 3 and 7 days of preculture with 0.75M sucrose. If shoot-tips cannot tolerate direct preculture with 0.75M sucrose, survival of dehydration and cooling in liquid nitrogen can be ensured by progressive increase of sucrose concentration in the preculture medium (0.3M, 0.5M, 0.75M and 1M). This procedure has been employed for shoot-tips of grape (Plessis et al., 1991).

At room temperature and humidity, best results are generally obtained after 4 to 6 hours of dehydration and 19-20% residual water. With such water content, differential scanning calorimetry thermograms of somatic embryos and encapsulating material display glass transitions during cooling and rewarming (10°C/min). Sigmoid curves corresponding to glass transitions occur between -50°C and -70°C, during cooling as well during rewarming.

With the species tested, this procedure allows regrowth to occur by reactivation of apical meristems without callus formation. For carrot somatic embryos at the torpedo stage, the procedure also allows recovery of hypocotyl tissues and direct recovery of plantlets (Dereuddre et al., 1991a).

Dehydration may also be performed without encapsulation (Uragami et al., 1990). Single nodes of *Asparagus* plantlets grown *in vitro*, pretreated with 0.7M sucrose, are collected on nylon mesh and subjected to dehydration for several hours in Petri dishes containing dry silica gel. The critical water content of single node segments is about 20%, very close to that of encapsulated shoot tips. This water content is also similar to that determined for artificial seeds.

DISCUSSION

Two procedures may be proposed to replace the conventional procedure used for cryopreservation of plant cells and organs: dehydration by evaporation with or without encapsulation, and vitrification in liquid medium.

In both cases, when the water content of encapsulating material or cryoprotective medium sucrose solutions drops to 25% or less, cooling and warming thermograms display glass transitions without any peak of crystallization or melting. Similar cryobehaviour has been observed with encapsulated somatic embryos and shoot-tips (Dereuddre et al., 1991b; Tannoury et al., 1991). In synthetic seeds, the residual water needed for a glass transition is similar to that required for aqueous solutions of sucrose (Dereuddre and Kaminski, 1991). It may be compared to liquid water, which remains unfrozen after cooling to -40°C, a temperature generally required for prefreezing and ensuring maximum survival of *Catharanthus* cells (Chen et al., 1984). With such vitrification solutions or dehydrated artificial seeds, survival becomes independent of the cooling and rewarming rates. With vitrification solutions containing higher water contents, rapid rewarming is generally required to avoid recrystallization and to obtain good survival rates (Uragami et al., 1989; Langis and Steponkus, 1989).

The resistance of plant cells to dehydration is generally induced after preculture in medium supplemented with sucrose. The exact role of sucrose is still unknown. Like other cryoprotective agents (DMSO, polyols), it seems that sucrose modifies the structure of water by the formation of hydrogen bonds and increases the viscosity of solutions, promoting glass transition during cooling and subsequent rewarming (Franks, 1985). With potato shoot-tips (Fabre, 1991), the resistance of shoot-tips to dehydration and cooling in liquid nitrogen appears to depend on the regression of plasmolysis during preculture. Hence, if cells can tolerate the osmotic stress needed for the glass transition to occur, they should be protected: phase transitions, and consequently the coexistence of the two phases, crystalline and liquid, can be avoided.

The encapsulation-dehydration process displays several advantages: easier handling of organs, simplification of cryoprotective media, elimination of costly programmed freezers, independence of survival from cooling rates. It increases the mass of the tissues which survive exposure to liquid nitrogen: primordial leaves of pear shoot-tips which are killed by the conventional procedure but survive after dehydration (Dereuddre et al., 1990a and b). This technique may be useful in practice for cryopreservation of shoot-tips and somatic embryos which are sensitive to freezing of cryoprotective medium.

Cryopreservation in liquid cryoprotective medium appears to be the more appropriate technique for long-term storage of plant cell suspensions. The encapsulation-dehydration technique has been applied succesfully to meristems and somatic embryos, and may also allow storage of plant organs at higher temperatures than that of liquid nitrogen without risk of recrystallization (Dereuddre et al., 1991a).

The vitrification and dehydration procedures have advantages compared with conventional procedures for tissue and organ cryopreservation. However, research is still needed to decrease the toxicity of vitrification solutions and to extend the results to other species. Progress may be possible through improved understanding of the behaviour of solutions in relation to the cytological and biochemical responses of cells to dehydration.

REFERENCES

Chen HH , Kartha KK, Constabel F, Gusta LV (1984) Freezing characteristics of cultured *Catharanthus roseus* (L.) G. Don cells, treated with dimethylsulfoxide and sorbitol in relation to cryopreservation. Plant Physiol 75: 720-725

Dereuddre J, Kaminski M (1991), Applications of thermal analysis in cryopreservation of plant cells and organs. J Thermal Analysis, in press

Dereuddre J, Blandin S, Hassen N (1991a) Resistance of alginate-coated somatic embryos of carrot (*Daucus carota* L.) to freezing in liquid nitrogen: 1. effects of preculture. Cryo-Letters 12: 125-134

Dereuddre J, Scottez C, Arnaud Y, Duron M (1990a) Effets d'un endurcissement au froid des vitroplants de Poirier (*Pyrus communis* L. cv Beurré Hardy) sur la résistance des apex axillaires à une congélation dans l'azote liquide. C R Acad Sci, sér III 310: 265-272

Dereuddre J, Scottez C, Arnaud Y, Duron M (1990b) Résistance d'apex axillaires de Poirier (*Pyrus communis* L.) à une déshydratation puis à une congélation dans l'azote liquide: effet d'un traitement au froid des vitroplants. C R Acad Sci, sér III 310: 317-323

Dereuddre J, Hassen N, Blandin S, Kaminski M (1991b) Resistance of alginate-coated somatic embryos of carrot (*Daucus carota* L.) to desiccation and freezing in liquid nitrogen: 2. thermal analysis. Cryo-Letters 12: 135-148

Fabre J (1991) Cryoconservation d'apex de Solanacées tubérifères après encapsulation et déshydratation. Etude des modifications cellulaires au cours du prétraitement. Thèse Univ. P. et M. Curie, 123 p.

Fabre J, Dereuddre J (1990) Encapsulation-dehydration: a new approach to cryopreservation of *Solanum* shoot-tips. Cryo-Letters 11: 413-426

Franks F (1985) Biophysics and biochemistry at low temperatures. Cambridge Univ Press, Cambridge 210 p.

Langis R, Steponkus PL (1990) Cryopreservation of Rye protoplasts by vitrification. Plant Physiol 92: 666-671

Langis R, Schnabel B, Earle ED, Steponkus PL (1989) Cryopreservation of *Brassica campestris* L. cell suspensions by vitrification. Cryo-Letters 10: 421-428

Plessis P, Leddet C, Dereuddre J (1991) Résistance à la déshydratation et à la congélation dans l'azote liquide d'apex enrobés de vigne (*Vitis vinifera* L. cv Chardonnay). C R Acad Sci, sér III, 313: 373-378

Sakai A, Kobayashi S, Oiyama IJ (1991) Survival by vitrification of nucellar cells of navel orange (*Citrus sinensis* var. *brasiliensis* Tanaka) cooled to - 196°C. J Plant Physiol, 137: 465-470

Sakai A, Kobayashi S, Oiyama IJ (1990) Cryopreservation of nucellar cells of navel orange (*Citrus sinensis* var. *Brasiliensis* Tanaka) by vitrification. Plant Cell Report 9: 30-33

Uragami A, Sakaï A, Nagaï M (1990) Cryopreservation of dried axillary buds from plantlets of *Asparagus officinalis* L. grown *in vitro*. Plant Cell Reports 9: 328-331

Uragami A, Sakai A, Nagai M, Takahashi T (1989) Survival of cultured cells and somatic embryos of *Asparagus officinalis* cryopreserved by vitrification. Plant Cell Reports 8: 418-421

Tannoury M, Ralambosoa J, Kaminski M, Dereuddre J (1991) Cryoconservation par vitrification d'apex enrobés d'oeillet (*Dianthus Caryophyllus* L.) cultivé *in vitro*. C R Acad Sci in press.

Towill LE (1990) Cryopreservation of isolated mint shoot tips by vitrification. Plant Cell Reports 9: 178-181

Yamada T, Sakaï A, Matsumura T, Higushi S (1991) Cryopreservation of apical meristems of white clover (*Trifolium repens* L.) by vitrification. Plant Science 78: 81-87

Use of Cryopreservation in Breeding Programs*

K.K.Kartha
Plant Biotechnology Institute
National Research Council
Saskatoon, Saskatchewan
Canada

Cultivated crop species, all over the world, are being improved continuously in order to meet the demand for more food, feed and fibre. Traditionally, crop improvement is based on conventional plant breeding strategies and this approach has resulted in the production and release of superior varieties of all major food crops. The breeding priorities are, however, dictated by the immediate needs of the consumer, with emphasis placed on developing crop species with increased yield, nutritional quality and resistance to pests, pathogens, and adverse environment. In order to ensure that new varieties of crop species are available to meet future demands, it becomes imperative that a high priority be accorded now for the acquisition and preservation of all valuable germplasm even if these germplasms are not required at this point in time. Additionally, with the emphasis placed globally on sustainable agriculture, preservation of plant genetic resources has become a high priority agenda.

Preservation of germplasm is to be viewed from the context of the type of plant species which need to be preserved. Storage of germplasm in the form of seeds is the most economical way of preservation of those types of crop species which are classified as 'orthodox' (Roberts 1973) and whose seeds have long longevity. However, here again, there are several reports which indicate that

* NRCC No. 33516

heritable point mutations, mostly recessive in nature and appear in the second generation, are induced in seeds as a consequence of ageing of seeds in storage (for reviews, see D'Amato 1986; Roberts 1988). As an alternative to conventional seed storage, cryopreservation has been attempted with seeds of over 250 species (Stanwood 1985). In contrast, there is another group of plant species known as 'recalcitrant' whose seeds cannot be dried, to facilitate storage, without immediate injury and viability loss. This group includes several economically important species such as coffee, cacao, coconut, rubber, oil palm, mango, etc. At the extreme end of the scale, one could also find a number of crop species which do not produce viable seeds, thus making preservation efforts impossible employing conventional means.

Amongst the vegetatively propagated species, some exhibit a high degree of heterozygosity and do not produce seeds. Such plants are clonally propagated through various vegetative means such as, tubers or cuttings which generally possess very limited life span. In addition to being labor intensive and expensive, this practice also increases the accompanying risks associated with field maintenance such as, exposure to pathogens, pests, climatic pertubations and human error.

As a pre-requisite for crop improvement programs in general, an area which requires immediate attention would then be the preservation of all the above 'problem' species. While conventional approaches have either proven to be ineffective or inadequate, the development of <u>in vitro</u> approaches involving organ cultures coupled with development of cryopreservation strategies should offer an acceptable solution in this regard. Fortunately, during the last 2-3 decades, organ culture techniques have advanced to a level of practical application. Similarly, newer and better methods of cryopreservation applicable to a wide range of culture systems are being developed in various laboratories. The purpose of this article is neither to review the whole area of cryopreservation nor to discuss its principles but rather to briefly project the applications of cryopreservation specifically to plant breeding programs.

Cryopreservation in the current context

Cryopreservation of plant cells and organs involves the viable freezing of these materials, their subsequent storage at ultra-low temperature preferably at that of liquid nitrogen (-196°C) and their eventual retrieval in a genetically unaltered condition. While the technology is influenced by a number of cryobiological and in vitro cultural parameters, the ultimate value of application of cryopreservation to plant breeding would depend, to a great extent, on the choice of the material to be preserved vis-à-vis their value to plant breeding programs. Naturally, preservation of meristems or shoot-tips, zygotic or somatic embryos and pollen assumes greater importance. Cryopreservation of whole seeds which are dessication-and liquid nitrogen-tolerant offers immense potential for long-term storage and this area of research has been discussed earlier in detail (Stanwood 1985) and hence, not included here. Similarly, cryopreservation of callus, cell and protoplast cultures are also excluded from coverage in this brief article despite the fact that a case could be made for their role in plant breeding programs.

Cryopreservation of meristems or shoot-tips

Ever since plant tissue culture has become a reliable technology, the potential offered by meristem culture in rapid clonal propagation and eradication of viral pathogens has been well recognized. Consequently, a number of commercial firms are employing this technique for the mass propagation of rare and elite plants as well as both ornamental and fruit tree species (George and Sherrington 1984). Subsequent to the pioneering discovery by Morel and Martin (1952) that virus-free plants could be produced from infected mother plants employing meristem culture, there occurred an explosion of similar activities resulting in the elimination of viral pathogens from an array of plant species (Kartha 1986). It was also well recognized that since the constituent cells of shoot apical meristems are genetically stable, the regenerants derived from meristem culture would also exhibit this genetic stability provided the plant regeneration process is direct and neither mediated by an intervening callus phase nor accompanied by adventitious shoot proliferation. Now if we consider the three major attributes of meristems, i.e., the potential for mass propagation, production of virus-free plants, and retention of genetic stability, it becomes very obvious that cryopreservation of meristems as a means of long-term germplasm preservation has a leading role to play in the arena of any plant breeding programs. The greatest application of meristem cryopreservation would be for the genetic improvement of crop species which are clonally propagated, which do not produce viable seeds, or whose seeds are recalcitrant to conventional storage strategies. Ensuring the availability of germplasm of such species for breeding applications would eventually necessitate cryopreservation efforts.

One of the most important factors to be considered before cryopreservation of meristmes is attempted is the development of meristem culture technology which facilitates plant regeneration from cultured meristems at very high frequency since the process of freezing and thawing is bound to result in some degree of viability loss even under the best of conditions.

It was generally assumed that meristem explants are ideal for cryopreservation by virtue of the presence of small and densely cytoplasmic cells, but in actual practice this turned out to be not the case. A limited number of studies carried out to examine the extent of cell survival in meristem explants following cryopreservation revealed that the entire meristem dome does not survive the freezing and thawing stress (Haskins and Kartha 1980; Grout and Henshaw 1980). This situation could result in indirect regeneration of plants from the surviving cells and invite problems associated with somaclonal variation which from a germplasm preservation perspective should be avoided. While small and densely cytoplasmic cells withstand the freezing process theoretically, the main problem encountered in the viable freezing of meristmes had been the retention of symplastic and structural integrity of the explant during cryopreservation. Consequently, the progress made in cryopreservation attempts with meristems lagged that of cell cultures wherein certain sub-populations of the cells would always be in a state structurally and physiologically conducive to withstand freezing and thawing stress.

Over the years, development of cryopreservation methodology revolved around examining various freezing methods such as, slow or two-step freezing, rapid freezing, droplet-freezing, and vitrification resulting in the viable freezing of meristmes from a number of taxonomically divergent species (for details, see Kartha 1985; Withers 1985; Langis et al. 1990; Yamada et al. 1991). A number of cryoprotectants as well as various cooling rate-dependent freezing methods have been examined in the past in great detail. Similarly, a number of other parameters such as, preculture of meristmes on nutrient media alone or supplemented with cryoprotectants have also been studied to enhance post-freezing survival(Grout and Henshaw 1978; Kartha et al.

1979; Baginol and Engelmann 1991). The observation that species which are inherently capable of acclimating to low temperature in nature has a greater propensity to withstand freezing stress has prompted several investigators to use meristems from cold acclimated plants for cryopreservation. Such attempts have contributed to enhance the survival of cryopreserved meristms of carnation (Seibert and Wetherbee 1977), apple (Tyler and Stushnoff 1988; Kuo and Lineberger 1985) Rubus (Reed 1988) and Pyrus sp (Dereuddre et al. 1990). Plant breeding efforts generally would benefit considerably by establishing germplasm banks based on cryopreserved meristems. However, development of simple and reliable routine cryopreservation methodology applicable to a wide range of target species is a consideration which warrants immediate attention.

Cryopreservation of zygotic and somatic embryos and pollen

Cryopreservation of zygotic embryos offers excellent opportunities for preserving germplasm of those species whose seeds are recalcitrant to conventional storage or for rescuing otherwise inviable embryos resulting from wide crosses aimed at expanding the genetic base. The advances made in tissue culture technology has permitted the application of embryos culture to crop improvement. Immature zygotic embryos currently find applications, notably in gramineous crops, for the establishment of totipotent culture systems for their use in various biotechnological research including direct gene transfer (Vasil et al. 1991; Kartha et al. 1989; Chibbar et al. 1991). Preservation of immature embryos at a particular developmental phase at which regenerative callus is induced upon in vitro culture offers special advantages in using such embryos as gene recipient systems in genetic engineering efforts. Cryopreservation of zygotic embryos has been reported for a number of species including such important tropical crops as oil palm (Grout et al. 1983) and coconut (Chin et al. 1989; Assy-Bah and Engelmann 1992).

Somatic embryos are also gaining in importance not only for large scale propagation of crop species but also for the production of synthetic seeds. Somatic embryo-based cryopreservation systems are particularly attractive for species which possess long regeneration cycles as is the case with coniferous trees (Kartha et al. 1988). An excellent example of germplasm preservation using cryopreservation of somatic embryos is in the case of oil palm and this technology holds great promise for both propagation and preservation of this economically important recalcitrant species (Engelmann 1990).

From a plant breeding perspective, preservation of pollen is an attractive proposition especially when synchrony in flowering does not occur in nature for fertilization to occur or when parental species are geographically isolated. This storage requirement might vary from a few days for crops with short-lived pollen or longer, up to a year, for certain woody species when a late-flowering strain is used for the pollination of an early flowering strain. Preservation of pollen for extended periods is particularly valuable for species which exhibit erratic flowering habit. Cryopreservation and/or freeze- or vaccum drying of pollen have been successful for the storage of pollen (for a review, see Towill 1985).

Cryoselection strategy for crop improvement

The conventional cryopreservation protocols utilize such steps as cryoprotection, freezing to an intermediate temperature and storage of the samples in liquid nitrogen in order to retain maximum viability of the germplasm. In contrast, the cryoselection strategy developed in the author's laboraotry involves freezing and cryostorage of cells or tissues without the application of cryoprotectants. Using this approach, immature embryo-derived calli of spring wheat could be cryostored in liquid nitrogen with a survival rate of 10-15%. By employing recurrent cryoselection, the survival rate of the originally cryopreserved calli could be increased to 40-50%. In the seed progeny derived from regenerants obtained from the cryoselected callus, several lines exhibited enhanced freezing tolerance up to -12°C (Kendall et al. 1990). Subsequent controlled freezing carried out on the third seed generation further revealed that the selected trait is genetically stable (Kendall and Kartha, unpublished). Although the exact mechanism by which cryoselection imparts enhanced freezing tolerance remains unclear, this approach, if applicable to other crop species which are sensitive to low temperature, will have great potential for crop improvement.

In conclusion, the cryopreservation technology holds great promise in any plant breeding program since preservation of all valuable germplasm is an integral component of crop improvement and sustainable agriculture. Moreover, as the field of plant biotechnology matures further, new and exciting lines of germplasm will appear in the scene and, in that context, the question of preservation of proprietory material including transgenic germplasm will have to be addressed.

References

Assy-Bah B, Engelmann F (1992) Cryopreservation of immature embryos of coconut (Cocos nucifera L.). Cryo-Letters (in press)

Baginol S, Engelmann F (1991) Effects of pregrowth and freezing conditions on the resistance of meristems of date palm (Phoenix dactylifera L. var. Bou Sthammi Noir) to freezing in liquid nitrogen. Cryo-Letters 12: 279-286

Chibbar RN, Kartha, KK, Leung N, Qureshi J, Caswell K (1991) Transient expression of marker genes in immature zygotic embryos of spring wheat (Triticum

aestivum) through microprojectile bombardment. Genome 34: 453-460
Chin HF, Krishnapillay, B, Hor YL (1989) A note on the cryopreservation of embryos from young embryos of coconuts (Cocos nucifera var. Mawa). Pertanika 12: 183-186
D'Amato F (1986) Spontaneous mutations and somaclonal variation. In: Nuclear techniques and in vitro culture for plant improvement. PP 3-10, IAEA, Vienna
Dereuddre J, Scottez C, Arnaud Y, Duron M (1990) Effects d'un endurcissement au froid des vitroplantes de poirier (Pyrus communis L. cv. Beurré Hardy) sur la résistance à une congélation dans l'azota liquide. CR Acad Sci Paris 310: Ser III. 265-272
Engelmann F (1990) Oil palm somatic embryo cryopreservation: progress and prospects. Bull Soc Bot Fr 137: Actual Bot (in press)
George EF,Sherrington PD (1984) Plant propagation by tissue culture - Hanbook and directory of commercial laboratories. Eastern Press Reading
Grout BWW, Henshaw GG (1978) Freeze-preservation of potato shoot-tip cultures. Ann Bot (London) 42: 1227-1229.
Grout BWW, Henshaw GG (1980) Structural observations on the growth of potato shoot-tip cultures after thawing from liquid nitrogen. Ann Bot (London) 46: 243-248
Grout BWW, Shelton, K, Pritchard HW (1983) Orthodox behaviour of oil palm seed and cryopreservation of the excised embryo for genetic conservation. Ann Bot (London) 52: 381-384
Haskins RH, Kartha KK (1980) Freeze-preservation of pea meristems: cell survival. Can J Bot 58: 833-840
Kartha KK (ed) (1985) Cryopreservation of plant cells and organs. CRC Press Inc Boca Raton Florida
Kartha KK (1986) Production and indexing of disease free plants. In: Withers LA, Alderson PG (eds) Plant tissue culture and its agricultural applications, pp 219-238, Butterworths, London
Kartha KK, Leung NL, Gamborg OL (1979) Freeze-preservation of pea meristems in liquid nitrogen and subsequent plant regeneration. Plant Sci Lett 15: 7-15
Kartha KK, Fowke LC, Leung NL, Caswell KL, Hakman I (1988) Induction of somatic embryos and plantlets from cryopreserved cell cultures of white spruce (Picea glauca). J Plant Physiol 132: 529-539
Kartha KK, Chibbar RN, Georges F, Leung N, Caswell K, Kendall E, Qureshi J (1989) Transient expression of chloramphenicol acetyl-transferase (CAT) gene in barley cell cultures and immature embryos through microprojectile bombardment. Plant Cell Repts 8: 429-432

Kendall EJ, Qureshi JA, Kartha KK, Leung NL, Chevrier N, Caswell K, Chen THH (1990) Regeneration of freezing-tolerant spring wheat (Triticum aestivum L.) plants from cryoselected callus. Plant Physiol 94: 1756-1762

Kuo CC, Lingberger RD (1985) Survival of in vitro cultured tissue of 'Jonathan' apples exposed to -196° C. HortScience 20: 764-767

Langis R, Schnabel B, Earle ED, Steponkus PL (1990) Cryopreservation of carnation shoot tips by vitrification Cryobiology 27: 657

Morel G, Martin C (1952) Guerison de dahlias atteints d'une maladie a virus. CR Acad Agric Fr 41: 471-474

Reed BM (1988) Cold acclimation as a method to improve survival of cryopreserved Rubus meristems. Cryo-Letters 9: 166-171

Roberts EH (1973) Predicting the storage of seeds. Seed Sci Technol 1: 499-514

Roberts EH (1988) Seed aging: The genome and its expression. In: Nooden LD, Leopold AC (eds) Senescence and aging in plants, pp 465-498 Academic Press, New York

Seibert M, Wetherbee PM (1977) Increased survival and differentiation of frozen herbaceous plant organ cultures through cold treatment. Plant Physiol 59: 1043-1046

Stanwood PC (1985) Cryopreservation of seed germplasm for genetic conservation. In: Kartha KK (ed) Cryopreservation of plant cells and organs pp 199-226 CRC Press Inc Baco Raton, Florida

Towill LE (1985) Low temperature and freeze-/vaccum-drying preservation of pollen. In: Kartha KK (ed) Cryopreservation of plant cells and organs pp 171-198 CRC Press Inc Boco Raton Florida

Tyler NJ, Stushnoff C (1988) Dehydration of dormant apple buds at different stages of cold acclimation induce cryopreservability in different cultivars. Can J Plant Sc 68: 1169-1176

Vasil V, Brown SM, Re D, Fromm ME, Vasil IK (1991) Stably transformed callus lines from microprojectile bombardment of cell suspension cultures of wheat. Bio/Technology 9: 743-747

Withers LA (1985) Cryopreservation of cultured plant cells and protoplasts. In: Kartha KK (ed) Cryopreservation of plant cells and organs pp 243-267 CRC Press Inc Baco Raton, Florida

Yamada T, Sakai A, Matsumura T, Higuchi S (1991) Cryopreservation of apical meristems of white clover (Trifolium repens L.) by vitrification. Plant Sci 78: 81-87

Genetic Diversity and Reproductive System

M. T. Clegg and B. K. Epperson

Department of Botany and Plant Sciences
University of California
Riverside, CA 92521 U.S.A.

AND

A. H. D. Brown

CSIRO Division of Plant Industry
P.O. Box 1600
Canberra, ACT 2501 AUSTRALIA

INTRODUCTION

The reproductive system determines the way in which gametes are brought together to form new zygotes and thereby exerts the primary level of control on genotypic frequency distributions in populations. Those who seek to improve crop plants genetically through the implementation of various selection strategies begin their task by manipulating genotypic frequency distributions. The success of an improvement program is, in turn, limited by the store of genetic variability available to the breeder. The reproductive system is also a primary determinant of the level of genetic variability within species. As a consequence, the reproductive system interacts with genetic diversity at two levels: first it controls the distribution of genetic diversity in genotypes, and second it determines the extent of genetic diversity through its interaction with the various forces of evolution. The latter interaction involves the integration of evolutionary forces acting over relatively long periods of time. Our purpose here is to explore these interactions.

Recent developments in molecular biology have provided a wealth of new and powerful tools for characterizing genetic diversity within and between species. This high resolution view of genetic differences has altered our understanding of mutational processes. For instance, molecular analyses have revealed that transposable elements play a major role in the origin of new mutations in many plant and animal species. In addition, molecular methods have allowed the characterization of genetic diversity at the ultimate level of resolution - that of individual nucleotide differences. With the advent of the polymerase chain reaction (PCR) method and direct primer mediated sequencing of PCR products, it is possible to determine the complete DNA sequences of a number of copies of a gene sampled from a population or species. This allows the population geneticist to study identity by descent directly rather than similarity in state. The famous French population geneticist, Malécot, pioneered the use of probability methods to study identity by descent among genes within populations. He then used this powerful concept to analyze genetic relatedness in space. Thus the conceptual framework introduced by Malécot (1948) allows us to describe the relationships among genes in both spatial and temporal dimensions. Our second purpose is, therefore, to consider how molecular analyses of diversity have enriched our understanding of the processes that shape patterns of genetic diversity within species.

SYSTEMS OF MATING AND GENETIC DIVERSITY

A large number of isozyme surveys of genetic diversity in plant species have been reported over the past quarter century. These combine to form a broad quantitative picture of the extent of plant genetic diversity as a function of reproductive system. Hamrick and Godt (1990) have recently summarized allozyme data from about 470 taxa where genetic diversity could be classified by reproductive system. The results show that, on average, inbreeding species have fewer polymorphic loci, a reduced number of alleles per locus and a reduced level of genetic diversity. The generalization that reduced levels of genetic variation are associated with inbreeding at the species level appears to be on very firm ground, owing to the large data base analyzed. It is important to add, however, that the variance in genetic diversity among species with similar breeding systems can be large, so the ability to predict the extent of genetic diversity based on a knowledge of the reproductive system alone is limited.

Reproductive system is just one of several variables that strongly influence genetic diversity. Hamrick and Godt (1990) also found that species with larger geographic ranges had higher average levels of total diversity. This pattern holds for all measures of diversity, including percentage polymorphic loci, and average number of alleles per locus. Lesser differences exist between averages of groups of species with different seed dispersal mechanisms, or with different modes of reproduction, successional status, climatic zone distribution, or differences in life form. Major taxonomic divisions also differ considerably in average levels of genetic diversity. Gymnosperms and monocots have greater diversity than do dicots. Interestingly, species with substantial clonal reproduction and purely sexual species have similar levels of genetic diversity.

Recently the RFLP (restriction fragment length polymorphism) technique has also been employed to estimate levels of plant genetic diversity. The advantage of this method is that an almost unlimited number of genetic markers can be obtained and this permits the high density mapping of plant genomes. The disadvantage is that the RFLP assay is more demanding of time and resources than is the case for isozymes. As a result, few detailed surveys of RFLP diversity based on adequate samples within plant species have been reported. In a recent study that is an exception to this trend, Miller and Tanksley (1990) surveyed 156 plants from eight species of *Lycopersicon* that included both selfing and self-incompatible representatives. They found a much higher level of variation among the self-incompatible forms relative to the selfing forms. Miller and Tanksley (1990) estimated that the likelihood that an accession would add new alleles to the sample was two-fold higher for the self-incompatible than for the selfing forms. Moreover, a single accession of the self-incompatible species *L. peruvianum* was twenty-times more likely to add new alleles than a single accession of cultivated tomato (a predominantly self-fertilizing form). Indeed, cultivated tomato is so depauperate in genetic diversity it has been necessary to rely on interspecific crosses to provide polymorphic RFLP markers.

Barley (*Hordeum vulgare*), in striking contrast to tomato, is a self-fertilizing species with relatively high levels of genetic variation (Brown 1992). Graner *et al.* (1990) have reported a RFLP analysis of barley diversity that was based on 23 single-copy probes. This analysis revealed that 43% of probes were polymorphic and when the barley accessions were partitioned into Spring and Winter forms the levels of polymorphism were 34% and 30% respectively. The reduction associated with the Spring/Winter barely partition probably reflects differentiation within subgroups associated with selection for different attributes.

Outcrossing plant species also differ from one another in levels of RFLP polymorphism. For example, RFLP analyses of maize (*Zea mays*) inbred lines, based on 82 genomic DNA clones, reveals substantial RFLP polymorphism (Melchinger *et al.* 1990) even through the inbreds do not span the full range of diversity in maize. Cultivated avocado also possesses high levels of RFLP polymorphism (about 90% of single-copy probes are polymorphic, unpublished data; Furnier *et al.* 1990). In contrast, studies of pearl millet (*Pennisetum glaucum*), based on ribosomal DNA (rDNA) and *Adh1* probes, suggest lower levels of polymorphism (Gepts and Clegg, 1989).

Knowledge of genetic diversity at the complete DNA sequence level is even more limited. To date, only one investigation has attempted to contrast complete DNA sequence diversity between an inbreeder and an outbreeder. Shattuck-Eidens *et al.* (1990) PCR amplified and sequenced several anonymous fragments (based on random clones of unknown function) from the genome of maize (*Zea mays*, an outcrosser) and melon (*Cucumis melo*, an inbreeder). The maize sample was based on seven inbreds, all of which trace back to the yellow dent foundation of US germplasm. Eight melon cultivars were sampled that span the world collection of melon (including muskmelon, Casaba, and Crenshaw). Three regions of the melon genome and four regions of the maize genome were sequenced from each sample. The resulting nucleotide substitution data were analyzed by estimating $\theta = 4N\mu$. The weighted mean estimate of θ for maize is more than 30-fold greater than the estimate for melon, indicating a vastly greater level of nucleotide sequence polymorphism, despite the rather narrow sample of maize lines, as compared to the melon lines. In addition, Shattuck-Eidens *et al.* (1990) tested the individual θ estimates for heterogeneity and found good evidence of heterogeneity over maize clones. (Homogeneous estimates provide support for the hypothesis that the evolutionary dynamics of the regions sequenced are accounted for by random genetic drift and mutation). Heterogeneity in θ is strong evidence that the level of sequence diversity at some of the cloned regions has been affected by selection or that mutation rates vary among regions.

A second aspect of the study by Shattuck-Eidens *et al.* (1990) is the occurrence of addition/deletion polymorphisms (indels). No indels were found in the melon regions, however, indels were common in the maize regions sequenced, yielding a weighted mean θ estimate of 0. 010 (expressed on a per site basis). Many of the indels were associated with inverted or direct repeat structures that are thought to be signatures of transposon activity. While no direct evidence of transposon activity was obtained, the authors speculate that transposons are an important mechanism in the origin of genetic diversity in maize. They also note that transposon activity is less often observed in inbreeders and this may account for the absence of indels in melon.

SPATIALLY STRUCTURED POPULATIONS

Interpopulation diversity. The fact that the level of genetic diversity is only partially correlated with breeding system is not surprising, because other factors such as effective population size and population history strongly influence genetic diversity within species. In addition, plant genetic diversity has a spatial dimension and the mating system interacts with spatial processes, such as seed dispersal and pollination, to shape the spatial pattern of genetic diversity. There are a number of approaches to measuring spatial patterns of genetic diversity. Perhaps the simplest approach is to sample population units in space and then to partition the total diversity into within and between population components.

One common measure of interpopulational diversity, introduced by Nei (1975) and denoted G_{st}, is the ratio of among population variation to the total variation. Hamrick and Godt (1990) found that selfing species have greater diversity or variation between populations (average $G_{st} \approx 0.50$), compared to outcrossed wind-pollinated species (average $G_{st} = 0.10$). Mixed selfing and outcrossing species are intermediate. Geographic range has little effect on G_{st}, but it is worth noting that G_{st} tends to be greater in plants that are annuals.

The magnitude of G_{st} is determined by the effective population size (N) and the extent of migration between populations (m). When the product $Nm > 1$, relatively small values of G_{st} are expected. Pollen migration in predominantly self-fertilizing plants is necessarily limited by the selfing mating system. Most genetic migration in these species is likely to be through seed dispersal. Even continuous plant populations tend to be fractured into many genotypic patches that reflect the limited pollen and seed dispersal capabilities of plant species (Epperson 1990a,b). Our spatial autocorrelation studies of continuous populations of morning glory (*Ipomoea purpurea*) reveal distinct genotypic patches even in this predominantly outcrossed species (Epperson and Clegg, 1986). The autocorrelation view of population genetic diversity is important for systems where migration (or mixing) is primarily between populations that are adjacent, or at least spatially proximal. In these so-called "stepping stone" type systems, spatially proximal populations are highly consanguinous (Malécot 1948) and have highly correlated gene frequencies (Kimura and Weiss 1964). The degree of spatial autocorrelation increases with larger values of m, but does not depend on the value of N.

Effective population size and breeding system. The effective population size of a purely self-fertilizing species is expected to be reduced by a factor of two relative to random mating species with an equivalent population number. Estimates of effective population size for a number of self-fertilizing and out-crossing plant species are consistent with this expectation (Schoen and Brown, 1991). The distribution of estimates of N over populations within species indicates much greater variability within inbreeding species than within out-breeding species. However, estimates of N from some inbreeding populations overlap the distribution of N for outbreeding species; thus a reduced effective population size is not always associated with self-fertilizing populations. Schoen and Brown (1991) suggest that inbreeding populations with low effective sizes are often characteristic of weedy or colonizing species that experience frequent population bottlenecks.

HISTORICAL INFLUENCES ON GENETIC DIVERSITY

Population bottlenecks. The level of genetic diversity within a plant species is determined by the evolutionary history of the species. For selectively neutral variants, the probability of drawing two copies of a gene at a locus that differ in primary nucleotide sequence, is approximately $F = 1 / 4N\mu + 1 = 1/\theta + 1$, where μ is the probability of mutation, and N is the effective population size (Crow and Kimura, 1970). For $N\mu > 1$, this probability rapidly approaches 1. Estimates of μ are roughly 10^{-9} per nucleotide site, which leads to estimates of $N \approx 5 \times 10^6$ over the evolutionary history of maize, based on the mean estimate of θ obtained by Shattuck-Eidens *et al.* (1990). To apply these calculations, a mutation-drift equilibrium must be assumed. Equilibrium is approached very slowly in mutation-drift models and many crop plant species have probably experienced recent population bottlenecks, owing to their history of domestication. Domestication almost certainly included many unpredictable factors, having to do with the culture of the human domesticators, opportunities for seed exchange, differential adaptation to local environments etc. These factors probably account for the large variance in genetic diversity within both inbreeding and outbreeding crop species noted above.

Population bottlenecks can have a strong effect on genetic diversity as shown in the mathematical analyses of Nei *et al.* (1975). These analyses showed that when the population is reduced to a very small size (*e.g.* N=2) and then experiences logistic growth, with mutation, the loss in average heterozygosity is moderate (about a 30% reduction) for high growth rate species. The loss in allelic diversity can be much greater, because in extreme cases (*e.g.* N=1 or N=2) only a very small number of genes are sampled. For low growth rate species the loss in heterozygosity can reach 90%, because the population sizes remain small for a longer period of time during the recovery phase, which allows more time for genetic drift to reduce variation. The restoration of genetic diversity through mutation occurs on a much slower time scale and roughly 10^6 generations are required for the equilibrium to be reestablished.

Some plant species that reproduce through predominant self-fertilization are also opportunistic colonizers. Such species exist in highly unstable environments and local populations experience a high rate of extinction and recolonization. Often a single propagule may found a new population and self-fertilization and high fertility are essential for successful establishment. Such species may exhibit reduced levels of genetic diversity owing to this adaptive strategy. Studies of isozyme variation in self-fertilizing species like the wild oat (*Avena barbata*) in California (Clegg and Allard 1972) are consistent with this scenario.

The common morning glory (*Ipomoea purpurea*) is a predominantly outcrossed self-compatible annual species that colonizes disturbed habitats and agricultural fields in the southeastern US. Despite an outcrossing habit, self-pollination is readily achieved in the morning glory with little or no loss in seed fertility. Very little isozyme variation is found in US populations of morning glory, consistent with its weedy population biology (Epperson and Clegg 1986). Hence outcrossing plant species may also employ a colonizing strategy, although levels of genetic diversity are likely to be reduced through frequent bottleneck events.

Selection during domestication. Historical factors have also been important in the domestication of morning glory. The plant was probably domesticated in central Mexico as a horticultural plant. It has showy flowers with a remarkable diversity of color polymorphisms (Epperson and Clegg 1988). The Mexican populations have higher levels of isozyme polymorphism than US populations (unpublished data), where the species was probably introduced in the 16th and 17th centuries. It appears that bottlenecks associated with the introduction of the plant to the southeastern US also reduced isozyme diversity, although flower color diversity remains high. Finally, the history of selection associated with domestication probably further reduced the store of genetic diversity in morning glory.

Selection acts to reduce the number of lineages in a population, thereby reducing the effective population size. Many crop plants have probably been subjected to very strong artificial selection, sometimes applied to relatively small base populations. The combination of these historical factors can greatly reduce genetic diversity. While the history of tomato domestication is obscure, it is tempting to invoke these kinds of explanations when attempting to account for the very low levels of genetic diversity found in the cultivated tomato. Similarly, the very low levels of diversity found in melon are likely to reflect historical events associated with domestication. It is also important to note that there is little correlation between morphological variation and molecular variation in melon and in morning glory. Both are quite variable for some phenotypic attributes (*e.g.* flower color in morning glory and fruit characteristics in melon). The high level of morphological diversity is probably also the result of selection by man for desirable phenotypes.

Complete DNA sequence data from a sample of genes within species may provide a means of detecting the cumulative effects of selection at a locus. We have already alluded to the importance of knowing the identity by descent of genes within populations. Complete DNA sequence data provides a means of inferring identity by descent because it is possible to infer the genealogical history of a sample of sequences. There is a rich theoretical framework associated with genealogical histories known as coalescence theory (Hudson, 1991). Coalescence models ask when in the past a sample of alleles (separate lineages) coalesce into a single ancestral lineage. The expected time of the coalescence event is a function of the mutation rate, the effective population size and selection parameters. Coalescence theory for neutral genes tells us that the average time back to the coalescence of all sample lineages into a single ancestral lineage (T) quickly approaches 4N (as the number of gene sequences in the sample increases). The standard deviation of T approaches 2.1N, so we expect considerable stochastic error from locus to locus in estimates of T.

In the case of a gene locus that has been subject to a recent selective sweep, where a favorable mutation has replaced a preexisting allele, the expected time to fixation for a selectively favored gene is relatively short. Population genetic calculations show that the expected time to fixation for favored genes in finite populations, given that fixation will occur, is on the order of a few hundred generations. The expected time to fixation for neutral genes, given that fixation will occur, is approximately 4N generations (Crow and Kimura 1970), which can be as long as 10^5 to 10^6 generations for many species. All polymorphic nucleotide sites within several thousand bases of the favored mutation are expected to be carried to fixation as the result of hitchhiking selection. The consequence of a recent selective sweep is a substantial reduction in nucleotide polymorphism in the neighborhood of the selected nucleotide relative to neutral genes. (The neighborhood size depends on the strength of selection and recombination, but is likely to be several thousand base pairs in length.) A very low estimate of θ for a sequenced region may reflect a recent selective sweep, and therefore, may indicate alleles that harbor adaptively important mutations. One maize clone in the above described study of Shattuck-Eidens (1990) fits this pattern.

Coalescence theory may also provide a framework for recognizing regions that have been held polymorphic by some form of balancing selection. The basic argument is that balancing selection will lead to the retention of two or more alleles in the population well beyond the expected 4N coalescence time. Other polymorphic nucleotide sites within a few hundred base pairs of the selected site are also expected to be retained for a time greater than 4N. As a consequence, levels of nucleotide polymorphism are expected to be elevated in the neighborhood of the selected site. The *Adh* data from *D. melanogaster* do indicate a higher level of polymorphism around the presumed site of selection. Detailed molecular studies may aid in identifying the location of adaptively important mutations.

There have been no thorough attempts to apply coalescence models to plant sequence diversity. The data base does not exist to provide good estimates of θ across a sequenced region because adequate population sampling has not been undertaken. The only estimate of the coalescence time for a plant gene is for alcohol dehydrogenase 1 in maize, where T is estimated to be about 2.6 million years for the alleles with the fast and slow electrophoretic forms (Gaut and Clegg, 1991). Interestingly, this value of T suggests a slightly more recent coalescence time than would be predicted based on the mean θ estimate from the Shattuck-Eidens *et al.* (1990) data.

MATING SYSTEM AND GENOME STRUCTURE

The effect of inbreeding on recombination and multilocus organization. The mating system exerts a major influence on the recombinational potential of a species. Effective recombination occurs in heterozygotes and inbreeding reduces the population frequency of heterozygotes. As a consequence, we expect recombination to be greatly reduced in self-fertilizing plant species. This expectation has been confirmed by a number of empirical studies of predominantly self-fertilizing plant species, where estimates of linkage disequilibria are often near their maximal values (Allard, 1975). A consequence of high levels of linkage disequilibrium is that selection at any particular locus is likely to effect much of the genome, including genes that are unlinked to the locus that is the target of selection (Clegg *et al.*, 1978). In contrast, linkage disequilibrium in random mating species tends to involve only tightly linked genes (typically within 2 or 3 map units) and most pairs of genes are effectively independent in their transmission (Clegg, 1984). These differential effects of mating system on multilocus organization constitute a fundamental distinction between outbreeding and inbreeding species and may profoundly influence levels of genetic diversity. As an example, consider a new favored mutation that has rapidly swept through the population to fixation in a random mating population. We asserted that polymorphic nucleotides within a few thousand base pairs of the favored mutation, would also be carried to fixation, thereby reducing the value of θ in the neighborhood of the selected site. In an inbreeding species where the opportunity for recombination is greatly reduced, the chromosomal region carried to fixation would be much larger, perhaps by as much as two orders of magnitude. It follows that we might expect to see large tracts of DNA sequence without polymorphism in inbreeding species that have had a recent history of selection as might be the situation in the melons discussed above.

Transposons in inbreeding species. We noted earlier that transposons seem to be rare in self-fertilizing species. Transposons gain a transmission advantage through a semi-autonomous system of replication and through the ability to infect chromosomes not previously infected. A semi-autonomous system of replication means that the transposon can increase its representation in the genome and therefore grow in numbers until some system to regulate the number of elements comes into play. In an outbreeding species, each genome is reconstituted every generation through a random draw of gametes. If some chromosomes in the population had not been previously infected with the transposon, the opportunity for new infection is presented in every sexual generation. The situation is quite different in a self-fertilizing species where the same chromosomal complement is transmitted within a lineage for many generations. In this situation the opportunity for infection is greatly reduced by the lack of genetic exchange. Moreover, such lineages are expected to be selected against owing to a higher frequency of trnasposon insertion causing deleterious mutations. This may account for the observation that transposons are rare in inbreeders and it may also help to explain the absence of indels in the melon sequences described above. If transposons are rare in inbreeders, then the total mutation rate is probably also reduced in inbreeders, because it is estimated that transposon generated mutations account for a substantial fraction of all mutations in outbreeders like maize.

CONCLUSIONS

The interaction between reproductive systems and genetic diversity is complex. It involves both a proximal effect on genotypic frequency distributions and the integration of the various evolutionary forces over time. To a first approximation, inbreeding species have on average less genetic diversity than do outbreeders and a wider range of genetic diversity among their populations. However, the variance in level of diversity within each major group is large and this makes prediction in specific cases uncertain.

The powerful tools of molecular biology are increasingly being applied to the study of genetic diversity. These provide a direct means of studying genealogies within the conceptual framework developed by Malécot nearly half a century ago. New theoretical developments in population genetics, such as coalescence theory, also provide an important framework for the interpretation of genetic diversity. It seems probable that when viewed at a microscopic scale (that of DNA sequences) genetic diversity in inbreeders will be much more restricted than in outbreeders. This follows from the way inbreeding restricts the recombination potential of populations. The restricted recombinational potential interacts with other forces such as selection to purge inbreeders of variation. It may also interact with transposon dynamics to diminish this important source of mutational variation. Clearly the reproductive system influences genetic diversity at many levels.

ACKNOWLEDGEMENTS: Supported in part by NIH grant GM 45144 and NSF grant BSR-9006984.

LITERATURE CITED

Allard RW (1975) The mating system and microevolution. Genetics 79: 115-126

Brown AHD (1992) Genetic variation and resources in cultivated barley and wild *Hordeum*. In: *Sixth International Barley Genetics Symposium*. Barley Genetics 6: Vol. II.

Clegg MT (1984) Dynamics of multilocus genetic systems. Oxford Surveys in Evol Biol 1: 160-183

Clegg MT, Allard RW (1972) Patterns of genetic variation in the slender wild oat, *Avena barbata*. Proc Natl Acad Sci USA

Clegg MT, Kahler AL, Allard RW (1978) Estimation of life cycle components of selection in an experimental plant population. Genetics 89: 765-792

Crow JF, Kimura M (1970) *An Introduction to Population Genetics Theory*. Harper and Row, New York, pp 591

Epperson BK (1990a) Spatial autocorrelation of genotypes under directional selection. Genetics 124: 757-771

Epperson BK (1990b) Spatial patterns of genetic variation within plant populations. In: *Plant Population Genetics, Breeding and Genetic Resources*, (Brown AHD, Clegg MT, Kahler AL, Weir BS, eds) Sinauer Associates, Sunderland, MA, pp 229-253

Epperson BK, Clegg MT (1986) Spatial autocorrelation analysis of flower color polymorphisms within substructured populations of morning glory I. *Ipomoea purpurea*) Am Natur 128: 840-858

Epperson BK, Clegg MT (1988) Genetics of flower color polymorphisms in the common morning glory (*Ipomoea purpurea*) J Heredity 79: 64-68

Furnier GR, Cummings MP, Clegg MT (1990) Evolution of the avocados as revealed by DNA restriction fragment variation. J Heredity 81: 183-188

Gaut BS, Clegg MT (1991) Molecular evolution of alcohol dehydrogenase 1 in members of the grass family. Proc Natl Acad Sci USA 88: 2060-2064

Gepts P, Clegg MT (1989) Genetic diversity in pearl millet (*Pennisetum glaucum* [L] R Br) at the DNA sequence level. J Heredity 80: 203-208

Graner A, Siedler H, Jahoor A, Herrmann RG, Wenzel G (1990) Assessment of the degree and the type of restriction fragment length polymorphism in barley (*Hordeum vulgare*). Theor Appl Genet 80: 826-832

Hamrick JL, Godt MJ (1990) Allozyme diversity in plant species. In: *Plant Population Genetics, Breeding and Genetic Resources*, (Brown AHD, Clegg MT, Kahler AL, Weir BS, eds) Sinauer Associates, Sunderland, MA, pp 43-63

Hudson RR (1991) Gene genealogies and the coalescent process. Oxford Surveys Evol Biol 7: 1-44

Kimura M, Weiss GH (1964) The stepping stone model of population structure and the decrease of genetic correlation with distance. Genetics 49: 561-516

Malécot G (1948) *Les Mathématiques de l'hérédité*. Paris: Mason

Melchinger AE, Lee M, Lamkey KR, Hallauer AR, Woodman WL (1990) Genetic diversity for restriction fragment length polymorphisms and heterosis for two diallel sets of maize inbreds. Theor Appl Genet 80: 488-496

Miller JC, Tanksley SD (1990) RFLP analysis of phylogenetic relationships and genetic variation in the genus *Lycopersicon*. Theor Appl Genet 80: 437-448.

Nei M (1975) Molecular Population Genetics and Evolution. Amsterdam: North Holland

Nei M, Maruyama T, Chakraborty R (1975) The bottleneck effect and genetic variability in populations. Evolution 29: 1-10

Schoen DJ, Brown AHD (1991) Intraspecific variation in population gene diversity and effective population size correlates with mating system in plants. Proc Natl Acad Sci USA 88: 4494-4497

Shattuck-Eidens DM, Bell RN, Neuhausen SL, Helentjaris T (1990) DNA sequence variation within maize and melon: Observations from polymerase chain reaction amplification and direct sequencing. Genetics 126: 207-217

Reproductive Systems and Dynamic Management of Genetic Resources

R. W. Allard
University of California
Davis, California 95616 USA

My approach to the above topic will be based on Dobzhansky's well-known dictum that "nothing in biology makes sense except in the light of evolution." My thesis is that genetic resource conservation, as well as the deployment of genetic resources in plant breeding, are evolutionary processes and that understanding of the evolutionary mechanisms responsible for the genetic progress that has occurred over time will provide the most certain guide to future progress. I turn now to empirical data to examine the genetic changes that have occurred in three species groups in response to many generations of natural and man-guided selection for high performance in agricultural environments. These species groups are: cultivated barley and its conspecific wild ancestor *Hordeum vulgare*, ssp. *spontaneum*; corn, *Zea mays* and its wild ancestor teosinte (*Zea* species); and the tetraploid slender wild oat, *Avena barbata* and its diploid ancestor, the *A. hirtula-A. wiestii* complex. The *Hordeum* and *Avena* species groups are heavily self-pollinated whereas the *Zeas* are outcrossers.

Wild and Cultivated Barley

Table 1 gives the numbers of alleles observed in 26 collections of wild barley and 18 landraces from the Middle East, two Californian experimental populations, and nine modern Californian cultivars of barley. The two experimental populations were synthesized from intercrosses among cultivars representing all of the major barley growing areas of the world: Composite Cross II (CCII) was synthesized in 1928 from 28 superior cultivars whereas CCXXI was synthesized using male-sterile assisted hybridization among 6,200 accessions from the U.S.D.A. world barley collection. Both experimental populations were grown annually in plots large enough to avoid genetic drift, harvested in bulk without conscious selection, and the next generation seeded from a random sample of seeds from

Table 1. Numbers of alleles in wild barley (*H.s.*) and cultivated barley (*H.v.*)

Item	No. loci	No. alleles	Mean no. alleles/locus	Relative no. alleles/locus
H.s. (26 locations)	20	103	5.15	100
H.v. (18 landraces)	25	55	2.75	53
H.v. (CCII, F_7)	26	41	1.58	31
H.v. (CCII, F_{53})	26	37	1.42	28
H.v. (CCXXI, F_4)	25	42	1.68	33
H.v. (CCXXI, F_{22})	25	38	1.52	30
H.v. (9 Calif. cultivars)	25	36	1.44	28

the previous harvest. The main thing to note from this table is that the total numbers of alleles decreased by nearly one-half from wild barley to the Middle-Eastern landraces but that the decreases in numbers of alleles were much smaller from the Middle Eastern landraces to the Californian experimental populations and elite cultivars and from the early to the late generations of the experimental populations, including broadly-based CCXXI. This raises an important question--which alleles survived and which were eliminated in the progression from wild barley to the modern cultivars? The answer to this question, which has been provided by studies of the dynamics of population change in barley and other species (reviews in Allard 1988, 1990), can be summarized as follows. Alleles that are present in moderate to high frequency in wild barley nearly always survive into primitive landraces and also into modern cultivars. In contrast, alleles that are infrequent in wild barley rarely survive into primitive cultivars; the population biology of such alleles indicates that their effects on survival vary from mildly deleterious to subvitals to lethals. However, a high proportion of alleles present in landraces survive into modern barley cultivars.

The studies of dynamic change have revealed another feature of the evolution of cultivated barley, a feature that appears to explain how populations that are so closely similar in allelic composition often differ widely in adaptedness and performance. This feature is that alleles of many loci interact epistatically at the 2-locus, 3-locus, and multilocus level and that increases in adaptedness and performance are correlated with the development of multilocus clusters of alleles that affect a variety of morphological and physiological characteristics (reviews in Allard 1988, 1990). It appears, therefore, that the improved performance of modern elite cultivars of barley has featured the assembly of relatively few alleles with favorable individual effects into favorably interacting complexes that enhance performance in specific agricultural environments.

Corn and Teosinte

It is widely accepted that Mexico was the cradle of domestication of corn and that North American corn was derived from Mexican landraces. Doebley *et al.* (1985) reported a total of 163 alleles, a mean of 7.1 alleles/locus, in 94 collections representing 34 Mexico landraces. Table 2 shows that 18 (11%) of the 163 alleles were present in virtually all of the 94 Mexican collections at overall frequencies (f) between 0.78 and 0.99. All 18 of these ubiquitous and frequent alleles survived into the six inbred lines that occur in the parentage of virtually all U.S. hybrid corn varieties. The survival ability of 97 rare alleles, present in $f<0.01$, was the opposite: only one of these alleles survived into the six most advanced inbreds. The 48 remaining alleles (29%) were present in the Mexican races in intermediate frequencies. The survival of these alleles was erratic and none became monomorphic in the elite inbreds. The most frequent allele of 20 of the 23 loci in the Mexican races was also the most frequent allele in the elite inbred

Table 2. Frequency of alleles of 23 allozyme loci in 94 Mexican collections of corn and in the six most widely used inbreds in U.S. hybrid varieties (From Goodman and Stuber 1980; Doebley *et al.* 1985).

Allelic freq. class in Mexican collections	% of collections in which observed	No. of alleles in frequency class	No. alleles that survived into 6 elite inbreds	No. alleles monomorphic in 6 elite inbreds
0.78-0.99	100%	18 (11%)	18 (100%)	14 (78%)
0.21-0.68	85%	10 (6%)	6 (60%)	0 (0%)
0.05-0.19	36%	13 (8%)	5 (39%)	0 (0%)
0.01-0.05	12%	25 (15%)	4 (16%)	0 (0%)
<0.01	2%	97 (60%)	1 (1%)	0 (0%)
Total		163 (100%)	34 (21%)	14 (9%)

lines, and the second or third most frequent allele of the Mexican races became the most frequent allele in the elite inbreds for the three remaining loci. Thus, as in barley, allelic frequencies are highly correlated with survivability in corn: frequent alleles survive through large numbers of generations of natural and man-directed selection whereas infrequent alleles do not survive.

The data of Table 2 show that substantial reorganization occurred at the single-locus level during the breeding of the elite corn hybrids and that this reorganization ran counter to expectations based on the hypothesis that heterosis is due to advantage of heterozygotes over homozygotes. Under this hypothesis it is expected that the parental inbreds of successful hybrids will diverge in allelic composition such that a maximum number of loci will be heterozygous in the F_1 hybrids. However, 14 of the 23 loci (61%) had become monomorphic (fixed) for the most frequent allele (denoted allele *1*) in both parental inbreds and, as a result, all nine hybrids among the three Lancaster-type and three Reid-type inbreds are homozygous (homoallelic) for these 14 loci (Table 3). Maximum divergence occurred for only one (4%) of the 23 loci (*Glu1*); the three Lancaster-type inbreds all carry allele *1* of this locus and the three Reid-type inbreds all

Table 3. Allozyme genotypes in the nine F_1 hybrids among the three Lancaster-type and three Reid-type inbred lines.

	Loci[b]										23-locus
Hybrid[a]	1	2	3	4	5	6	7	8	9	10	genotypes
1	*11*	*12*	*11*	*11*	*12*	*12*	*11*	*11*	*12*	*11*	I
2	*11*	*12*	*11*	*11*	*12*	*12*	*11*	*11*	*12*	*11*	II
3	*11*	*12*	*11*	*11*	*11*	*11*	*12*	*12*	*11*	*33*	III
4	*11*	*12*	*12*	*12*	*11*	*11*	*11*	*11*	*12*	*13*	IV
5	*11*	*12*	*11*	*11*	*12*	*12*	*11*	*11*	*12*	*11*	II
6	*11*	*12*	*11*	*11*	*11*	*11*	*12*	*12*	*11*	*33*	III
7	*11*	*12*	*12*	*12*	*11*	*11*	*11*	*11*	*12*	*12*	I
8	*11*	*12*	*11*	*11*	*12*	*12*	*11*	*11*	*12*	*11*	II
9	*11*	*12*	*11*	*11*	*11*	*11*	*12*	*12*	*11*	*33*	III

[a] 1, C103 x B73; 2, C103 x B37; 3, C103 x A632; 4, Oh43 x B73; 5, Oh43 x B37; 6, Oh43 x A643; 7, Mo17 x B73; 8, Mo17 x B37; 9, Mo17 x A632.

[b] 1, 14 monomorphic loci, 2, *Glu1*, 3, *Idh1*; 4, *Mdh3*; 5, *Pgd1*; 6, *Pgm1*; 7, *Pgd2*; 8, *Phi1*; 9, *Est8*; 10, *Acp1*.

carry allele *2* so that all nine F_1 hybrids among the Lancaster x Reid inbreds are *12* heterozygotes (Table 3). The six elite inbreds are polymorphic for the eight remaining loci (35%) so that some of the F_1 hybrids are homoallelic and others are heteroallelic for these eight loci (Table 3). The number of different single-locus genotypes (207) in the nine F_1 hybrids of Table 3 is vastly smaller than in the Mexican races. Also many more of the single-locus genotypes of the F_1 hybrids are homoallelic (171/207 = 83%) and many fewer are heteroallelic (36/207 = 17%) than in the Mexican races.

The data of Table 3 show that substantial reorganization also occurred at the 23-locus level during the progression from the Mexican races to the elite hybrids. During this progression the number of different 23-locus genotypes was reduced from very large numbers in the Mexican landraces to only four (genotypes I, II, III, and IV of Table 3). Among these 23-locus genotypes Genotype I is heteroallelic for five loci, II for four loci, and III and IV for only three loci. Thus heterosis in corn is associated far more frequently with homoallelism (genetic uniformity) than with heteroallelism (genetic diversity) and also with few genotypes (genetic uniformity) rather than with large numbers of genotypes. Again, as in barley, the improved performance of modern elite cultivars appears to have been accomplished by assembling relatively few alleles into favorably interacting multilocus complexes. Stated another way, the quality of allelic variability appears to be more important than the quantity of allelic variability in promoting heterosis.

The Slender Wild Oat Group

Avena barbata (abbreviated *Ab*) is a tetraploid grass (2n=28 chromosomes) derived by polyploidization from the diploid (2n=14 chromosomes) *A. hirtula* (*Ah*)-*A. wiestii* (*Aw*) complex. *Ab* and *Aw* are indigenous to the Mediterranean Basin where they are usually found in sparse and disjunct stands. *Ab* is much more widely distributed: it is common throughout the Mediterranean Basin across the Middle East to Nepal. It thrives under a wide range of environmental conditions, often occurring in dense stands of millions of plants in undisturbed areas and as a weed in cultivated fields. *Ab* is clearly much more widely adapted and much more successful in covering vast areas more or less continuously than its diploid progenitors.

Cytogenetic studies have shown that *Ab* regularly forms 14 bivalents at meiosis, which suggests that it is a fully diploidized tetraploid. This hypothesis is strongly supported by recent studies of segregation patterns of allozymes which have established that pairing is fully preferential within each of the two genomes (Hutchinson *et al.* 1983). It has often been postulated that the successes of polyploids such as *Ab* are due to heterosis associated with ability to breed true for a highly "heterozygous" state in which intralocus allelic diversity is fixed through chromosome doubling, *i.e.*, to "fixed-heterozygosity."

To illustrate the relationships among allelic frequencies, allelic diversity, and adaptedness in the diploid and tetraploid wild oat group, I turn to data for 14 allozyme loci from samples of ~100 individual plants collected from each of 10 Spanish diploid and 48 Spanish tetraploid populations (García *et al.* 1989, Pérez de la Vega *et al.* 1991, García *et al.* 1991), and from several earlier Californian studies cited in García *et al.* 1989. To simplify tables and discussion, I have abbreviated allelic codes to a single digit; thus the code for allele 100 (the first and usually most frequent allele to be described for each locus) is *1* and the codes for the second, third, ... alleles to be discribed are *2*, *3*, ... All genotypes for the diploid are denoted by a single digit in the diploid and by two digits in the tetraploid, *e.g.*, *1* denotes the *100 100* homoallelic duplex genotype of the diploid whereas *12* denotes the *100 100 101 101* heteroallelic quadriplex genotype of the tetraploid. The mating system of both diploid and tetraploid is one of nearly complete selfing (~99%) and almost all individuals in all populations are homozygous at all loci. Although heterozygotes play an important role in generating the novel genotypes on which adaptive improvement depends, they are too infrequent to have much effect on immediate adaptedness; the rare heterozygotes that were observed were consequently not included in analyses of the relationship between genetic diversity and adaptedness.

Various standard statistics calculated from allelic and genotypic frequency data indicate that genetic diversity is substantially greater in *Ab* than in its diploid ancestor. However, the 14 loci differ widely in amount and kinds of genetic variability so that a locus-by-locus examination is required to identify the mechanisms that promote superior adaptedness for each locus. The 14 loci will be discussed in three groups, those that form only homoallelic quadriplexes, those that form only heteroallelic quadriplexes, and those that form both kinds of quadriplexes.

Locus *Pgm1* is monomorphic for the *1* duplex in the diploids and monomorphic for the *11* quadriplex in Spain and throughout the entire range of both species (Table 4). The patterns of distribution for three additional loci, *Got1*, *Got2*, and *Mdh2*, are virtually identical to that of *Pgm1*. This pattern, which is common in plants (Allard 1988), indicates that a single allele (Allele *1*) of each of these four loci is a generalist that confers superior adaptedness in all genetic backgrounds, diploid or tetraploid, and in all environments in which the diploid and tetraploid are found.

Table 4. Frequencies (f) of representative duplexes or quadriplexes in Spanish *Avena* diploids and tetraploids.

	Diploid		Tetraploid			
Locus	Duplex	f	Quadriplex	f	No. populations in which most frequent	fixed
Pgm1	*1*	1.00	*11*	1.00	48	48
Mdh1	2	1.00	12	1.00	48	48
	1	0.00				
Prx1	2	0.72	12	0.46	25	12
	1	0.17	11	0.26	12	4
	4	0.06	13	0.14	6	1
	5	0.05	14	0.12	4	1
			22	0.02	1	0

Locus *Mdh1* (Table 4) is monomorphic for a single allele (allele *2*) in all diploid populations whereas all tetraploid populations are monomorphic for the *12* quadriplex. Thus the diploid is monomorphic for the *2* duplex and, although allele *1* has not been found in the diploid, it is always present with allele *2* in the *12* heteroallelic quadriplex. Clearly alleles *1* and *2* interact favorably with each other in ways that enhance adaptedness in all environments in which *Ab* is found. Superior adaptedness for locus *Mdh1* is thus associated with genetic uniformity in the form of the *2* duplex in the diploid and also in the form of the monomorphic *12* quadriplex in the tetraploid; however, the *12* quadriplex is heteroallelic so that superior adaptedness in the tetraploid is associated with intralocus allelic diversity (but not heterozygosity) as well as with genotypic uniformity for the monomorphic *12* quadriplex.

The eight remaining loci (*Pgi1*, *Mdh3*, *Pgd2*, *Acp2*, *Pgd1*, *Acp1*, *Prx1*, *Lap1*, *Est1*, represented by *Prx1* in Table 4) are genetically diverse in several ways. Populations of the diploid are often polymorphic for two or more homoallelic duplexes and populations of the tetraploid are often polymorphic for two or more homoallelic and two or more heteroallelic quadriplexes. In all 80 different quadriplexes were observed in Spanish *Ab*, among which 18 heteroallelic and 13 homoallelic quadriplexes were present in f>0.10. Detailed analyses of the *Ab* data (García *et al.* 1991; Pérez de la Vega 1991) have shown that the superior adaptedness of the tetraploid is due in substantial part to fixed intraallelic diversity (not heterozygosity) and also in substantial part to epistatic interactions among alleles of different loci.

Summary and Conclusions

What insights concerning the management of genetic resources emerge from the barley, corn and *Avena* data? The main finding relevant to managing genetic resources is that the frequencies of alleles affecting discretely recognizable morphological, disease resistance, allozyme, and restriction fragment variants are highly correlated with adaptedness and productivity. Although the effects of single alleles on various attributes of adaptation, yield or quality of product are difficult to measure directly, it is clear that selection recognizes those attributes that contribute to survivability and that it causes alleles that govern such attributes to increase in frequency over generations. Stated another way, what matters in evolutionary terms is the frequency at which alleles of a given locus are represented in future generations. If an allele is frequent in many different populations from a given ecogeographical region, and particularly if it is frequent in several ecogeographical regions, prospects are good that this allele will be useful in regions in which it is not present. Fortunately, frequencies of alleles governing discretely recognizable morphological, allozyme, and restriction fragment variants can be determined quickly and relatively inexpensively in accessions in germplasm collections, landraces and older once popular cultivars, genetically enhanced populations, and in locally adapted elite cultivars. This is also the case for resistance to damaging races of pathogens, particularly if specific pathotypes are available for gene-for-gene tests of host resistance *vs.* pathogen virulence. The rapidly growing arrays of restriction-fragment variants, including random amplified polymorphic DNA variants (*e.g.*, Williams *et al.* 1990), appear to offer outstanding opportunities for identifying promising alleles for introgression into breeding stocks. When an allele is frequent somewhere, but is not present in the local stocks of another area, prudent genetic resource management by the breeder responsible for the latter area calls for introgressing such alleles into his breeding stocks. If an allele is infrequent everywhere, it is unlikely to be useful anywhere, and cost-effective managers of germplasm will focus their efforts elsewhere.

The task of managing germplasm at the genotypic level is unfortunately much less straightforward and much more difficult than at the allelic level. This is because having superior alleles in breeding stocks is not enough in itself--superior alleles must be assembled into superior combinations at the intralocus level to exploit favorable heteroallelic within-locus interactions and also at the interlocus level to exploit favorable interactions among alleles of different loci (Allard 1988). These are substantial complications because numbers of genotypes increase exponentially with numbers of loci and numbers of alleles/locus; the consequences are that large numbers of cycles of segregation and recombination, carried out in large populations, are required merely to produce the most useful genotypes. Furthermore, laborious and expensive testing is required over many years at many locations to determine the real value of potentially useful novel genotypes (Bradley *et al.* 1988). Despite such complexities and complications it is clear that a great deal of genetic progress has been made in improving adaptedness, productivity, and quality during the thousands of years our major crops have been grown in agricultural environments--landraces of our major crops perform their agricultural tasks for better than their wild progenitors and modern cultivars are still more successful in this respect than the landraces they replaced. Prospects for rapid and long-continued advances appear to be even better in the future as we take advantage of modern technological advances to introgress more and more favorable alleles into local breeding stocks and assemble these alleles into increasingly superior interactive combinations adapted to specific agricultural environments.

Literature Cited

Allard RW (1988) Genetic changes associated with the evolution of adaptedness in cultivated plants and their wild progenitors. J Hered 79:225-238

Allard RW (1990) The genetics of host-pathogen coevolution: Implications for genetic resources conservation. J Hered 81:1-6

Bradley JP, Knittle KH, Troyer AF (1988) Statistical methods in seed corn product selection. J Prod Agric 1:34-38

Doebley JF, Goodman MM, Stuber CW (1985) Isozyme variation in races of corn from Mexico. Amer J Bot 72:629-639

García P, Morris MI, Sáenz-de-Miera LE, Allard RW, Pérez de la Vega M, Ladizinsky G (1991) Genetic diversity and adaptedness in tetraploid *Avena barbata* and its diploid ancestors, *A. hirtula* and *A. wiestii*. Proc Nat'l Acad Sci USA 88:1207-1211

García P, Vences FJ, Pérez de la Vega M, Allard RW (1989) Allelic and genotypic composition of ancestral Spanish and colonial Californian gene pools of *Avena barbata*: evolutionary implications. Genetics 122:687-694

Goodman MM, Stuber CW (1980) Genetic identification of lines and crosses using isoenzyme electrophoresis. Annual Corn Sorghum Res Conf Proc 35:10-31

Hutchinson ES, Price SC, Kahler AL, Morris MI, Allard RW (1983) An experimental verification of segregation theory in a diploidized tetraploid: esterase loci in *Avena barbata*. J Hered 74:381-383

Pérez de la Vega M, García P, Allard RW (1991) Multilocus genetic structure of ancestral Spanish and colonial Californian populations of *Avena barbata*. Proc Nat'l Acad Sci USA 88:1202-1206

Williams JGK, Kubelik AR, Livak KJ, Rafalski JA, Tingey SC (1990) DNA polymorphisms amplified by arbitrary primers are useful as genetic markers. Nucleic Acids Res 18:6531-6535

Pollen Competition: Genetics and Implications for Plant Breeding

Carla Frova and Mirella Sari-Gorla
Department of Genetics and Microbiology
University of Milano
Via Celoria 26, 20133 Milano, ITALY

Introduction

Pollen competition can take place during all stages of the male gametophytic generation, both during development within the anthers and during function, i.e. germination, tube growth and fertilization. As a consequence, non-random fertilization will occur with the fittest grains siring a larger proportion of seeds: this fact can affect the genetical structure of populations and can be exploited for practical purposes.

Here we shall consider the phenomenon in both natural populations and cultivated crops, its genetic control and its implications for plant breeding.

Evidence

Pollen competitive ability (PCA) is the result of pre-shedding components (variability and competition within the anther) and of post-shedding components (variability of the pollen grain during the independent stage, differential pollen germination, pollen tube growth rate, pollen-style interaction and fertilization ability). The evidence for PCA will here be reviwed in relation to the different stages of male gametophyte development.

Development in the anther Pollen competition within the anther is particularly important since in this phase the sporophyte (mother plant) equally affects all pollen grains. Therefore fitness variability among individual grains can derive only from segregation and expression of pollen genes. Although direct proof of PCA during this stage is difficult to obtain, a large body of indirect evidence has been found. A first set of data regards the amount of post-meiotic gene expression of genes controlling basic metabolic functions (detected by isozyme analysis) or with unknown function (revealed by mRNA analysis). Several studies have shown that a very large proportion of the genome is expressed during the pollen formation phase. The estimates range from 55 to 80% according to plant species (see Ottaviano et al 1991 and Mascarenhas 1990 for review).

A second kind of evidence can be obtained from the analysis of response to selection applied during pollen maturation. That male gametophytic selection (MGS) can affect the sporophytic generation has been demonstrated in many plant species for a variety of traits (see Ottaviano and Mulcahy 1989 for review). The biological basis of the phenomenon lies in the large extent of genetic overlap, i.e. in the large number of genes expressed in both phases of the life-cycle of higher plants. When genes contributing to pollen fitness are expressed also in the diploid phase and affect sporophytic characteristics, variability in PCA at any male gametophytic stage can be inferred from a positive response to selection applied at that stage. Although most of the results reported in the literature refer to selection applied during pollen function. improvement of sporophytic fitness in the progeny via pre-shedding MGS has been obtained for tolerance to abiotic stresses (Sari et al 1989).

And finally, indications of pollen fitness variability and competition within the anther can be deduced from the analysis of mutations which affect grain development. Such mutants are usually difficult to detect because the within species morphological variability of pollen is poor and efficient selection methods for viable gametophytic mutants are not easy to set up. One of the best known cases is that of Rf3 in maize, a fertility restoration factor specific for cms-S cytoplasm (Laughnan and Gabay-Laughnan 1983): heterozygous Rf3/rf3 plants produce 50% sterile pollen grains, easily identifiable by their empty apppearance, and 50% normal viable grains.

Mutants can also be detected by an indirect approach, i.e. by selecting sporophytic or endosperm mutants and then testing for their gametophytic expression which is revealed by distortion from the expected mendelian segregation. In maize the analysis of 33 independent mutants affecting endosperm development has revealed gametophytic expression in 65% of them (de-ga) and also indicated the phase in which the gene is expressed: in 12 cases the expression could be attributed to pre-shedding stages (Ottaviano et al 1988a).

Mature grain. Expression at the time of shedding and during the early steps of germination has been known to occur for a number of genes. Classical examples are adh-1, ae, wx in maize and B-gal in Brassica. The pollen population from heterozygous +/- plants is composed of 50% plus and 50% null grains. In at least one case, that of wx, a difference of pollen fitness in relation to the allele expressed in the grain has been observed: pollen grains carrying the recessive allele (wx) are more resistant to storage.and germinate better in these conditions than Wx grains (Pfahler and Linskens 1972).

In the group of de-ga mutants mentioned above, the recessive allele was shown to affect grain viability, germinability and early tube growth in 4 cases.

Tube growth. Most evidence of pollen competition relates to this stage, to the extent that the term is often used to indicate competition between pollen tubes growing in the style to achieve fertilization. One of the reasons is that also in natural populations the most frequent situation is that of many grains germinating and growing at the same time through individual stigmas and styles. Post-pollination PCA can be measured quite easily by modulating the intensity of selection in two basic ways: i) by regulating pollination intensity, i.e. the amount of pollen deposited on the stigma and, ii) by varying the distance between the stigma and the ovule. With the first technique overall differences in PCA are detected, while the second approach is very efficient in revealing variability in tube growth rate. These methods, coupled with a mixed pollination technique, have revealed the existence of a large variability of post-pollination PCA in many plant species (For a review see Ottaviano and Mulcahy 1989).

Pollen-style interaction. Pollen germinability and tube growth rate are determined not only by the genetic makeup of the pollen itself and of the paternal sporophyte, but also by the stylar environment. The most extreme example is given by gametophytic self incompatibility, but the phenomenon has been observed in several other cases. For instance in maize, there are at least nine "gametophytic factors" whose effect can range from 0 to complete inhibition of tube elongation, depending on which Ga allele is carried by the stylar tissue (Bianchi and Lorenzoni 1975). Moreover, even in the absence of known gametophytic factors, the influence of the stylar genotype on tube growth rate, first reported by Pfahler (1967), has been demonstrated in a comparative study of four inbred lines: their relative pollen tube growth varied significantly in relation to the genotype of the lines used as female plant (Sari Gorla et al 1976).

Genetics

The several components of pollen fitness can be genetically determined by both the diploid (paternal plant and stylar genotypes) and the haploid genome, but their relative importance has yet to be clearly established. Also, the main components of PCA display the characteristics of quantitative traits (Zamir 1983,

Wendel et al 1987, Ottaviano et al 1988, Sarr et al 1988, Rajora 1989), but little is known about the number, function and localization of the genes involved.

With regard to the first question it is important to evaluate the amount of post-meiotic gene expression correlated with pollen PCA. The estimate of the genetic overlap obtained by electrophoretic analysis of a large sample of multimeric enzymes (see Ottaviano et al 1991 for review) refers to genes controlling basic metabolic pathways (they are mainly involved in carbohydrate metabolism), common to pollen and sporophyte. Thus it can reasonably be assumed that the genetic constitution (with regard to these genes) of individual pollen grains contributes to some extent to their fitness.

A second group of genes expected to affect pollen fitness are those involved in tolerance to abiotic stresses. Known examples are those referring to tolerance to low temperature or salt in tomato, to heavy metals in Mimulus Guttatus and Silene dioica (for review see Ottaviano and Mulcahy 1989), to herbicides (Sari et al 1989, 1991) and to heat stress (Frova unpubl.) in maize. In three cases (herbicides, salt, heavy metals) within plant gametophytic selection produced a significant tolerance increase in the resulting sporophytic progeny. This indicates unequivocally that for these characters genetic variability of haploid origin is expressed in the pollen population and that the same genes are expressed also in the sporophytic phase.

In the case of heat tolerance in maize, a positive response to selection applied during microspore development has been verified in the next gametophytic generation in terms of germination ability under stress (Frova unpubl.). A correlated effect on the sporophyte has so far been observed only when the selective pressure was applied in post-pollination stages in maize (Petolino 1992) and cotton (Rodriguez-Garay and Barrow 1988). Taken together, these data suggest that the trait is controlled by a large number of genes, some expressed during pollen formation and probably gametophyte specific, and some during pollen tube growth, which show haplo-diploid transcription. There are several indications in the literature that genes coding for heat-shock proteins could be part of these complexes (Frova et al 1989, 1991, Lin et al 1984).

At dehiscence the pollen is in a dehydrated state and shows a very low level of metabolism. Cellular activity and gene expression are resumed upon rehydration on a stigmatic surface or in an artificial medium and are quickly followed by germination. At this particular time variability in pollen fitness can be evaluated in terms of germination ability also in relation to environmental factors, such as storage conditions and duration, high or low temperature, drought, humidity, pollutants. In general, however, this PCA component appears to be mostly under sporophytic genetic control (paternal plant genotype). This has been clearly demonstrated in maize in a study of MGS carried out according to a recurrent selection scheme (Ottaviano et al. 1982, 1988b). The experimental procedures take advantage of the peculiar structure of the maize ear in which the silk (style) length varies according to the position of the flower on the ear, increasing from the top to the base. Selection experiments exploit the fact that, since several grains germinate on the same stigma and their growing tubes compete within the same silk, the intensity of gametophytic selection increases from the apex to the base of the ear. Consequently, basal ovules are fertilized after more intense gametophytic competition than those at the apex. Gametophytic competitive ability can be measured in precise quantitative terms by using a mixture of two pollen types, distinguished by a detectable marker such as an aleurone color gene. This procedure allows estimation of: i) relative pollen germinability and viability (measured as the proportion of uncoloured kernels in the apical section of the ear) and ii) relative pollen tube growth rate, expressed by the regression of the proportion of uncoloured kernls on ear segments. The results have demonstrated that MGS has little effect on the first component, viability and time of germination, the wide variability of which appears to be mostly sporophytically controlled. On the contrary, with regard to the second component, a clear response to selection has been observed: under high selection intensity, the parameters describing tube growth rate had a significantly higher value than those observed with low selection intensity. The amount of genetical variability released by selection exceeded 20% (Ottaviano et al 1988b). Moreover, correlated responses were obtained for sporophytic traits relating to general plant vigour, in particular mean kernel weight, indicating that pollen competitive ability due to tube growth rate and kernel development is controlled, to a considerable extent, by genes expressed in both tissues.

Similar results have been obtained in *Cucurbita pepo* (Stephenson et al 1988), *Dianthus chinensis* (Mulcahy and Mulcahy 1975), *Anchusa officinalis* (McKenna 1986), and *Hibiscus moscheutus* (Snow and Spira 1991)

Gametophytic control of pollen tube growth rate was detected also in six de-ga mutants (Ottaviano et al 1988a). The F2 progeny obtained by selfing the entire set of de/+ F1s was analyzed according to the same criterion, i.e. the regression of the proportion of de kernels on different sectors of the ear. In this case the effect is obtained in response to within plant MGS and is therefore clearly attributable to post meiotic gene expression.

In some plant species gametophytic factors have been described: the alleles of these genes confer very different competitive abilities to the gametophytes carrying the dominant (Ga) or the recessive (ga) allele. In maize the system has been characterized and 5 out of 9 such factors have also been mapped (Bianchi and Lorenzoni 1975). Difference in competitive ability between competing Ga and ga pollen is expressed on GaGa or Gaga silks: in the absence of the competitor or on gaga styles, ga pollen performs normal fertilization. The system leads to preferential self-fertilization in Ga plants. The opposite situation is found in the case of self-incompatibility, where the growth of incompatible tubes, bearing the same S allele as the style, is completely blocked.

In these last two cases, pollen performance is due to single gene segregation, but in general, as indicated by the body of evidence reported above, pollen tube growth appears to be under the control of a number of quantitative trait loci (QTLs). The genetic dissection of pollen competitive ability can be approached by means of RFLP analysis of its variability. A recent study in maize (Sari et al 1992) was based on a set of 44 Recombinant Inbreds derived from the F1 between two divergent lines and characterized for about 200 restriction loci (Burr and Burr 1988). The pollen tube growth rate of each line was measured by the pollen mixture technique. The chromosome segments carrying loci affecting the trait variability are detected by simple regression analysis. The results show that: i) the character has a typical quantitative pattern and a large portion of the variability is due to genetic effects (h^2=0,77); ii) the variability of the character can be explained on the basis of 6 QTLs, accounting for about 60% of the genetic variability; iii) three of them are found on chromosomal segments in which also Ga factors have been mapped. As Ga alleles with major genetic effects do not segregate in the RI set, the observation suggests that in a maize population the Ga loci may be present as isoalleles having small phenotypic effects not detectable by standard genetic analysis.

Implications for plant breeding

A first consideration concerns the effect of pollen competition on the genetical structure of populations, particularly with regard to the mating system. Variability of PCA, in particular of its main components, germination and tube growth rate, can be explained by the expression of at least two categories of genes: i) genes specific for pollen development, grain germination and tube growth; ii) genes controlling basic metabolic pathways, common to pollen and sporophyte.

In maize Ga factors belong to the first category, enzyme coding genes or genes affecting both pollen and endosperm development to the second. According to which genetic system is operating, selfing or outcrossing should be favoured. In fact, a multigenic system of Ga even with minor effects, would lead to preferential fertilization by self pollen, thus producing inbreeding. On the other hand, the complex of genes controlling basic metabolic processes should favour adaptive allelic combinations for pollen and sporophyte, thus leading to heterosis. These expectations have been tested in maize (Ottaviano et al 1983) in experiments in which PCA in fertilizing the self or an unrelated female was compared using a set of 5 standard inbreds not carrying any known Ga allele, and their F1s. The results showed an overall competitive advantage of self pollen both from inbred lines and F1s, indicating the presence of many gametophytic factors with minor phenotypic effects, the combined action of which might origin a quantitative phenotypic expression. On the other hand, the self combination was not always the best one. This points to the existence of associations in which physiological complementation between pollen and style favours non self pollen tube growth. Pollen style complementation at the nutritional level has been observed in maize (Linskens and Pfahler 1977) and Pisum (Wolff 1975), where the high content of a specific amino acid in the pistil seems to have a role in promoting the tube growth rate in pollen grains with a low content of that amino acid. It would be intersting to investigate the relative competitive ability of the two pollen types (high and low amino acid content).

Finally, when considering the great self pollen advantage in the case of Ga factors with major effects. a substantial increase in Ga allele frequency is to be expected in open pollinated populations. This, however, is not observed, suggesting that a contrasting mechanism favouring outcrossing is active at the same time.

The second consideration regards the opportunities offered by MGS for the improvement of useful agronomical traits in the sporophyte. The large extent of genetic overlap predicts that selection during pre- or post-anthesis stages should have a large effect on the resulting sporophyte. Furthermore, MGS can be more efficient than sporophytic selection because of the haploid state and the large population size. The number of generations needed to produce a given change in gene frequency under equal selection pressures in the two phases of the plant life-cycle hase been calculated and is considerably smaller in the case of gametophytic selection. The large pollen population size also allows a very strong selection pressure to be applied without a drastic reduction in sporophytic fitness, that is at a low cost. With regard to complex traits, a large amount of genetic load is produced by recombination at each generatioin: the removal of the unfitted genotypes can be carried out more efficiently and at a lower cost in the pre-zygotic stage. In fact, the elimination of even a substantial portion of the pollen produced does not affect the size of the sporophytic generation. since the remaining pollen is sufficient to ensure normal pollination.

On these theoretical grounds, which are largely confirmed by experimental data, improvement of pollen competitive ability by means of MGS is of particular value in plant breeding for two reasons. In the first place, pollen quality is a major factor in plant fertility and thus also in seed production. In maize, for instance, both grain viability and germination ability show a large variability in response to high temperature stress, and this has consequences with regard to the seed set (Herrero and Johnson 1980, Shoper et al 1986). Second, it has been shown that a correlation exists between PCA due to pollen tube growth rate and sporophytic traits representing plant vigour, and that it is at least in part attributable to the effects of genes expressed in both pollen and sporophytic tissues.

As previously stated, a number of studies in the last decade have provided evidence that MGS can be efficiently exploited for the improvement of specific useful traits, such as tolerance to physical or chemical stresses or resistence to diseases. The results suggest that MGS is a very promising tool to select for traits that are expressed at the cellular level. In this respect it can be compared to methods based on in vitro selection of cell and tissue cultures. It offers the advantages of in vitro selection procedures, but also other favourable conditions, mainly haploidy, genetic variability, absence of need for regeneration processes. A comparison of the results obtained by selection based on in vitro culture of cell and plant tissues and those of MGS, reveals that most of the characters that can be selected in vitro are also suitable for MGS (Ottaviano et al 1990). The more numerous results obtained by means of in vitro selection surely reflects the far greater use of this technology by both private and public institutions. In any case, the results of this comparative analysis also suggest a strategy for the definition of the most promising applications of MGS: characters which reveal genetic variability in vitro are likely to be suitable for the MGS approach.

Acknowledgements. This research work was supported by the Ministero Agricoltura Foreste: a special grant "tecnologie avanzate applicate alle piante".

References

Bianchi A, Lorenzoni C (1975) Gametophytic factors in *Zea mays*, In Mulcahy DL(ed.) Gamete Competition in Plants and Animals. Elsevier, Amsterdam, pp 257-264.

Burr B, Burr F A, Thomson KH, Albertson CM, Stuber, CW (1988) Gene mapping with recombinant inbreds in maize. Genetics 118: 519-526

Frova C, Taramino G, Ottaviano E (1991) Sporophytic and gametophytic heat shock protein synthesis in *Sorghum bicolor*. Plant Science 73: 35-44

Frova C, Taramino G, Binelli G (1989) Heat-shock proteins during pollen development in maize. Developmental Genetics 10: 324-332

Herrero MP, Johnson RR (1980) High temperature stress and pollen viability of maize. Crop Sci 20: 796-800

Laughnan JR, Gabay-Laughnan SJ (1983) Cytoplasmic male sterility in maize. Annu Rev Genet 17: 27-48

Lin CY, Roberts JK, Key JL (1984) Acquisition of thermotolerance in soybean seedlings: synthesis and accumulation of heat shock proteins and their cellular localization. Plant Physiol 74: 152-160

Linskens HF, Pfahler PL (1977) Gametophytic effects on the aminoacid relationships in maize (Zea mays L.) pollen and style. Theor Appl Genet 50: 173-177

Mascarenhas JP (1990) Gene activity during pollen development. Annu Rev Plant Physiol Plant Mol Biol 41: 317-338

McKenna MA (1986) Heterostyly and microgametophytic selection: The effect of pollen competition on sporophytic vigor in two distylous species. In: Mulcahy DL, Bergamini Mulcahy G, Ottaviano E, (eds.) Biotechnology and Ecology of Pollen. Springer-Verlag, New York, pp 443-448

Mulcahy DL, Mulcahy GB (1975) The influence of gametophytic competition on sporophytic quality in Dianthus chinensis. Theor Appl Genet 46: 277-280

Ottaviano E, Sari-Gorla M, Mulcahy DL (1990) Pollen selection, efficiency and monitoring. In: Ogita ZI, Markert CL (eds) Isozymes: Structure, Function and Use in Biology and Medicine. Wiley-Liss, Inc, New York, pp 575-588

Ottaviano E Mulcahy DL (1989) Genetics of Angiosperm Pollen. Advances in Genetics 26: 1-64

Ottaviano E, Pe' ME, Binelli G (1991) Genetic manipulation of male gametophytic generation in higher plants. In: Biswas BB, Harris J R (eds) Subcellular Biochemistry 17, Plant Genetic Engineering. Plenum Press, New York, pp 107-142

Ottaviano E, Petroni D, Pè E (1988a) Gametophytic expression of genes controlling endosperm development in maize. Theor Appl Genet 75: 252-258

Ottaviano E, Sari-Gorla M, Pe' E (1982) Male gametophytic selection in maize. Theor Appl Genet 63: 249-254

Ottaviano E, Sari-Gorla M, Villa M (1988b) Pollen competitive ability in maize: Within population variability and response to selection. Theor Appl Genet 76. 601-608

Ottaviano E, Sari-Gorla M, Arenari I (1983) Male gametophyte competitive ability in maize. Selection and implications with regard to the breeding system. In: Mulcahy DL, Ottaviano E (eds) Pollen: biology and implications for plant breeding. Elsevier, New York, pp 367-373

Petolino JF, Cowen NM, Thomson SA, Mitchell JC (1992) Gamete selection for heat stress tolerance in maize. In: Ottaviani E, Mulcahy DL, Sari-Gorla M (eds) Angiosperm pollen and ovules. Basic and applied aspects. Springer Verlag, New York, in press

Pfahler PL (1967) Fertilization ability of maize pollen grains. II: Pollen genotype, female sporophyte and pollen storage interaction. Genetics 57: 513-521

Pfahler PL, Linskens HF (1972) In vitro germination and pollen tube growth of maize (*Zea mays* L.) pollen. IV. Combined effects of storage and the alleles at the Waxy (*wx*), Sugary (su_1) and Shrunken (sh_2) loci. Theor Appl Genet 42: 136-140

Rajora OP (1989) Pollen competition among *Populus deltoides* Marsh., *P. nigra* L.and *P. maximowiczii* Henry in fertilizing *P. deltoides* ovules and siring its seed crop. Sex Plant Reprod 2: 90-96

Rodriguez-Garay B, Barrow JR (1988) Pollen selection for heat tolerance in cotton. Crop Science 28: 857-859

Sari-Gorla M, Bellintani R, Ottaviano E (1976) Competitive ability of maize pollen. Interaction between genotypes of pollen and stylar tissues. Maydica 21: 77-88

Sari-Gorla M, Ottaviano E, Ferrario S, Gianfranceschi L, Villa M (1991) Herbicide (thiocarbamate) tolerance in maize. Genetics and pollen selection. In: Ottaviano E, Mulcahy DL, Sari-Gorla M (eds)Angiosperm Pollen and Ovules. Basic and Applied Aspects. Springer-Verlag, New York, in press

Sari-Gorla M, Ottaviano E, Frascaroli E, Landi P (1989) Herbicide-tolerant corn by pollen selection, Sex Plant Reprod 2: 65-69

Sari-Gorla M, Mulcahy DL, Pè E, Ottaviano E (1992) Genetic dissection of pollen competitive ability in maize. In preparation

Sarr A, Sandmeier M, Pernes J (1988) Gametophytic competition in pearl millet, *Pennisetum typhoides* (Stapf and Hubb). Genome 30: 924-928

Schoper JB, Lambert RJ, Vasilas BL (1986) Pollen viability, pollen shedding and combining ability for tassel heat tolerance in maize. Crop Sci 27: 27-31

Snow A, Spira TP (1991) Pollen vigour and the potential for sexual selection in plants. Nature 352: 796-797

Stephenson AG, Winsor JA, Schlichting CD (1988) Evidence for non-random fertilization in the common zucchini, *Cucurbita pepo.* In: Cresti M, Gori P, Pacini E (eds) Sexual Reproduction in higher Plants. Springer-Verlag, Berlin, pp 333-338

Wendel JF, Edwards MD, Stuber CW (1987) Evidence for multilocus genetic control of preferential fertilization in maize. Heredity 58: 297-301

Wolff G. (1975) Investigations on pollen tube growth of Pisum sativum and some of its mutants. In: Mulcahy DL (ed) Gamete competition in plant and animals. Elsevier, Amsterdam, pp 161-175

Zamir D (1983) Pollen gene expression and selection: applications in plant breeding. In: Tanksley SD, Orton JJ (eds) Isozymes in plant genetic and breeding. Elsevier, Amsterdam, pp 313-330

Relationship Between Genetic Expression and Hybrid Vigor in Maize

D. de Vienne, A. Leonardi and C. Damerval
Station de Génétique Végétale, INRA/CNRS/Université Paris Sud
La Ferme du Moulon
91190 Gif-sur-Yvette
France

Whether heterosis is explained by dominance or overdominance a positive relationship is expected between heterozygosity and hybrid performance. Thus kinship coefficient or genetic distance estimated from neutral markers (allozymes or molecular) distributed throughout the genome provide, in principle, a way to predict heterosis between related lines. However the power of neutral markers to predict heterosis in the range of genetic distances explored by maize breeders seems rather low (Frei *et al.* 1986, Lee *et al* 1989, Melchinger and Lee 1990). Actually, the only relevant loci in terms of phenotypic expression of heterosis are those affecting the quantitative traits (QTLs, quantitative trait loci). Since the number of QTLs with significant effect for any given character is limited, the prediction could be improved by selecting markers in linkage disequilibrium with the QTLs. However such a disequilibrium is not generally expected among unrelated genotypes. Increasing the number of random markers will necessarily result in decreasing the correlation by increasing the number of noninformative markers (Charcosset *et al.* 1991).

Parameters derived from the variability of genome expression could provide a source of nonneutral markers. Since regulatory and structural mutations were first distinguished in the early 1960s, various lines of evidence led one to ask whether mutations that change the concentration (as opposed to the structure) of specific proteins play a major role in adaptation and evolutionary changes. In particular Wallace (1963) proposed that the polymorphism of the genes controlling the amount and the appearance of the proteins during development was responsible for heterosis. From that time numerous examples have been described in which the concentration of proteins, or the activity of enzymes, vary between or within species (reviewed in Wilson *et al.* 1977 and Paigen 1986). However, even though most biologists are certainly predisposed to accept that the variability of genetic expression can be a source of nonneutral polymorphisms, the unambiguous demonstration of phenotypic or adaptive consequences of such variations is experimentally difficult (*e.g.* Leary *et al.* 1984, Potter *et al.* 1985, Flint *et al.* 1986).

In this paper we summarize various results obtained in our laboratory concerning the variability of genetic expression in maize revealed by high resolution two-dimensional electrophoresis (2DE), and its relationships with hybrid vigor.

Protein variations revealed by two-dimensional electrophoresis

When two inbred lines are compared by 2DE (O'Farrell 1975), the protein variation can appear as two forms :

- quantitative, when the intensity of a polypeptide spot is different between the two lines (Fig. 1a). Such variations, which seem to be consistently found in various species (Klose 1982, Zivy *et al.* 1983, Bahrman *et al.* 1985, etc.), reveal the polymorphism of genetic factors affecting protein amounts. This polymorphism will be referred to as PAP, for Protein Amount Polymorphism.

- qualitative, when two close polypeptide spots with similar staining, and differing usually in their isoelectric point, are mutually exclusive between the two lines (Fig. 1b). Such variations correspond usually to allelic variation of the gene coding for the polypeptide, rather than to post-translational modifications (Coyne *et al.* 1979). As shown by segregation analyses in maritime pine (Bahrman and Damerval 1989), barley (Zivy *et al.* 1992) and maize (in preparation), the position shift criterion is reliable to score the cases of allelism of putative structural loci.

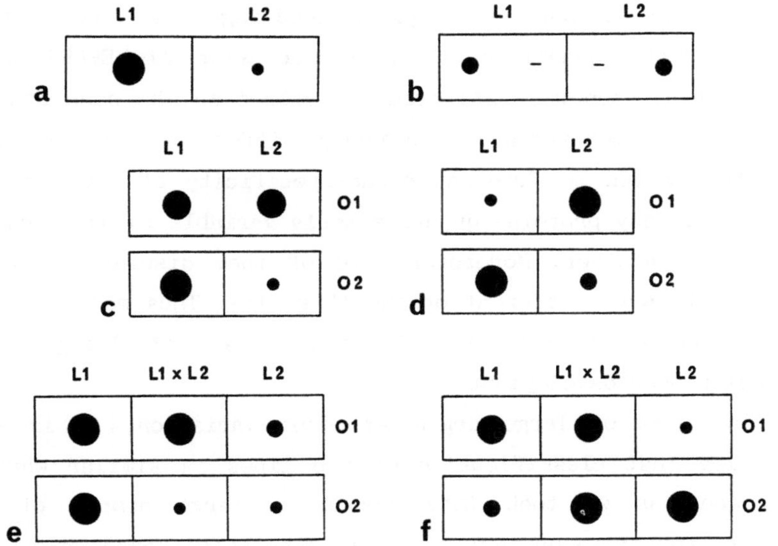

Figure 1. a and **b**, protein variations between two inbred lines L1 and L2 ; a, quantitative variation; b, position shift of two allelic polypeptides. **c** and **d**, examples of organ-specific quantitative variations ; 01 and 02, arbitrary organs ; c, variation only detected in 02 ; d, inversion of the variation between 01 and 02. **e** and **f**, examples of inheritance of quantitative variations ; e, inversion of the direction of nonadditivity between 01 and 02. f, inversion of the variation between 01 and 02, with the same inheritance in both cases.

Organ-dependence of protein amount polymorphism

A stricking feature of the PAP is the influence of the organ or physiological condition on its expression. For example, in an experiment including 5 inbred lines (F2, F1254, F1772 [all three flint type], F252 and W117 [both dent type]), we examined protein variations in three different organs or physiological conditions : the coleoptile and the blade of the first leaf of 8 day-old green seedlings, and the coleoptile of 8 day-old etiolated seedlings. For two lines (W117 and F1254), the sheath and the blade of the second leaf and the mesocotyl of 3 week-old plantlets were also analyzed. A visual survey revealed that the percentage of proteins quantitatively variable between two lines ranged from about 2 % to 8 % for the blades of first or second leaf, and from about 9 % to

15 % for the coleoptiles of green seedlings (de Vienne *et al.* 1988). Automatic quantification performed using the KEPLER package of LSB (Washington DC, USA) produced values in the same order of magnitude (in preparation). Percentage differences between organs are actually due to frequent organ-specificity of the variation (Fig. 1c) : many proteins quantitatively variable in one organ are constant in another. Morerover some of them display a reversed variation between different organs (Fig. 1d). Thus maize displays high levels of polymorphism of the factors controlling genetic expression over development.

In spite of large organ-dependent variation, it is worth mentioning that classification of the lines is similar whatever the organ. As we took into account a large number of gene products, any organ properly reflects genotypic characteristics of the individuals analyzed (de Vienne *et al.* 1988).

Heterosis for protein expression

The inheritance of individual protein amounts was examined for the 10 possible hybrids between the 5 lines mentioned above, in the same organs or physiological conditions. Hybrid patterns were compared to parental line patterns and to coelectrophoreses of the parental samples, these latter representing the expected hybrid patterns under the hypothesis of additivity of protein amounts. The inheritance was considered as nonadditive only when the hybrid spot differed from the coelectrophoresis spot, *and* was either similar to one of the parental spots, or more intense than the most intense, or fainter than the faintest parental spot (Leonardi *et al.* 1988).

Combining all organs/conditions and genotypes, we found 3117 cases where the inheritance could be assessed, *i.e.* where the parental lines displayed spot intensity differences (including absence of one spot), or, less frequently, where the parental lines displayed a similar spot intensity different from the hybrid spot intensity.

In all cases where a polypeptide spot was detected in one or both parents, it was detected in the hybrid. Nonadditive inheritance was found in 128 cases (4 %). Interestingly, most of the nonadditive situations, for a given spot, concern only one organ or condition. The consequence is that the more numerous the organs studied will be, the more numerous will be the proteins displaying nonadditive inheritance. Moreover our estimates of nonadditivity are minimal values, because when hybrid spot intensity seemed to be half-way between "theoretical" hybrid intensity and either parental spot intensity, we preferred to pool these cases with additive ones. Computer-assisted 2D gel analysis allows intermediary situations to be scored. In a cross between two related lines, whereas only 3 situations of nonadditive inheritance were found, automatic quantification allowed the detection of 8 additional cases, all intermediary (in preparation).

Another relevant result concerns the existence of heterosis for the relative amounts of polypeptides. First, in 14 cases the hybrid spot intensity was higher than in either parent. In 12 out these 14 cases, both parents displayed similar spot intensity. Using the QTL mapping approach (Beckman and Soller 1983), we recently showed that oligo- or polygenic determinism of protein amounts is common, which could easily explain these observations without involving overdominance (in preparation). Second, considering the 114 remaining nonadditive cases, a significant asymmetry was observed, since in 91 cases (80 %) the hybrid spot was as intense as the most intense parental spot (Leonardi 1989). In a hybrid between the maize lines B73 and Mo17, Romagnoli *et al.* (1990) also found "positive" nonadditivity in the primary root tips of young seedlings. Thus the relative protein amounts are, in average, higher in hybrids than in their parental inbred lines, and this appears to be all the more significant as more organs/conditions are taken into account. Considering the whole developmental cycle of the plant, this supports the existence of heterosis for genetic expression in maize. The asymmetry of nonadditivy might well be a general phenomenon, as suggested by

Leonardi (1989) from a review of the literature about genetics of enzyme amount or activity in various species.

Moreover various original situations were found. For example, a polypeptide showed an inversion of the direction of the nonadditivity between two organs displaying the same interline difference, *i.e.* the hybrid spot intensity was not similar to that of the same parental line in both organs (Fig. 1e). For a polypeptide whose interline difference was reversed between two organs, the large amount was in both cases observed in the hybrid (Fig. 1f). These examples demonstrate that hybrid and mid-parental patterns of protein expression during development can be very different from each other.

Relationships between protein amount polymorphism and hybrid vigor

To examine the possible ability of PAP to predict hybrid vigor, we used a set of 8 inbred lines (the 5 ones previously mentioned, plus F188 [flint-dent type], F220 and F254 [both dent type] and their hybrids. The 28 possible pairs of lines were characterized for :

- the number, *H*, of structural loci displaying different alleles in the two lines (therefore *H* is also the number of heterozygous structural loci in the hybrid between the two lines, in the samples of loci revealed by 2DE). In the 8 lines, 32 protein loci were found polymorphic. The *H* values ranged from 8 to 24 according to the hybrid.

- a distance index, *PAD* (Protein Amount Dissimilarity), equal to the ratio of the number of spots displaying quantitative variation to the number of common spots between two lines.

- the hybrid values for total and ear insertion heights and fresh forage yield, measured in four different year/locations (in this experiment hybrid values were tightly correlated to heterosis).

Moreover, the number, *NA*, of cases of nonadditive inheritance found in coleoptiles of etiolated and green seedlings for the subset of the 10 hybrids analyzed by 2DE was determined.

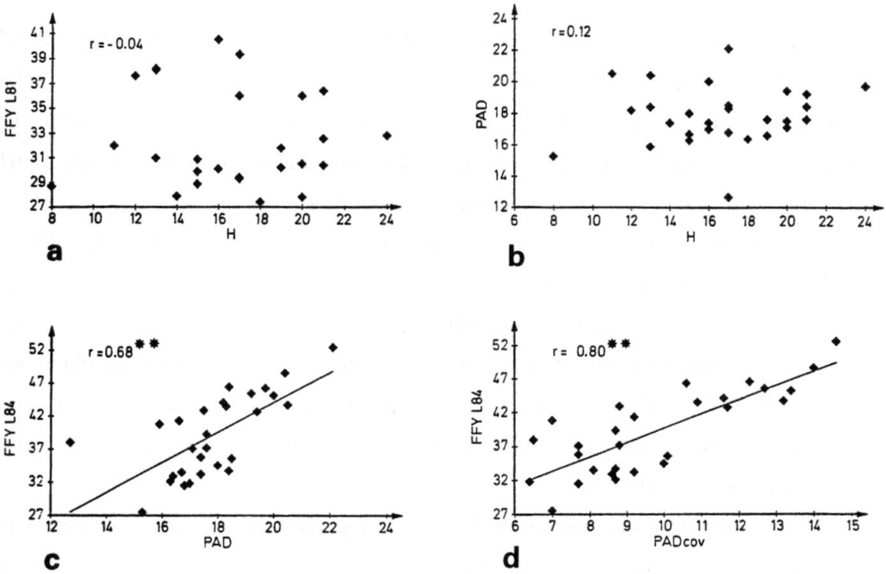

Figure 2. a, correlation between *H*, heterozygosity of the hybrid, and its fresh forage yield measured in Lusignan (France) in 1981 (FFY L81). **b**, correlation between *H* and *PAD* (protein amount dissimilarity index). **c**, correlation between *PAD* and the fresh forage yield measured in Lusignan (France) in 1984 (FFY L84). **d**, correlation between *PADcov* (computed as *PAD* on the groups of covariable spots alone) and FFY L84. The correlation is higher than in **c**. Conversely, when *PAD* is computed after removal of groups of covariable spots, there is no more significant correlation for any variable (Leonardi *et al.* 1991).

The heterozygosity of the sample of structural loci (*H*) failed to be predictive of hybrid values for any agromorphological trait (Leonardi *et al.* 1991) (Fig. 2a). Such a result shows that the polymorphism of the structural genes studied does not have any effect on quantitative trait variation, and that, in our material, the level of linkage disequilibrium between structural loci and loci controlling the agromorphological characters under study (QTLs) was limited, if present at all. It is worth noting that our sample of polymorphic structural loci covers a large part of the genome, as shown by mapping studies (in preparation). This result is consistent with the idea that random, supposedly neutral markers would not be the most relevant markers to predict heterosis.

As *H* was related neither to *PAD* index (Fig. 2b), nor to *NA* index, the same conclusions can be put forward, namely the structural loci are neutral with regard to PAP, and there is limited (if any) linkage disequilibrium between the structural loci and the loci affecting protein amounts.

On the other hand, *PAD* appeared correlated to *NA* (r = 0.61*), and both were correlated to hybrid values of agronomical traits (Fig. 2c). It is worth noting that the correlations between *PAD* and hybrid values appear to be due to a limited number of groups of covariable polypeptides (Fig. 2d), as if only a few genes with multiple effects were responsible for the relation (Leonardi *et al.* 1991).

The correlations between *PAD* and performances of the hybrids mean that loci affecting protein amounts either (i) are in strong linkage disequilibrium with the QTLs, or (ii) are themselves QTLs. Under the first hypothesis the force that would maintain such a specific disequilibrium remains to be explained, since we did not find linkage disequilibrium with structural loci. Actually the major role of a limited number of polypeptide groups is in favor of the second hypothesis. Moreover *PAD* is correlated to *NA*, which cannot, in any case, be explained in terms of linkage disequilibrium. Finally, it is interesting to compare our graphs of correlations with the one published by Lee *et al.* (1989), which also involved 8 inbred lines covering a similar range of diversity. For the upper half of their RFLP-based distances, no relation exists with F1 yield : high RFLP distances could correspond to low as well as high yield, while in our case, high values of *PAD* corresponded exclusively to high F1 yield. This result is actually expected if loci controlling protein amounts are not QTL-linked markers but are themselves QTLs.

In conclusion, our results do not allow to explain heterosis, but provide two relevant pieces of information : (i) there is, in average, heterosis for relative protein amounts ; a consequence might be a possible compensation of limiting gene products. (ii) protein amount polymorphism could allow the definition of predictors of heterosis based on functional rather than genetic relationships.

Literature references

Bahrman N and Damerval C (1989) Linkage relationships of loci controlling protein amounts in maritime pine (*Pinus Pinaster* Ait.). Heredity 63:267-274

Bahrman N, Vienne D de, Thiellement H and Hofmann JP (1985) Two-dimensional gel electrophoresis of proteins for genetic studies in Douglas fir (*Pseudotsuga menziesii*). Biochem Genet 23:247-255

Beckman JS, Soller M (1983) Restriction fragment length polymorphisms in genetic improvement : methodologies, mapping and costs. Theor. Appl. Genet. 67:35-43

Charcosset A, Lefort-Buson M and Gallais A (1991) Relationship between heterosis and heterozygosity at marker loci : a theoretical computation. Theor Appl Genet 81:571-575

Coyne JA, Eanes WF and Lewontin RC (1979) The genetics of electrophoretic variation. Genetics 92:353-356

Flint J, Hill AVS, Bowden DK, Oppenheimer SJ, Sill PR, Serjeantson SW, Bana-Koiri J, Bhatia K, Alpers MP, Boyce AY, Weatherall DJ and Clegg JB (1986) High frequencies of α-thalassaemia are the result of natural selection by malaria. Nature 321:744-750

Frei OM, Stuber CW and Goodman MM (1986) Use of allozymes as genetic markers for predicting performance in maize single cross hybrids. Crop Sci 26:37-42

Klose J (1982) Genetic variability of soluble proteins studied by two-dimensional electrophoresis on different inbred mouse strains and on different mouse organs. J Mol Evol 18:315-328

Leary RF, Allendorf FW and Knudsen KL (1984) Major morphological effects of a regulatory gene : *Pgm1-t* in rainbow trout. Mol Biol Evol 1:183-194

Lee M, Godshalk EB, Lamkey KR and Woodmann WW (1989) Association of restriction fragment length polymorphism among maize inbreds with agronomic performance of their crosses. Crop Sci 29:1067-1071

Leonardi A, Damerval C and Vienne D de (1988) Organ-specific variability and inheritance of maize proteins revealed by two-dimensional electrophoresis. Genet Res 52:97-103

Leonardi A. (1989) Variabilité et hérédité des quantités de protéines chez le maïs : analyse par électrophorèse bidimensionnelle et relations avec la variabilité agromorphologique. PhD thesis, University of Paris VII, 166p

Leonardi A, Damerval C, Hebert Y, Gallais A and Vienne D de (1991) Association of protein amount polymorphism (PAP) among maize lines with performances of their hybrids. Theor Appl Genet 82:552-560

Melchinger AE and Lee M (1990) Use of RFLPs for investigating relationships among inbreds and predicting heterosis in maize. In : Proc 15th Congress of Eucarpia, June 4-8, 1990, Baden, Austria

O'Farrell P (1975) High resolution two-dimensional electrophoresis of proteins. J Biol Chem 250:4007-4021

Paigen K (1986) Gene regulation and its role in evolutionary processes, pp 3-37. In Karlin S and Nevo E, eds : Evolutionary theory and processes. Academic Press Orlando

Potter J, Ho MW, Bolton H, Furth AJ, Swallow DM and Griffiths B (1985) Human lactase and the molecular basis of lactase persistence. Biochem Genet 23:423-439

Romagnoli S, Maddaloni M, Livini C and Motto M (1990) Relationship between gene expression and hybrid vigor in primary root tips of young maize (*Zea mays* L.) plantlets. Theor Appl Genet 80:769-775

Vienne D de, Leonardi A and Damerval C (1988) Genetic aspects of variation of protein amounts in maize and pea. Electrophoresis 11:742-750

Wallace B (1963) Genetic diversity, genetic uniformity and heterosis. Can J Genet Cytol 5:239-253

Wilson AC, Carlson S and White T (1977) Biochemical evolution Annu Rev Biochem 46:573-639

Zivy M, Thiellement H, Vienne D de and Hofman JP (1983) Study on nuclear and cytoplasmic genome expression in wheat by two-dimensional gel electrophoresis. 1. First results on 18 alloplasmic lines. Theor Appl Genet 66:1-7

Zivy M, Devaux P, Blaisonneau J, Jean R and Thiellement H (1992) Segregation distorsion and genetic linkage studies in microspore derived doubled haploid lines of *Hordeum vulgare L.* Theor Appl Genet (in press)

Prediction of Heterosis

A. Charcosset
I.N.R.A. - U.P.S.
Station de génétique végétale du Moulon
Ferme du Moulon
91190 Gif sur Yvette
France

Summary

Efficiency of hybrid breeding could be greatly enhanced by reliable methods of prediction of F1 hybrids values or heterosis. Use of parents per se values through different approaches (including distance computation) should be of poor efficiency for species that exhibit high heterotic effects, such as Maize. In such a situation, testcross designs lead to efficient predictions of F1 hybrids values, as was illustrated by two experimental studies on Maize. When using a classical approach based on general combining ability (GCA) estimation, the most important discrepancies between predicted and actual values of the hybrids were obtained for overestimated crosses (actual value poorer than predicted). It was demonstrated, both theoretically and experimentally, that the use of parents * testers specific effects (SCA) could enhance the prediction efficiency without additional measurements. However this method still leads to misleading conclusions for some hybrids between "original" lines (poorly related to any tester). RFLP assays should provide a powerful way to detect such situations through the estimation of parents similarity and a better knowledge of germplasm structuration.

Introduction

The type of variety to be chosen for a given species depends on the genetic basis of the variation of the traits of economical (agronomical) value and the ability to control pollination within that species. If the species exhibits heterosis (i.e. a given hybrid is superior to his best parent) and pollination can be controlled on a large scale, F1 hybrids (crosses between homozygous lines) will optimize the use of genetic variation (Shull, 1909).

Pollination control through castration, male sterility or gametocides has allowed the development of hybrid varieties in many species such as Maize, Sunflower, Rapeseed, Sugarbeet and more recently Wheat. Hybrid breeding requires the isolation of inbred lines (homozygous genotypes). These lines can be considered as "reproducible gametes" and are used to produce large quantities of identical heterozygous individuals of superior agronomical value.

Once inbred lines have been developed according to a given selection scheme (see Hallauer, 1990) breeders test them in numerous single-cross combinations.

The primary goal of these crosses is the detection of interesting hybrid combinations. The second objective is to estimate the combining ability of the new lines (i.e. their potential as hybrid parents) which will help in the determination of the next crosses to be tested. Testing combinations is so costly and time consuming that the choice of the single-crosses to be tested has a great incidence on breeding efficiency.

Usually, crossing designs depend on what is known about lines origin. If a line can clearly be attributed to a given heterotic group (i.e. a group of lines with related origins), it will be generally crossed to elite lines of a "complementary group". Complementary groups are such that crosses between elite lines of both groups generally lead to highly producing hybrids. When little is known about the origin of a line, or if it is generated from a cross between lines of different groups, it is generally crossed to elite lines that represent the different heterotic groups.

The efficiency of crossing designs to detect hybrids of superior agronomical value can be enhanced through (i) methods that would help in lines classification and (ii) methods that could lead to reliable predictions of the value of a given cross (or its heterosis). Basically, predicted values can be computed from information on the parents. Several types of information may be used, such as morphological information on the parents, values of the crosses already tested (if both parents have already been involved in crosses), marker information (isozymes, RFLP and other techniques)... Predicted values can be computed using different statistical models such as distance computation, estimation of parents general combining ability ... Thus, the aim of this study is to present several approaches of this prediction, and discuss their relative efficiencies.

I Morphological and yield data on the parents

Several studies on Maize have reported a significant relationship between the per se values of inbred lines (or populations) and the yield of the crosses involving these lines. However, the authors generally concluded that this relationship was not tight enough to be actually reliable, and that poor lines may lead to highly producing hybrids (Hallauer and Miranda, 1981). However, this conclusion depends on the genetic basis of the variation, and the relationship is expected to be strong when the magnitude of dominance effects is relatively small when compared to additive effects (see Smith, 1986 for a theoretical discussion). Thus, this kind of prediction might be interesting for some species: for instance, Lefort-Buson and Dattée (1985) found a significant relationship in a study on 140 Rapeseed hybrids evaluated for yield. In the case of winter Wheat, Oury et al. (1990a) found a significant relationship for thirty five hybrids between lines with a broad genetic base. However, Oury et al. (1990b) did not find any significant relationship in a study in which solely lines of superior per se value were considered.

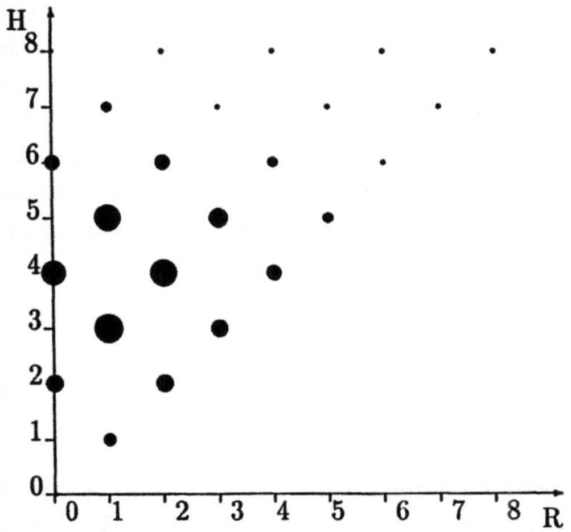

Fig. 1 Relationship between the distance (R) between two lines computed for a single trait and the heterosis of the hybrid (H). A simple genetic model is assumed: eight biallelic loci with equal effects on trait variation and dominance effects. In this situation H is proportional to the number of loci for which the two parent lines differ. The size of the disks is related to the frequency of the hybrids at a given position, estimated numerically in the case where allelic frequencies are 0.5.

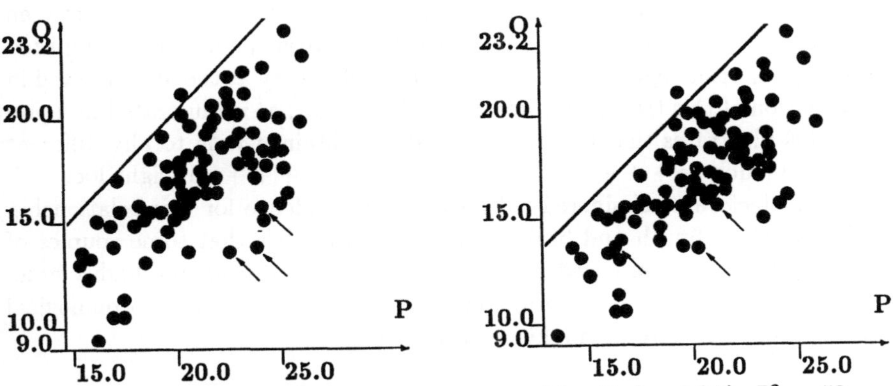

Fig. 2a (model 3) , $R_p^2 = .43$

Fig. 2b (model 7), $R_p^2 = .53$

Fig. 2 Plots of observed silage yield (O) ($t.ha^{-1}$) vs. predicted yield (P) by models (3) and (7). The line is the P=O axis, R_p^2 is the phenotypic coefficient of determination. Arrows indicate hybrids for which predicted values by models (3) and (7) show the most important differences.

Several studies have focused on distances computed from morphological data on the parents (see Goodman, 1972 for a discussion on distance computation) as a tool to predict heterosis of the crosses. Quantitative genetics theory (Falconer, 1981) shows that the heterosis expressed in a cross (computed as the difference between the value of the cross and midparent value), either between lines or populations, is a function of the differences in the allele frequencies of the parents at loci that exhibit dominance. If biallelism is assumed:

$$H_{ij} = \sum_{l=1}^{Nl} d_l (f_l^i - f_l^j)^2 \quad (1)$$

where f_l^i and f_l^j are the frequencies of a given allele at locus l in populations i and j (note that in the case of inbred lines frequencies will take values 0 or 1). d_l is the dominance effect at locus l.

So, any indirect measurement of genetic diversity is a potential tool to predict heterosis. However, experimental studies (see Lefort-Buson and de Vienne, 1985; Charcosset et al., 1990b for reviews) have generally lead to disappointing conclusions.

Relationship between heterosis and this type of distances depends on several parameters. First the loci involved in distance computation may not be involved in heterosis values, as is true for loci exhibiting no dominance effects. Thus, the relationship will depend on complex linkage disequilibria between both type of loci. Moreover, the relationship between the distance computed for a single morphological trait and the genetic diversity at the loci involved in the variation of this trait is not linear, but triangular, as is illustrated in figure one. This illustrates that a same value can be obtained with totally different genotypes (such as $++++----$ and $----++++$ if eight loci with identical effects are considered). Experimental evidence for this relationship is illustrated by Smith and Smith (1989). This implies that some couples of lines may be very close based on their morphology and lead to a highly heterotic cross. So discarding hybrid combinations because of the morphological similarity of their parents might be hazardous.

II Testcross designs

We have seen in the introduction that breeders generally test newly released lines in single-cross combinations with elite lines, which are widely called testers. Designs that involve crosses between lines and testers are usually called testcrosses. If several testers are considered, statistical analysis of the data can be performed using the classical model for factorial designs:

$$Y_{it} = \mu + \alpha_i + \beta_t + \phi_{it} \quad (2)$$

where Y_{it} is the value of the cross between line i and tester t, α_i and β_t are the main effects of line i and tester t, respectively, and ϕ_{it} is the interaction effect.

If the testers are "representative" of the tested germplasm, lines main effects (α) can be considered as an estimate of lines general combining ability (GCA), whereas ϕ_{it} can be considered as the specific combining ability (SCA) of the cross between line i and tester t (Sprague and Tatum 1942). Lines main effects can be used to predict the value of the crosses using the following formula:

$$\tilde{Y}_{ij}^{gca} = \mu + \alpha_i + \alpha_j \quad (3)$$

This approach can be also used when a single tester is considered, and in this situation α_i will be the genotype effect of the hybrid between line i and the tester.

Two studies have been conducted in our laboratory to estimate the efficiency of this approach in the case of Maize hybrids in Northern France. The first study has been described in Charcosset et al (1990b) and the second study (Charcosset, 1990) has been built in a similar way: fifty eight inbred lines of different origins have been crossed to four tester lines during year 1987 and field trials were carried out in Gif/Yvette during year 1988, using replicated block trials (3 replications), to estimate silage yield. Prediction efficiency was tested through a set of 88 hybrids representative of the possible crosses (each line was crossed to approximately three other lines). Experimentation of these hybrids was carried out as for testcrosses. A similar design was built for grain yield; it involved 53 lines (subset of the 58 lines), 4 testers (the same as for silage yield) and seventy-one lines * lines crosses. Among lines * lines hybrids, fifteen hybrids for silage study and thirteen for grain study were generated from crosses between related lines. In this case, the maximum coancestry coefficient (Malecot 1969) was 0.5.

The efficiency of a given predictor was estimated by the coefficient of determination ($R2p$) between predicted and observed values for lines * lines hybrids. Moreover, a genetic coefficient of determination ($R2g$) was computed, by analogy to the computation classically done in the case of random effects.

The conclusions of these studies were consistent with the conclusion of Hull (1945) that the best tester should carry the largest amount of recessive alleles: correlations between predicted and observed values of the 88 hybrids were higher for low productive testers (per se value) than for high productive testers. It was also demonstrated, both theoretically and experimentally that the use of several "complementary" testers improves prediction efficiency, this was consistent with experimental results of Green (1948).

Results are given in table 1 in the case considering the 4 testers. All the correlations were highly significant, and $R2g$ estimates lead to the conclusion that predicted value explain roughly 50 % of the genetic variation observed between crosses. Figure 2a illustrates the relationship between predicted and observed values in the case of silage yield. Most important deviations are due to overestimated crosses (predicted value superior to observed value). These deviations were mostly observed for crosses between related lines. However, two lines, although from a similar origin were not related according to the pedigree information we had, and lead to an overestimated cross.

Table 1. Phenotypic coefficient of determination (R_p^2) and genotypic coefficient of determination (R^2g) between predicted and observed values using model (3) for all hybrids and hybrids between unrelated lines (unr.). The number of hybrids involved in each study is given within parenthesis.

	Silage yield 58 lines		Grain yield 53 lines	
	R_p^2	R_g^2	R_p^2	R_g^2
all hybrids	.433 (88)	.573	.472 (71)	.542
unr. hybrids	.557 (73)	.767	.667 (58)	.781

To look for the incidence of relatedness between lines on the relationship between predicted and observed values, a similar study was carried out for hybrids between unrelated lines (table 1). In this situation $R2g$ values were 0.767 and 0.781 for silage and grain yield, respectively. This result illustrates that a large amount of the variation of SCA effects in Maize should be due to genetic relatedness between lines. However, the exact amount of SCA variation that should be due to other reasons is difficult to estimate since relatedness between lines is not always accurately known because of lacks in pedigree information.

To go further in the use of testcross data, we studied the potential use of line * testers specific effects to increase prediction efficiency. The first idea was derived from Hanson and Casas (1968) results that in case of a diallel mating design between populations, heterosis was highly correlated to a distance (R2) computed from populations * populations specific effects (specific combining ability). Such a distance can be computed in the case of top-cross designs as follows (T beeing the total number of testers):

$$R_{ij} = \sqrt{\sum_{t=1}^{T}(\phi_{it} - \phi_{jt})^2} \quad (4)$$

This distance has a potential advantage on other distances in that solely loci showing dominance effects are considered in its computation. It was related significantly to the difference between observed values and predicted values (using model 3) in the first study (Charcosset et al., 1990b). In the second study, if four testers were considered the relationship had $R2$ values of .09 and .07 for grain and silage yield, respectively (both values were significant at the $\alpha = 0.05$ risk level).

Thus using this distance may improve significantly prediction efficiency. However, when two testers were considered, relationship between heterosis and its predicted value on the distance computation is triangular, as discussed in chapter one. Moreover, it is not easy to establish *a priori* the coefficients to give to lines GCA effects and distance to establish a synthetic prediction formula.

These drawbacks lead us to look for alternative ways to take lines * testers specific effects into account. A basic idea was to build a model that could explain the variation of lines * testers specific effects, estimate its parameters, and then use this model to predict the values of lines * lines specific effects. When a single trait is considered, this can be achieved through multiplicative models (Mandel 1971). In the case of testcross designs, the interaction between line i and tester t can be written (Charcosset et al., 1990a):

$$\phi_{it} = \sum_{m=1}^{M} c_m \gamma_i^m \gamma_t^m \tag{5}$$

Were γ_i^m is the parameter for line i at the mth multiplicative term, c_m is the parameter specific to the mth multiplicative term. M is the total number of multiplicative terms in the model.

Predicted value of the specific effect of the cross between line i and line j will be:

$$\tilde{\phi}_{ij} = \sum_{m=1}^{M} c_m \gamma_i^m \gamma_j^m \tag{6}$$

And so the predicted value of the hybrid will be:

$$\tilde{Y}_{ij} = \mu + \alpha_i + \alpha_j + \sum_{m=1}^{M} c_m \gamma_i^m \gamma_j^m \tag{7}$$

In (5) γ characterises both testers and lines so testers * testers crosses have to be included in the analysis to make the comparison of lines and testers possible. When testers per se values are included in the analysis, simple solutions can be found for parameters estimation if T-1 multiplicative terms are involved in (5). Other situations have to be treated with an alternate least squares procedure (Denis, 1991), using for instance INTERA program (Decoux and Denis, 1990).

This way of using specific effects has theoretical advantages on distance computation since relationship between predicted and observed SCA is no longer triangular (Charcosset, 1990). An experimental study on the previously described silage yield data was conducted (Charcosset et al 1990a). Testers * testers hybrids and testers per se values used for that study were estimated in the same conditions as lines * testers hybrids. When four testers were considered, $R2p$ between predicted values (including specific effects) and observed values was 0.534. $R2g$ value was estimated (unpublished result) as 0.757. Although no statistical test could be performed these values are superior to those obtained using model (3) ($R2p = .433$ and $R2g = .573$) Fig 2b illustrates this relationship. It can be seen that some crosses that were overestimated by the first method were much better estimated. For instance this was the case for the cross between unrelated lines previously described. However some crosses remain overestimated. It has been theoretically demonstrated that the efficiency of this approach will decrease when the number of independent loci implied in SCA variation increases. Moreover, theoretical analysis and pedigree information show that this method will be very efficient for crosses involving at least one line clearly related to a given tester but will not improve prediction efficiency when both lines of a given cross are not clearly related to any tester. However, in this situation predicted SCA value should be close to 0 and so, the use of this method should not decrease prediction efficiency when compared to the method (3).

III Markers information

On the opposite of polygenic traits, markers provide a mean to estimate directly, for a given set of loci, the number of loci for which two lines are different. More generally, they can be used to compute distances between populations directly from allelic frequencies (see for instance Rogers, 1972):

$$MRD_{ij}^2 = \frac{\sum_{a=1}^{Na}(f_a^i - f_a^j)^2}{2Nl} \tag{8}$$

where Na is the total number of alleles, Nl is the total number of marker loci, f_a^i and f_a^j are the frequencies of allele a in populations i and j, respectively.

This formula has the same type as (1), with the difference that in (8) all alleles coefficients are fixed to 1. So, under the assumption that marker loci provide a reliable sampling of the genome, distances between parents computed from markers information will be related to heterosis.

Several studies on Maize have analyzed this relationship using enzymatic markers. Correlations between distances and either heterosis or grain yield were generally low and non significant (Hunter and Kannenberg 1971; Heidrich-Sobrinho and Cordeiro 1975; Gonella and Peterson 1978; Hadjinov et al. 1982; Price et al. 1986; Lamkey et al. 1987; A. Charcosset, M. Lefort-Buson and

M. Grenèche, unpublished results). Similar results were found in the case of rice (*Oryza sativa* L.) by Peng et al. (1988). However, some studies showed significant relationships. In Maize, Frei et al. (1986) established a significant relationship for hybrids between related lines but no relationship for hybrids between unrelated lines. A similar result was established by Strauss (1986) in *Pinus attenuata*: correlation between heterozygosity at enzymatic loci and basal area of the trees was higher for inbred trees than for crossbreds (although both correlations were significant). Smith and Smith (1989) established a significant relationship ($R2 = 0.36$) between heterozygosity for 31 isozyme loci and yield within a set of more than 100 Maize hybrids derived from 37 inbred lines. This set of hybrids included crosses between lines that were highly related by pedigree.

Recently, RFLP provided a mean to conduct studies with a much higher number of marker loci. Results of the previous study (Smith and Smith, 1989) showed that correlation between marker diversity (distance) and yield was much higher for RFLP (230 probe/restriction enzyme combinations) than for isozymes ($R2$ were respectively 0.87 and 0.36). Positive results were also found by Lee et al (1989), with 33 DNA probes and 5 restriction enzymes in the case of a diallel mating design.

We have seen in chapter two that testcross designs could lead to efficient predictions of hybrid values. To test if marker loci information could lead to a significant improvement of prediction efficiency, we have assayed the lines involved in the study described in chapter two for their RFLPs using 44 probe / restriction enzyme combinations. Probes were part of the "core map" set provided by E.H. Coe and D.A. Hoisington (MNL; pers. comm.). DNA was extracted by salting out method (derived from Miller et al., 1988). 8 μg of DNA was run in 0.7 % agarose gels and transferred on charged nylon membranes. DNA probes, radioactively labelled by random priming, were hybridised using conditions described by Church and Gilbert (1984). Only 51 lines (out of 58) were assayed for silage device, and 46 (out of 53) were assayed for grain device. The number of lines * lines hybrids reduced subsequently to 71 and 57 (59 and 47 if only crosses between unrelated lines were considered).

Modified Rogers'distances (MRD^2) between lines were computed (Rogers, 1972). In the case of purely inbred lines, MRD^2 is reduced to the percentage of loci for which two lines differred. Correlation between MRD^2 and yield of the hybrids was first computed. To test if distance could lead to significant improvement of the prediction based on model (3), the partial correlation between MRD^2 and yield was computed with $\tilde{Y}_{gca}$ being fixed. Since partial correlation is the correlation between the distance and the error term of prediction by Yagc, this correlation is equivalent to the correlation between distance and SCA effect of a cross. Moreover the multiple correlation between yield and the variables $\tilde{Y}_{agc}$ and MRD^2 was computed, and the significancy of the increase due to introduction of MRD^2 in prediction formula was tested using a T test.

If the entire set of hybrids is considered, it can be seen from table 2 that distances are significantly correlated to hybrid values. Partial correlation, with $\tilde{Y}_{gca}$ being fixed, is higher than simple correlation. This observation is consistent with results of Lee et al (1989), Melchinger et al (1990a and 1990b), Dudley (1991) who found that distances were better related to SCA effects than F1 hybrid value.

Table 2. Use of MRD^2 to increase prediction efficiency, for all hybrids and hybrids between unrelated lines (unr.). $R^2_p(Y;\tilde{Y}_{gca})$ and $R^2_p(Y;MRD^2)$ are the phenotypic coefficient of determination of F1 hybrid value (Y) by $\tilde{Y}_{gca}$ (model 3) and MRD^2 (model 8). *, ** and *** indicate the levels of significance (5 %, 1 % and 0.1 % respectively). $R^2_{par}(Y;MRD^2/\tilde{Y}_{gca}fixed)$ is the partial coefficient of determination of Y by MRD^2 once $\tilde{Y}_{gca}$ has been introduced in the model. $R^2_{mul}(Y;\tilde{Y}_{gca},MRD^2)$ is the multiple coefficient of determination of Y by $\tilde{Y}_{gca}$ *and* MRD^2. +++ indicate that the increase of R^2 due to MRD^2 introduction in the regression was significant at the 0.1 % level. ns indicates that a relationship was not significant at the 5 % level.

	Silage yield (51 lines)		Grain yield (46 lines)	
	all hybrids	unr. hybrids	all hybrids	unr. hybrids
number of hybrids	71	57	57	47
$R^2_p(Y;\tilde{Y}_{gca})$	.431***	.592***	.454***	.680***
$R^2_p(Y;MRD^2)$	.061*	ns	.097*	ns
$R^2_{par}(Y;MRD^2/\tilde{Y}_{gca}fixed)$	.085+++	ns	.117+++	ns
$R^2_{mul}(Y;\tilde{Y}_{gca},MRD^2)$	.516***	.595***	.571***	.683***

Introduction of MRD^2 as a predictive variable in addition to $\tilde{Y}_{gca}$ significantly increases the amount of yield variation ($R2$) explained by the model (8.5 % and 11.7 % for silage and grain yield, respectively).

However, efficiency of MRD^2 is no longer significant when only crosses between unrelated lines are involved in the analysis. This result is consistent with a study of Godshalk et al (1990) who found no significant relationship between MRD^2 distance and hybrid value for a set of hybrids between lines and unrelated testers.

Published studies and our results lead apparently to the following conclusions:

- For similar studies, distances computed from numerous RFLP data appear more efficient than distances involving a restricted number of isozyme loci.

- In diallels, correlations are higher for SCA effects than for F1 hybrid value.

- For a given technique, results appear very dependent on the material which is considered. Moreover, a relationship exists between prediction efficiency (using distances) and the presence of crosses between related lines in the studied hybrids.

Variation in prediction efficiency may be due to some extent to a variation in the magnitude of linkage disequilibria between markers and QTL (quantitative trait loci) exhibiting dominance effects. A theoretical computation (Charcosset et al., 1991) has illustrated the fact that in the case of random mating populations, relationship between heterozygosity at marker loci and heterosis could decrease very quickly unless markers are very close to the QTLs. Markers which do not mark (are not in linkage disequilibrum with) any QTL implied in heterosis variation lead to a decrease of prediction efficiency. Markers selection previous to distance computation should in this situation enhance prediction efficiency. Markers techniques that consider gene expression variation such as revealed by two-dimensional gel electrophoresis should also provide efficient information (Léonardi et al., 1991).

Results also depend on the genetic effects implied in traits variation. Melchinger et al (1990b), have illustrated the occurrence of additive * additive epistatic effects. SCA variation due to this type of effects cannot be achieved through distance computation. Moreover, according to our study, variation due to SCA effects appears to be extremely low compared to the variation explained by GCA effects when only crosses between unrelated lines are considered.

Conclusions

The choice of a method to predict heterosis or hybrid value depends on the type of information that can be collected on the potential parents of hybrids and the genetic basis of the variation of the trait(s) considered.

In the case of Maize yield, efficient methods can be proposed using testcross designs information. According to our results, predictors involving parents GCA estimates are very efficient. The largest errors in hybrid value estimations were observed in the case of overestimated hybrids (predicted value superior to the actual value). We theoretically demonstrated that the use of lines * testers specific effects through multiplicative models should improve the accuracy of prediction (without any additional measurements). This was experimentally supported by results from grain and silage yield data sets. However the efficiency of this approach will depend on the genetic structure of the tested germplasm and should be evaluated for several environments. Theoretical and experimental results both illustrate that it will be very efficient for hybrid combinations in which one line at least is very close to a given tester. In other cases some hybrids will still have overestimated values.

Some overestimated hybrids can be discarded based on pedigree information. However some crosses might not be discarded in case of incomplete pedigree information. In such a situation marker loci should provide a useful complementary information since it has been demonstrated in several studies that RFLP provide a mean to determine the origin of a given line (see for instance Melchinger et al., 1991), and discard crosses between lines that show important similarities.

Once this type of hybrids have been discarded, only deviations of relatively small magnitude between the actual value of a cross and it's predicted value using lines GCA estimates are expected. For instance Hallauer and Miranda (1981, p 345) have concluded that in the case of variety crosses, "only occasionally is the superiority of a given cross increased by significant specific effects".

It has been demonstrated that the use of RFLP data through distance computation should be of minor help in predicting these deviations. Several elements of explanations have been proposed to explain this result, they all lead to the conclusion that the improvement of prediction efficiency would require a better understanding of the genetic basis of such deviations. This knowledge should allow for instance the selection of markers introduced in distance computation, or the construction of alternative models (for instance in the case of epistatic effects).

ACKNOWLEDGEMENTS

I thank A. Gallais and M. Lefort-Buson for their helpful attention and encouragements throughout the work that is summarized here. I thank J.B. Denis for cooperating in the interaction models study. I thank C. Lariagon, H. Miteul, D. Pacault, F. Pfeiffer, S. Roger for their technical assistancy in collecting field data and J.P. Sampoux for his cooperation. I also thank all the persons who contributed to the marker analysis of the lines: M. Grenèche (from GEVES) for isozymic data, D. de Vienne (and his team), D. Brunel, P. Moreau, F. Hallard and especially E. Krecji and P. Dufour who both collected the RFLP data used in the present study. RFLP analysis was partly supported by French MRT and Promaïs association. I thank M. Causse for helpful corrections on the manuscript.

References

Charcosset A (1990) Etude de l'hétérosis chez le Maïs : prédiction de la valeur d'hybrides F1 sur la base d'informations agronomiques, biochimiques et moléculaires. I.N.A.P.G., Thèse de doctorat

Charcosset A, Denis JB, Lefort-Buson M (1990a) Performance prediction of Maize single cross hybrids. XVth International Biometric Conference, Budapest.

Charcosset A, Lefort-Buson M, Gallais A (1990b) Use of top-cross designs for predicting performance of maize single cross hybrids. Maydica 35:23-27

Charcosset A, Lefort-Buson M, Gallais A (1991) Relationship between heterosis and heterozygosity at marker loci: a theoretical computation. Theor Appl Genet 81:571-575

Church GM, Gilbert W (1984) Genomic sequencing. Proc Natl Acad Sci 81:1991-1995

Decoux G, Denis JB (1990) INTERA : logiciels pour l'interprétation statistique de l'interaction entre deux facteurs. INRA, Versailles 120p.

Denis JB (1991) Ajustements de modèles linéaires et bilinéaires sous contraintes linéaires avec données manquantes. in La Revue de Statistique Appliquée 39(2):5-24

Dudley JW, Saghai Maroof MA, Rufener GK (1991) Molecular markers and grouping of parents in Maize breeding programs. Crop Sci. 31:718-723

Falconer DS (1981) Introduction to quantitative genetics, 2nd edn. Longman, London

Frei OM, Stuber CW, Goodman MM (1986) Use of allozymes as genetic markers for predicting performance in maize single cross hybrids. Crop Sci 26:37-42

Godshalk EB, Lee M, Lamkey KR (1990) Relationship of restriction fragment length polymorphisms to single-cross hybrid performance of Maize. Theor Appl Genet 80:273-280

Gonella JA, Peterson PA (1978) Isozyme relatedness of inbred lines of Maize and performance of their hybrids. Maydica 23:55-61

Goodman MM (1972) Distance analysis in biology. Syst. Zool., 21:174-186

Green JM (1948) Relative value of two testers for estimating topcross performance in segregationg maize populations. J Am Soc Agron 40:45-57

Hadjinov MI, Scherbak VS, Benko NI, Gusen VP, Sukhorzhevskaya TP, Vorona LP (1982) Interrelationships between isozymic diversity and combinig ability in maize lines. Maydica 27:135-149

Hallauer AR (1990) Methods used in developing maize inbreds. Maydica 35:1-17

Hallauer AR, Miranda JB (1981) Quantitative genetics in maize breeding. Iowa State University Press, Ames 468 p.

Hanson HC, Casas E (1968). Spatial relationship among eight populations of Zea Mays L. utilizing information from a diallel mating design. Biometrics 24:867-880.

Heidrich-Sobrinho E, Cordeiro AR (1975) Codominant isoenzymic alleles as markers of genetic diversity correlated with heterosis in maize *(Zea mays)*. Theor Appl Genet 46:197-199

Hull HF (1945) Recurrent selection for specific combining ability in corn. J Am Soc Agron 37:134-145

Hunter RB, Kannenberg LW (1971) Isozyme characterisation of corn *(Zea mays)* inbreds and its relationship to single cross hybrid performance. Can J Genet Cytol 13:649-655

Lamkey KR, Hallauer AR, Kahler AL (1987) Allelic differences at enzyme loci and hybrid performance in maize. J Hered 78:231-234

Lee M, Godshalk EB, Lamkey KR, Woodman WW (1989) Association of restriction fragment length polymorphisms among maize inbred lines with agronomic performance of their crosses. Crop Sci 29:1067-1071

Lefort-Buson M, Dattée Y (1985) Etude de l'hétérosis chez le colza oléagineux d'hiver (*Brassica napus* L.). II. - Structure génétique d'une population de lignées. Agronomie 5:201-208

Lefort-Buson M, de Vienne D. (eds) (1985) Les distance génétiques, INRA, Paris

Léonardi A, Damerval C, Hébert Y, Gallais A, de Vienne D (1991) Association of protein amount polymophism (PAP) among maize lines with performances of their hybrids. Theor Appl Genet 82:552-560

Malecot G (1969) The mathematics of heredity. WH Freeman and Co., San Francisco

Mandel J (1971) A new analysis of variance model for non-additive data. Technometrics 13:1-18

Melchinger AE, Lee M, Lamkey KR, Hallauer AR, Woodman WL (1990a) Genetic diversity for restriction fragment length polymorphisms and heterosis for two diallel sets of maize inbreds. Theor Appl Genet 80:488-496

Melchinger AE, Lee M, Lamkey KR, Woodman WL (1990b) Genetic diversity for restriction fragment length polymorphisms: relation to estimated genetic effects in maize inbreds. Crop Sci 30:1033-1040

Melchinger AE, Messmer MM, Lee M, Woodman WL, Lamkey KR (1991) Diversity and relationships among U.S. Maize inbreds revealed by restriction fragment length polymorphisms. Crop Sci. 31:669-678 (1991)

Miller SA, Dykes DD, Polesky HF (1988) A simple salting out procedure for extracting DNA from human nucleated cells. Nucleic Acids Research 16:12-15

Oury FX, Brabant P, Pluchard P, Bérard P, Rousset M (1990a) Etude multilocale de blés hybrides : niveaux d'hétérosis et élaboration du rendement. Agronomie 10:735-748

Oury FX, Koenig J, Bérard P, Rousset M (1990b) Une comparaison entre des blés hybrides produits par voie chimique et leurs parents : niveaux d'hétérosis et élaboration du rendement. Agronomie 10:291-304

Peng JY, Glaszmann JC, Virmani SS (1988) Heterosis and isozyme divergence in indica rice. Crop Sci. 28:561-563

Price SC, Kahler AL, Hallauer AR, Charmley P, Giegel DA (1986) Relationship between performance and multilocus heterozygosity at enzyme loci in single-cross hybrids of maize. J Hered 77:341-344

Rogers JS (1972) Measures of similarity and genetic distance. Studies in Genetics 7. Univ. Tex. Publ. 7213:145-153

Shull GH (1909) A pure line method of corn breeding 5:51-59

Smith OS (1986) Covariance between line per se and testcross performance. Crop Sci. 26:540-543.

Smith JSC, Smith OS (1989) The use of morphological, biochemical, and genetic characteristics to measure distance and to test for minimum distance between inbred lines of maize *(Zea mays L.)*. UPOV Workshop Versailles, France, October 1989

Sprague GF, Tatum LA (1942) General vs. specific combining ability in single crosses of corn. J. Am. Soc. Agron. 34:923-932

Strauss SH (1986) Heterosis at allozyme loci under inbreeding and crossbreeding in *Pinus attenuata*. Genetics 113:115-134

The Potential of Inbred Lines in the Presence of Heterosis

M.J. Kearsey and H.S. Pooni
Plant Genetics Group
School of Biological Sciences
The University of Birmingham
Edgbaston
Birmingham B15 2TT, U.K.

1. Introduction

In most crop species it is possible to choose either F_1 hybrids or inbred cultivars as alternative objectives of a breeding programme. To a very large extent, the decision has been made in the past either on the reproductive biology of the crop, i.e. whether it is naturally an outbreeder or an inbreeder, or on commercial considerations, i.e. the word 'hybrid' is important in marketing. Genetical factors have generally been subservient to these and it is widely believed that hybrids are inately higher yielding and more stable than inbreds.

Over the past 15 to 20 years, however, a great deal of reliable and consistent genetical information relevant to such decisions has become available from a wide variety of sources. It is the purpose of this paper to summarise these data and to discuss their implications in the light of current theory and practice. In particular, we wish to explore the potential value of inbred cultivars which might be obtained from heterotic crosses.

2. Does the genetical nature of heterosis imply inbred inferiority?

Heterosis is most usefully defined as the superiority of an F_1 hybrid over its better (inbred) parent, P_1. Some authors refer to 'mid-parent heterosis' but this is of no commercial importance and so will not be considered further. For simplicity we will mainly consider positive heterosis, though arguments for negative heterosis will follow identical lines.

In model terms, following Mather & Jinks (1982):

$$\text{Heterosis} = F_1 - P_1$$

where $F_1 = m + [h] + [l]$

$P_1 = m + [d] + [i]$

$$\therefore \text{Heterosis} = \{[h] + [l]\} - \{[d] + [i]\}$$

where $[d] = r_d \Sigma d$

$[h] = \Sigma h$

$[i] = r_i \Sigma i$

$[l] = \Sigma l$

Thus heterosis occurs if [h] + [l] is greater than [d] + [i] which implies that the heterozygous (dominance) deviations from the mean together with their interactions exceed the combined homozygous additive effects of the best parent of the cross together with their associated additive by additive interactions. Both of the latter ([d] & [i]) are reduced relative to [h] and [l] by gene dispersion (r) in the parents i.e. when the two parents carry a mixture of increasing and decreasing alleles, which is typical of parents in a breeding programme. The most extreme superior inbred line (I_s) that could be obtained from this F_1 by inbreeding and selection would have a mean:-

$$I_s = m + \Sigma d + \Sigma i$$

Its potential relative to the F_1 would thus depend on the relative size of Σh to Σd and of Σl to Σi.

Whether or not there is epistasis, the relative sizes of [h] and [d] tell us little about the underlying genetical cause of heterosis since, even with unidirectional dominance as the formulae above show, [d] could be less than [h] either due to gene dispersion (r_d <1.0) or overdominance (heterozygote advantage). From classical, major gene genetics we would expect the heterozygote to look like one of the two homozygotes or be intermediate between them but not to be more extreme than either homozygote. Thus we might expect $-d_i < h_i < d_i$.

What is the available evidence on the degree of dominance and the nature of epistasis for quantitative traits?.

Dominance

Information about true dominance requires avoiding using parameters affected by gene dispersion and this implies looking at second degree statistics, variances and covariances, of generations derived from an F_1 (F_2, BX's, F_3 etc.) because, in the absence of linkage, the genes from an associated or dispersed cross will yield the same F_2. Thus we can use estimates of $\Sigma d^2(=D)$ and $\Sigma h^2(=H)$ from appropriate experimental procedures applied to the F_2 of a heterotic cross to explore the true average dominance ratio

$$f = \sqrt{\frac{H}{D}} = \sqrt{\frac{\Sigma h^2}{\Sigma d2}} = \frac{h}{d}$$

Most experimental breeding designs such as F_3 or full sib families give greater weight to D than H and hence might be considered unreliable for the estimation of the dominance ratio. However the North Carolina Design III (NCIII) (Comstock & Robinson 1952) and Triple Test Cross (Kearsey & Jinks, 1968) overcome this problem and estimate both with equal precision and so we will confine our attention to data from this experiment applied to F_2 populations.

Table 1 gives examples of dominance ratios for a range of characters in different species including both obligate inbreeders and outbreeders. In selecting data for this table, we have simply identified crosses which show heterosis for commercial traits and for which NCIII or Triple Test Cross data are available. This has invariably meant going back to original data in PhD theses etc as full information is seldom published.

With few exceptions, we can see that the dominance ratio is less than unity, indicating partial, incomplete dominance and that the size of the dominance ratio bears no obvious relationship either with the amount of heterosis or with the inbreeding, outbreeding nature of the species. We can develop these conclusions further by examining the information in Table 2 which summarises the available data on dominance ratios from

Table 2. Distribution of dominance ratios taken from triple test crosses in a wide range of traits in the species in Table 1.

Dominance ratio ($\sqrt{\frac{H}{D}}$	Number of cases	%	
0-0.2	7	10	
0.21-0.4	12	18	
0.41-0.6	18	27	84
0.61-0.8	9	13	
0.81-1.0	10	16	
1.01-1.2	7	10	16
1.21-1.4	4	6	
	67	100	

Source: Devey *et al* (1989), Kearsey *et al* (1987), Smith *et al* (1985), Thomas & Tapsell (1983), Werner *et al* (1989, 1990)

some 67 Triple Test Crosses in a wide variety of traits in the same range of species as Table 1. The great majority have dominance ratios less than one, with a modal value around 0.5 i.e. the heterozygote at a given locus has a phenotype halfway between the average of the two homozygotes and that of the best homozygote. The remaining 16% just exceed unity but in no case significantly so. Partial dominance is clearly the predominant pattern.

Of all the cases illustrated in Table 2, all but one of the few cases of apparent overdominance ($\sqrt{H/D}>1$) occurred in inbreeding species. However, these were mainly restricted to seed production traits which are intimately related to fitness and are thus expected to show strong directional dominance.

Could the absence of overdominance be due to biases in our estimation procedures?

In fact, if there is any bias, it is likely to be towards exaggerating the degree of dominance since the two most likely factors, epistasis and linkage of genes in dispersion in the F2, both inflate H and deflate D so overestimating dominance (Jinks & Pooni 1982, Kearsey 1985). Only coupling linkages could cause such a bias.

It is thus not unreasonable to conclude that dominance is generally less than complete and that evidence for overdominance is insubstantial if not absent i.e. $\Sigma h < \Sigma d$.

Epistasis

Epistasis or non allelic interaction has long been known to be closely associated with heterosis (Jinks & Jones 1958). If, as we argued earlier, Σh is not greater than Σd, could it be that it is $\Sigma l > \Sigma i$ which is a major cause of heterosis? Certainly on a priori grounds it would seem reasonable to expect complementation to be a likely cause of hybrid superiority, the genes from both parents 'nicking' or complementing in the F_1. Thus, in classical terms, we would be looking for complementary epistasis.

What do we find in practice?

Given a certain number of basic generations derived from a cross between two inbred lines (e.g. P_1, P_2, F_1, F_2, BX's, F_3 etc.) it is possible to estimate m, [d], [h], [i] and [l] and to test their significance and ability to explain the data. Table 3 illustrates a typical example taken from a cross in Barley.

Table 3. Model fitting for grain number/plant in barley cross Clipper x Ymer. (Tapsell & Thomas 1983; Tapsell 1984)

Generation	Mean	Model
P_1	23.70	m + [d] + [i]
P_2	12.85	m - [d] + [i]
F_1	18.80	m + [h] + [l]
F_2	19.80	m + 1/2[h] + 1/4[l]
B_1	22.30	m + 1/2[d] + 1/2[h] + 1/4[i] + 1/4[j] + 1/4[l]
B_2	17.30	m - 1/2[d] + 1/2[h] + 1/4[i] - 1/4[j] + 1/4[l]

Estimates

m	18.1***
[d]	5.2***
[h]	6.2***
[l]	-5.5***

P=0.86 i.e. model fits

We see from this Table that the dominance interaction term [l] is in fact opposite in sign to [h], i.e. it is reducing rather than reinforcing the effects of heterozygosity. This is not a unique example and the results of 18 other cases chosen at random from the literature are summarised in Table 4. In every case, irrespective of the sign of [h], [l] takes the opposite sign. Indeed we have yet to find a reported case from any species in which [h] and [l] take the same sign.

Table 4. Relative direction of sign of [h] and [l] from 18 randomly chosen data sets

		[h] +	[h] -
[l]	+	0	7
[l]	-	11	0

$\chi^2_1 = 18;\ P < .001$

This observation is very significant since complementary epistasis requires [h] and [l] to take the same sign. What we find is always the reverse, i.e. duplicate epistasis.

A further point concerns the relative sizes of Σl and Σi. It is normally difficult to obtain unbiased estimates of Σi because they are seriously reduced by gene dispersion in the parents. There are however a number of documented cases where divergent recombinant inbred lines have been derived by SSD or anther culture from a dispersed heterotic cross. Such lines should be, if not in complete association, certainly approximating to it, for the same genes which were dispersed in the original parents. They can thus be used to provide information on Σd and, more particularly, Σi. Table 5 presents data from two such crosses for a variety of quantitative traits but these results are typical of all cases. Again [h] and [l] have opposite signs but, more interestingly, so do [i] and [l] i.e. whereas [l] reduces heterosis, [i] increases and improves inbred performance.

Table 5. Components of generation means associated crosses in *Nicotiana rustica* (Pooni & Jinks 1981)

Parameter	Final Height	Flowering time	Height -3
m	98.73	71.11	29.51
[d]	42.22	11.61	32.22
[h]	70.44	32.32	48.44
[i]	19.38	14.40	20.32
[j]	21.13	7.02	9.53
[l]	-30.79	-23.84	-15.26
gene association (r)	0.63	0.55	0.52

Thus, the weight of available data from a wide range of species and traits supports the arguments that heterosis is caused by dispersed genes showing directional dominance and not to heterozygote superiority or to complementary epistasis. Indeed even with complete unidirectional dominance, the F_1 yield could be worse than the best inbred because [l] is reducing while [i] is reinforcing the gene effects in the F_1 and inbred respectively.

It therefore follows that inbred cultivars should potentially be able to outyield heterotic F_1's and that this should be true both for inbreeding and outbreeding species.

2. Probability of producing inbreds that outyield heterotic hybrids

If one takes a heterotic F_1 cross and inbreeds by single seed descent or produces double haploids a population of F_∞ inbred lines will be produced, the means of which should be normally distributed with overall mean 'm' and variance D i.e. N(m,D).

Using information from the early generations plus an F_3 or TTC it is possible to estimate both m and D for the cross and hence predict the distributional properties of the recombinant inbred lines that could be derived from it. Using this information, one can use the basic properties of the normal distribution to predict the proportion of inbred lines exceeding some specific target yield e.g. that of the heterotic F1, as shown in Table 6. This prediction is as reliable as the estimates of m, D and the target mean allow, but does indicate to the breeder the likelihood of success in a given case as well as the scale of the inbreeding and testing operation required. Some typical predictions quoted in the literature are shown in Table 7.

Table 6. An example to illustrate the prediction of the distributional properties of recombinant inbred lines

Generation	Mean	Parameter estimates
P_1	20	m = 23
P_2	26	D = 9
F_1	30	

want to know what proportion of SSD lines will outyield F_1 mean.

The SSD lines will have a mean m=23 and variance D=9.

∴ proportion of inbreds exceeding F1 is obtainable from

P of Z, where $Z = \dfrac{F1\text{-}m}{\sigma} = \dfrac{30\text{-}23}{\sqrt{9}}$

P=.01

The variability in these probabilities indicates that, though it may be possible to produce inbreds which outyield heterotic

F_1's, they may be difficult to obtain in some cases. The most likely reason for this is that in these cases, the heterosis arises from dominance at very many genes and it is difficult to accumulate all the 'good' alleles together in one parent. Several cycles of inbreeding, selection and repeated crossing may be required to obtain such inbreds. The potential exists but the rate of achieving it may be slow. However, in many cases predicted yields of good inbreds are high and, if accurate, easily achievable.

The best possible inbred (I_S) obtainable from a cross in the absence of epistasis, is:-

$$I_S = m + \Sigma d$$

Unfortunately, although we can estimate m, we do not know Σd, but only $[d] = r_d \Sigma d$. If we assume, however, that the dominance ratio is constant at all loci (i.e. $f = \frac{hi}{di}$ is a constant) then $\Sigma d = \Sigma h/f$. Both these parameters are estimable, since $\Sigma h = [h]$ and $f = \sqrt{H/D}$

$$\text{i.e. } I_S = m + [h]/\sqrt{H/D}$$

Unfortunately, estimates of $\sqrt{H/D}$ have large standard errors in practice even though m and [h] can be accurately established, so such predictions are often unreliable.

(3) <u>How do the predictions compare with trial values</u>?

The prediction methods described above have been extensively used and tested in recent years. Generally such tests have involved growing the recombinant inbred lines in the same trial as the predictor generations in order to avoid problems of g x e, but others have involved predictions and tests in different seasons. Table 7 compares the observed frequency of superior recombinant inbred lines with predictions for a range of traits and species.

As might be expected, agreement is better when predictions and tests are in the same season, but generally the predicted scale of success is confirmed by observation. If there is a consistent

Table 7. Predicted and observed frequencies of superior recombinant inbred lines

	Predicted %	Observed %
Onions		
Yield	0.11	0.33
Quality	5.53	7.25
Maturity	2.96	2.21
Brussel Sprouts		
Market weight	0.02	0.26
Quality	<0.01	0.08
Sprout number	0.23	1.90
Swedes		
Fresh weight	4.00	-
Dry weight	4.00	-
Dry matter %	78.00	-
Hardness	11.70	-
Barley	21	29
EE	8	16
NT	21	37
Rice		
NUP	16.4	
NPP	2.5	
DH	56.8	
TGW	0.7	

difference, it is that observed gains tend to exceed predictions with SSD. This is simply due to further recombination between linked genes which had not occurred in the generations used for predictions; i.e. predictions will tend to be conservative.

The major problem of accurate predictions is that of having to use necessarily small scale early generation trials in one season to predict yield in large scale plot trials in a different season. In the present context, however, it is sufficient to indicate that predictions that inbreds can outyield hybrids do work in practice; experiment has shown that such superior inbreds are obtained.

The fact that inbred lines can be produced which exceed the heterotic hybrid does not in itself imply that one cannot produce an even better hybrid from the same cross. Indeed if one selected the top two recombinant inbred lines, both of which exceeded the original F1, we would expect their F1 to be higher yielding still as long as the genes display only partial dominance.

For example if the initial parents differed at 10 loci for which $f = \frac{h}{d} = 0.5$ then the F_1 would have a score of m + (10 x .5) = m + 5. If we chose 2 superior inbreds which share 6 '+' alleles but still differed at 4, their mean scores would be m + 6 while the F1 obtained by intercrossing them would be m + 6 + (4 x .5) = m + 8. It is possible to predict the distribution of second cycle hybrids which can be obtained from the SSD derived recombinant inbreds in the same way as the distribution of the inbreds were predictable (Toledo *et al.* 1984). They will be approximately normally distributed with N(m + 1/2[h]; 1/2D + 1/4H).

4. Conclusion

Available data on the causes of heterosis for a wide variety of trials in outbreeding and inbreeding crop plants strongly opposes the idea of heterozygote superiority per se. Instead the original view that dispersed, partially dominant alleles are responsible is confirmed and that the effects of dispersion is enhanced by linkage and epistasis. Epistasis is of a duplicate and not a complementary type and as such is opposing instead of reinforcing dominance.

These two factors imply that it should always be possible to produce inbred lines which are as good as or better than F1 hybrids whether the species is normally inbreeding or outbreeding.

The ease of obtaining such lines is predictable and these predictions are confirmed by trials on derived lines. In many cases, however, the expected frequency of such lines is small following a single round of inbreeding and may be insufficient to warrant discarding F_1 hybrids in such cases.

Final decisions on short or medium term commercial objectives have to be based on the balance of costs of inbred v. hybrid production but genetical evidence predicts that inbreds could be produced which will outyield current hybrids no matter what the species or the traits to be improved.

5. References

Comstock RE, Robinson HF (1952) Estimation of average dominance of genes. In: Gowen JW (ed) Heterosis. Iowa State College Press, Ames, Iowa: 494-516.

Coombs DT (1980) Biometrical genetics of tobacco. PhD thesis University of Birmingham, U.K.

Devey FD, Hayward MD, Kearsey MJ, McAdam N, Eggleston MP (1989) Genetic analysis of productive characters in *Lolium* II Triple Test Cross analysis of drill and plant performance. Plant Breeding 103: 63-72.

Gardner CO (1963) Estimates of genetic parameteres in cross-fertilising plants and their implications in plant breeding. In: Hanson WD, Robinson HF (eds). Statistical Genetics and Plant Breeding. NAS-NRC, Washington DC Publication 982: 225-252.

Jinks JL, Jones RM (1958) Estimating the components of heterosis. Genetics 43: 223-234.

Jinks JL, Pooni HS (1982) Predicting the properties of lines extractable from a cross in the presence of linkage. Heredity 49: 223-234.

Kearsey MJ (1985) The effect of linkage on the additive genetic variance with inbreeding an F_2. Heredity 55: 139-143.

Kearsey MJ, Jinks JL (1968) A general method of detecting additive dominance and epistatic variation for metrical traits: I. Theory. Heredity 23: 403-409.

Kearsey MJ, Hayward MD, Devey FD, Arcioni S, Eggleston MP, Eissa MM (1987) Genetic analysis of production characters in *Lolium*. 1. Triple Test Cross analysis of spaced plant performance. Theor. Appl. Genet 75:66-75.

Mather K, Jinks JL (1982) Biometrical Genetics, 3rd Edn., Chapman and Hall, London.

Perera ALT (1985) Biometrical genetic studies of rice (*Oryza sativa*). PhD thesis, University of Birmingham, UK.

Pooni HS, Jinks JL (1981) The true nature of non-allelic interactions in *Nicotiana rustica* revealed by association crosses. Heredity 47: 253-258.

Ramsay LD (1991) Application of Biometrical Genetics to swede breeding. PhD Thesis, University of Birmingham, UK.

Setter AP (1987) Predicting the results of breeding programmes in bulb onions (Allium cepa) and Brussels sprout (*Brassica oleracea* var gemmifera) PhD thesis, University of Birmingham.

Smith BM, Rogers WJ, Kearsey MJ (1985). The feasibility of producing inbred rather than F_1 hybrid cultivars in Brussels sprouts: a preliminary genetical analysis and choice of material for inbreeding. Ann. Appl. Biol. 107:87-90.

Tapsell CR (1984) Cross prediction studies on Spring barley. PhD thesis, University of Birmingham, UK.

Tapsell CR, Thomas WTB (1983) Cross prediction studies in spring barley. Estimation of Genetical & Environmental control of Morphological and maturity characters. Theor. Appl. Genet. 64:345-352.

Thomas WTB, Tapsell CR (1983) Cross prediction studies in Spring barley. 2. Estimation of genetical and environmental control of yield and its component characters. Theor. Appl. Genet. 64: 353-358.

Toledo JFF de, Pooni HS, Jinks JL (1984) Predicting the properties of second cycle hybrids produced by intercrossing random samples of recombinant inbred lines. Heredity 53: 283-292.

Werner CP, Setter AP, Smith BM, Kubba J, Kearsey MJ (1989)

Performance of recombinant inbred lines in Brussel sprouts (*Brasica oleacea* var. gemmifera) Theor Appl Genet 77:527-5334.
Werner CP, Kearsey MJ, Crowther TC, Dowker BD (1990) Prediction of performance of inbred lines derived from a population cross in autumn sown onions (Allium cepa L.) Theor Appl Genet 79: 507-512.

The Potential of Hybrid Varieties in Self-Pollinating Vegetables

James Nienhuis and Gavin Sills
Dept of Horticulture
University of Wisconsin
1575 Linden Drive
Madison, WI 53706

The next generation of successful cultivars in many self-pollinated crops will be F_1 hybrids developed through a more systematic exploitation of the phenomenon of heterosis. Longer term, marketing and varietal protection issues aside, the most successful cultivars in terms of superior genotypes in self-pollinated crops may be inbred lines. Presently, commercial exploitation of hybrid vigor is limited to a relatively small number of vegetables despite evidence that heterosis above high-parent exists within many species (Table 1). Commercial development is limited to those crops in which the added value of heterosis is sufficient to justify the cost of hybrid seed production and development. The formula for successful application of heterosis is: added value > added cost of seed production. Much research has focused on the second half of the equation, *i.e.* the development of efficient pollination mechanisms for hybrid seed production (George, 1985). The objective of this paper is to focus an examination of the first half of the equation, *i.e.* maximizing the value added, by examining the genetic information relevant in the development of F_1 hybrid cultivars in self-pollinated species.

Earlier this century vegetable breeders quickly realized from the success of hybrid maize that F_1 hybrid vegetables would have a tremendous impact upon the seed industry. Nevertheless, most F_1 hybrid vegetables do not exploit heterosis *per se*, but rather combine desirable characteristics possessed by their parents. The general varietal attributes observed in F_1 hybrids include: 1) maximum performance under optimal growing conditions, 2) stability in stress environments, and 3) joint improvement of traits which are negatively associated (Mayo, 1987). In vegetables, the usual attributes of F_1 hybrids include: uniformity, earliness, combination of desirable dominant disease resistances, proprietary control of parents and reduced time for cultivar release. In many self-pollinated crops F_1 hybrids represent the quickest way available for a plant breeder to accumulate the maximum number of favorable dominant genes in one genotype (Gallais, 1989).

Table 1. Estimated percentage of selected self-pollinated vegetable crops planted to hybrid seed in regions of the United States.

Crop	Region	Percent hybrid
Tomato		
processing	California	50%
fresh market	USA	100%
Pepper		
sweet	California	15%
sweet	Florida	>85%
Eggplant	USA	>70%
Beans	USA	0%
Peas	USA	0%
Lettuce	USA	0%

A great deal of research has been conducted to investigate the potential of heterosis in crops. Among self-pollinated vegetable species significant heterosis above high parent can be found, but the range of values indicates that results are generally more erratic than in allogamous species (Table 2). The erratic expression of heterosis may be due in part because the superior crosses are identified by trial and error, rather than systematic procedures.

Table 2. Examples of heterosis in self pollinated vegetables.

	High parent heterosis (%)		
Crop/trait	Average	Range	Citation
Tomato			
days to fruit	-13	- 45 — 10	Peter and Rai, 1978
fruits/plant	31	- 55 — 174	Peter and Rai, 1978
fruit yield	41	- 59 — 168	Peter and Rai, 1978
fruit yield	34	40 — 24	Tikoo, 1987
soluble solids	-10	- 42 — 53	Peter and Rai, 1978
soluble solids	- 3	- 10 — 6	Khalil et al., 1987
Eggplant			
total yield	80	- 29 — 242	Thakur et al., 1968
yield/plant	30	- 79 — 358	Peter and Singh, 1974
Peppers			
days to flower	-11	- 48 — 9	Meshram and Mukewar,1986
fruits/plant	39	- 18 — 75	Meshram and Mukewar,1986
fruits/plant	13	- 16 — 64	Gill et al.,1973
fruit yield	53	- 22 — 158	Meshram and Mukewar,1986
fruit yield	15	- 16 — 52	Gill et al.,1973
Beans			
seed yield	29	- 38 — 146	Nienhuis and Singh, 1986
pods/plant*	-7	- 31 — 14	Dickson, 1967
Peas			
seed yield/plant	28	116 — 218	Gritton, 1975
seed yield/plant	16	1 — 41	Kumar and Das, 1975
pods/plant	-1	- 17 — 11	Kumar and Das, 1975
Lettuce			
yield	6	- 6 — 119	Langton et al., 1990

* Heterosis values based on performance of F_2 plants.

Self-pollinated agronomic crops, such as oilseed rape (Brandle and McVetty, 1989), wheat (Brears and Bingham, 1989), and rice (Yuan and Virmani, 1988), can also display significant heterosis above high parent. As yield potential of wheat parents increased, the already low hybrid advantage declined. The authors felt the commercial potential of hybrid wheat was limited. Hybrid rice in China, however, yields about 20% more than the best commercial varieties. From 8.43 million ha was harvested 26.4% of China's total rice harvest in 1985 (Yuan and Virmani, 1988). Hybrid seed production is facilitated by a cytoplasmic-genetic three-line breeding system.

Despite the presence of heterosis, F_1 hybrids are exploited in only a small number of self-pollinating species (Table 1) because of the prohibitive costs in terms of increased labor, materials and land resources required to produce F_1 seed. The relative magnitude of these cost factors is largely dependent upon the nature of the pollination mechanism and the various systems employed in F_1 seed production (Table 3).

Table 3. Strategies for producing F_1 seed.

1. genetic male sterility
2. cytoplasmic-genetic male sterility
3. monoecy
4. dioecy
5. gynoecy
6. self-incompatibility
7. manual emasculation
8. chemical alteration, *e.g.*, gametocides and plant growth regulators

Theories of Heterosis and Lines vs. Hybrids

Several genetic hypotheses have been proposed to explain heterosis, including , dominance, overdominance and epistasis (Moll and Stuber, 1974). Although there is some empirical evidence to support all three theories (Hallauer and Miranda, 1981), dominance is now widely considered the primary cause of heterosis (Houle, 1989).

The dominance hypothesis asserts that heterosis is due to decreased homozygosity for unfavorable recessive alleles in the hybrid as compared to its inbred parents (Davenport, 1908). Theoretically, plants homozygous for all favorable genes could be developed that would perform as well as genotypes heterozygous for these genes. Nevertheless, because of the large numbers of genes, numerous traits of interest, and linkage, the probability of accumulating all favorable genes in a single homozygous genotype is near zero. With dominance as the predominant mechanism of heterosis, the focus of the breeder should be inbred line development, using selection procedures that maximize opportunities for recombination.

The overdominance hypothesis (East, 1907; Shull, 1908) states that heterozygotes at single loci have greater values for desirable traits than either of the homozygotes for those alleles. Unequivocal examples of heterosis tied to a particular locus are very rare. Functional overdominance may be caused by the actual superiority of a heterozygote in a given environment, or it may be due to marginal overdominance (Wallace, 1968), where heterozygotes are superior only when averaged over environmental conditions. Thus, even if heterozygotes never exceed the range of corresponding homozygotes, functional overdominance may be detected. If an overdominant locus is tightly linked to a scored locus, the scored locus may be incorrectly considered overdominant. This associative overdominance is due to genetic disequilibrium between the two loci (Thompson, 1977). Apparent overdominance may be caused by dominance correlation heterosis (Houle, 1989) in which genotypic correlations occur between scored loci and loci segregating for deleterious recessive alleles. This is actually dominance which appears as overdominance due to gametic-phase disequilibrium.

In our laboratories, heterozygous RFLP genotypes with performance superior to that of either parent have been observed for individual loci associated with water use efficiency (Martin et al., 1989) and insect resistance (Nienhuis et al., 1987) in tomato (Table 4).

Overdominant effects of associated RFLP markers suggest that individual RFLP loci which contribute to heterosis may be identified, and thereby be selectable by the plant breeder. The heterotic effects listed in Table 4, as well as others, need to be further investigated as the effects may be due to statistical sampling or linkages (*i.e.* associative or dominance-correlation heterosis).

The genetic causes and importance of heterosis affect not only the type of cultivar, inbred or hybrid, but also choice of the optimal breeding methodology. In self-pollinated crops, the quickest method of accumulating favorable genes in a single genotype is through an F_1 hybrid. However, in self-pollinated crops parents are often developed using some form of pedigree selection based on performance *per se*, rather than selected for specific combining ability. Hence the erratic expression of heterosis in many self-pollinated crops may be due to the random contributions of overdominant-like effects. In the near term, breeders of self-pollinated crops will begin to better exploit breeding methodologies, such as recurrent selection for specific combining ability, which systematically select for heterosis. With any significant contribution of overdominance to heterosis, the best cultivar will always be a hybrid (Gallais, 1989), and selection procedures should focus on specific combining ability. Longer term, if functional overdominance is minor, selection methods which maximize opportunities for recombination will break repulsion phase linkages, and replace apparent overdominance with dominant loci.

Table 4. Analysis of variance and genotypic means for RFLP loci associated with apparent overdominant effects for water use efficiency and 2-tridecanone, respectively.

Source	Water Use Efficiency ‡ d.f.		loci: C56	2-tridecanone § d.f.		loci: Q90
			mean square			
Genotype	2			2		
Linear		1	125,601		1	4.25**
Quadratic		1	218,063*		1	3.19**
Error	92		191,162	16		0.40
Genotype			means			
esculentum homozygote			114			-25.61
heterozygote			525			-25.51
wild species homozygote			474			-27.11
Average effect of an allelic substitution†			60 ± 76			0.66 ± 0.25

‡ Nienhuis et al., 1986. § Martin et al., 1989.
*, ** Significant at the 0.05 and 0.01 probability levels, respectively
† Linear regression coefficient and standard error, respectively.

Thus, important plant breeding decisions and allocations of enormous amounts of plant breeding resources hinge on our understanding of the genetic causes of heterosis.

Relationship between genetic distance and heterosis

The ability to predict expression of heterosis, or specific combining ability, between a pair of genotypes has been as elusive goal of plant breeders. Studies by Moll et al. (1965) and others showed positive relationships between genetic distance, as measured by geographical distance, and yield and heterosis in maize. The studies showed that as genetic distance increased so did expression of heterosis; however, geographic origin may be an unreliable estimator of genetic distance. Restriction fragment length polymorphisms (RFLPs) are abundant in many self and cross-pollinated crops, and have been used to estimate genetic distance (Smith et al., 1990). In maize, the coefficients of determination (r^2) from regressing the coefficient of parentage, grain yield, and grain yield heterosis on Nei's (Nei and Li, 1979) measure of genetic distance based on RFLP data were 0.81, 0.87 and 0.77, respectively (Smith et al. 1990). In contrast, in a similar study Godshalk et al. (1990) found no relationship between RFLP-based genetic distance and hybrid performance. We conducted a similar experiment in a self-pollinated crop, dry beans, to measure the relationship between genetic distance, based on RAPD molecular markers, and heterosis. RAPDs are molecular markers observed by PCR amplification of arbitrary primers (Williams et al., 1990). In a previous study of a 9 x 9 diallel in dry beans Nienhuis and Singh (1986) observed heterosis for grain yield above the high-parent ranging from near zero to over 100 percent. When the values for heterosis were regressed on Nei's measure of genetic distance based on RAPD data the coefficients of determination (r^2) were 0.30 and 0.18 for mid- and high-parent heterosis, respectively (Figure 1).

Fig. 1. Plot of relationship between heterosis for yield in a 9 x 9 diallel of dry beans (*Phaseolus vulgaris*) and genetic distance based on RAPD data.

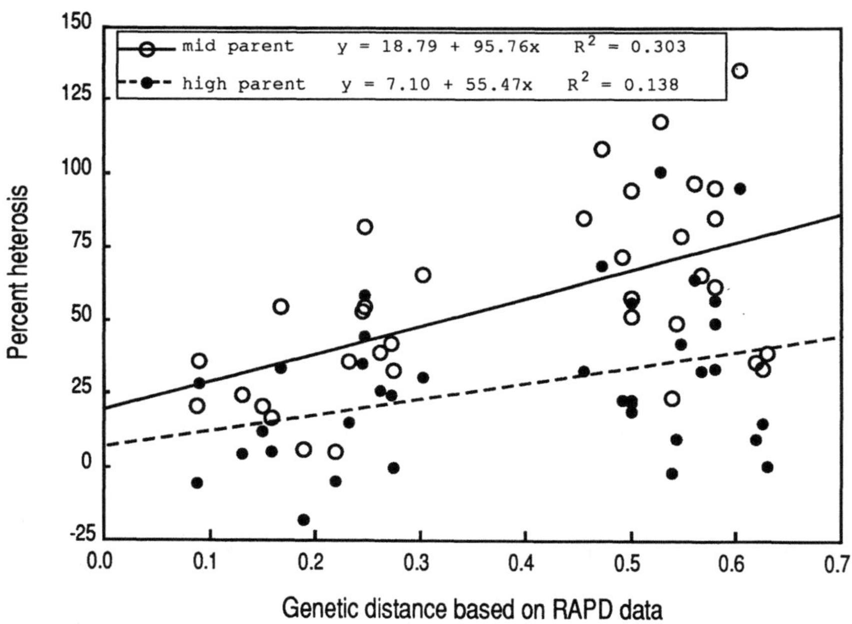

The relationship between genetic distance and heterosis in a self-pollinated crop, dry beans, is less than that observed for hybrid maize; nevertheless, the results from both crops suggest that genetic distance can predict hybrid performance. Moreover, as Smith et al. (1990) point out, prediction of hybrid performance aside, molecular markers can be used to identify individuals within heterotic pools from which superior breeding populations can be developed.

Breeding for Heterosis: Recurrent Selection

Recurrent selection is a breeding methodology designed to increase the frequency of favorable alleles. In all genotypic recurrent selection schemes the families tested are either inbred (S1, S2, etc.) or hybrid (half-sib, full-sib, or test cross). This fundamental difference in the types of families tested may have genetic consequences which will impact breeding for maximum heterosis if any form of overdominant gene action is present.

Ultimately, heterozygote performance determines the success of a breeding program designed to produce F_1 hybrids. Typically, in such a breeding program, selection is initially practiced for inbred performance, *per se*. Superior inbreds are then tested for their performance in hybrid crosses. Inbred line performance, however, is often only weakly correlated (*e.g.*, $r=0.09$; $p>0.05$) with performance of inbreds in hybrid combination (Gama and Hallauer, 1977). A critical question for plant breeders is whether selection for inbred family performance versus hybrid family performance influences which alleles change frequency in response to selection. Research in maize suggests that it does.

Recurrent selection for specific combining ability (RS-SCA) was proposed as a method of breeding for overdominance in test hybrids (Hull, 1945). This method of population improvement uses a narrow genetic base tester in the development of test hybrids for evaluation. Several studies in maize have concluded that RS-SCA improved general combining ability along with specific combining ability, and that overdominance was, therefore, relatively less important than partial or complete dominance (*e.g.*, Walejko and Russel, 1977). Others, however, have shown that population performance, *per se*, improved little or none over cycles of selection, while test hybrid performance did improve (*e.g.*, Eberhart et al., 1973). This could result from selection for certain alleles favorable only in hybrid combination with the tester.

The effects of selection for inbred versus hybrid performance on "allele" frequencies can be investigated via RFLP marker alleles. Directional changes in marker frequencies over cycles of selection, barring mutation and migration, will be due to either selection or drift. A breeding program in which RS-SCA and S1 family recurrent selection (RS-S1) could be initiated from the same cycle zero population. Cycle zero could be characterized for all RFLP marker alleles at each of many loci, and these marker alleles could be monitored over cycles of selection. A linear trend in marker frequencies can be estimated by use of weighted least squares methods (Schaffer et al., 1977) and tested for significant deviations form the null hypothesis of drift acting alone. Ultimately, the loci and alleles favored by the contrasting selection methods could be determined. If any loci change allele frequencies in opposite directions using the contrasting selection methods, then one can hypothesize that early generation testing of hybrid performance is necessary because different alleles contribute to inbred and hybrid performance. This suggests that selection for inbred performance *per se* may preclude opportunities to exploit heterosis because alleles contributing to heterosis may be lost as those best suited to inbred performance are selected instead.

A direct, systematic approach to breeding for heterosis, such as RS-SCA, above, would likely be a change of objectives for some vegetable breeders. Instead of simply viewing F_1 hybrids as a rapid, profitable way of combining in a uniform progeny the desirable characteristics possessed by two parents, F_1 hybrids would truly be valued for their heterotic performance. In allogamous species, where inbreeding depression can make an inbred line useless for commercial deployment, there is a great risk in selecting only for hybrid performance: inadequate seed yields of the inbred parent of the F_1 hybrid. In self-pollinating species, inbreeding depression resulting in low seed yield is not a concern. Therefore, an approach to breeding for heterosis such as RS-SCA should not be hampered in this regard.

Conclusion

The present level of knowledge regarding the genetic mechanisms involved in heterosis limits the ability of plant breeders to make informed decisions regarding both the "ideal" genotype, inbred or hybrid, and optimal selection strategies in many self-pollinated crops. The tools of molecular biology combined with carefully structured populations may lead to a clearer understanding of the importance of overdominance, and hence the importance of F_1 hybrid cultivars, in self-pollinated crops.

Literature references

Brandle JE, McVetty PBE (1989) Heterosis and combining ability in hybrids derived from oilseed rape cultivars and inbred lines. Crop Sci 29:1191-1194

Brears T, Bingham J (1989) Exploitation of heterosis in hybrid wheat using gametocides. Vortr. Pflanzensuchtg 16:397-409

Davenport CG (1908) Degeneration, albinism and inbreeding. Science 28:454-455

Dickson MH (1967) Diallel analysis of seven economic characters in snap beans. Crop Sci. 7:121-124

East EM (1907) Inbreeding in corn. Annu. Rep. Conn. Agric. Exp. Stn. for 1907. pp 419-428

Eberhart SA, Debela S, Hallauer AR (1973) Reciprocal recurrent selection in the BSSS and BSCB1 maize varieties and half-sib selection in BSSS. Crop Sci 13:451-456

Gallais A (1989) Lines versus hybrids - The choice of the optimum variety. Contribution of quantitative genetics and selection theory. Vortr. Pflanzenzuchtg 16:69-80

Gama EEG, Hallauer AR (1977) Relation between inbred and hybrid traits in maize. Crop Sci. 17:703-706

George RAT (1985) Vegetable Seed Production. Longman Group, Essex, England

Gill HS, Thakur PS, Thakur TC (1973) Combining ability in sweet pepper (*Capsicum annuum* L.) var. Grossum Scudt. Indian J. Agric. Sci. 43:918-921

Godshalk EB, Lee M, Lamkey KR (1990) Relationship of restriction fragment length ploymorphisms to single-cross performance in maize

Gritton ET (1975) Heterosis and combining ability in a diallel cross of peas. Crop Sci. 15:453-457

Hallauer AR , Miranda Fo JB (1981) Quantitative Genetics in Maize Breeding. Iowa State Univ. Press, Ames

Houle D (1989) Allozyme -associated heterosis in *Drosophila melanogaster*..Genetics 123:789-801

Hull F (1945) Recurrent selection for specific combining ability in corn. J. Am. Soc. Agron. 37:134-145

Khalil RM, Midan AA, Hatem AK (1987) Breeding studies of some characters in tomato (*Lycopersicon esculentum* Mill.) Acta Horticulturae 220:77-84

Kumar H, Das K (1975) Analysis of gene effects and heterosis in certain productive traits in pea. Indian J. Agric. Sci. 45:301-307

Langton FA, Smith JWM, Edmondson RN (1990) Heterosis in crisphead lettuce (*Lactuca sativa* L.) hybrids. Euphytica 49:15-23

Martin B, Nienhuis J, King G, Schaefer A (1989) Restriction fragment length polymorphisms associated with water use efficiency in tomato. Science 243:1725-1728

Mayo O (1987) The Theory of Plant Breeding (2nd Ed.) Clarendon Press, Oxford.

Meshram LD, Mukewar AM (1986) Heterosis studies in chilli (*Capsicum annuum* L.). Sci. Hortic. 28:219-225

Moll RH, Lonnquist JH, Fortuna JV, Johnson CE (1965) The relationship of heterosis and genetic divergence in Maize. Genetics 52:139-144

Moll RH, Stuber CW (1974) Quantitative genetics - empirical results relevant to plant breeding. In: Brady NC (ed) Advances in Agronomy. Academic Press, San Francisco, pp 277-313

Nei M, Li W (1979) Mathematical model for studying genetic variation in terms of restriction endonucleases. Proc. Natl. Acad. Sci. USA 76:5256-5273

Nienhuis J, Singh SP (1986) Combining ability analysis and relationships among yield, yield components and architectural traits in dry bean. Crop Sci. 26:21-27

Nienhuis J, Helentjaris T, Slocum M, Ruggero B, Schaefer A (1987) Restriction fragment length polymorphism analysis of loci associated with insect resistance in tomato. Crop Sci. 27:797-803

Peter KV, Rai B (1978) Heterosis as a function of genetic distance in tomato. Indian J. Genet. and Plant Breed. 38:173-178

Peter KV, Singh RD (1974) Compatibility, heterosis and analysis of phenotypic variation in brinjal. Indian J. Agric. Sci. 44:393-399

Schaffer HE, Yardley D, Anderson AA (1977) Drift or selection: a statistical test of gene frequency variation over generations. Genetics 87:371-379

Shull GH (1908) The composition of a field of maize. Rpt. Amer. Breeder's Assoc. 4:296-301

Smith OS, Smith JSC, Bowen SL, Tenborg RA, Wall SJ (1990) Similarities among a group of elite maize inbreds as measured by pedigree, F_1 grain yield, grain yield, heterosis and RFLPs. Theor. Appl. Genet. 80:833-840

Thakur MR, Singh K, Singh J (1968) Hybrid vigor studies in brinjal (*Solanum melongena* L.). Punjab Agric. Univ. J. Res. 5:490-495

Thompson G (1977) The effect of a selected locus on linked neutral loci. Genetics 85:753-788

Tikoo SK (1987) Breeding tomatoes for processing in India: present status and future prospects. Acta Horticulturae 200:73-81

Walejko RN, Russell WA (1977) Evaluation of recurrent selection for specific combining ability in two open-pollinated maize cultivars. Crop Sci. 17:647-651

Wallace B (1968) Topics in population genetics. Norton WW, New York

Williams JGK, Kubelik AR, Livak KJ, Rafalski JA, Tingey SV (1990) DNA polymorphisms amplified by arbitrary primers are useful as genetic markers. Nucleic Acids Research 18:6531-6535

Yuan LP, Virmani SS (1988) Status of hybrid rice research and development. In: Hybrid Rice, IRRI. Manila, Philippines, pp7-24

Developing a Genetic Understanding of Agronomic Traits with Complex Inheritance

Tim Helentjaris, Mary Ann F. Cushman, & Rod Winkler
Dept. of Plant Sciences
University of Arizona
Tucson, Arizona 85721
UNITED STATES[1]

The development of RFLPs[2] (Botstein *et al.*, 1980) and PCR-based marker systems, such as RAPDs (Williams et al., 1990), have given geneticists powerful new tools to use in the analysis of genes and their effects upon phenotypes. These genomic technologies are being applied to problems in plant genetics and maps based upon molecular markers have now been produced for a variety of crop plants, including corn (Helentjaris *et al.*, 1986), tomato (Helentjaris *et al.*, ibid.;Bernatsky and Tanksley, 1986), lettuce (Landry *et al.*, 1987), soybean (Keim et al., 1989) and rice (McCouch *et al.*, 1988). Several groups have subsequently examined their application to basic questions in plant breeding, including germplasm relationships (Walton and Helentjaris, 1987; Smith *et al.*, 1990), the mapping of single genes (McMullen and Louie, 1989), and the genetic basis of quantitative traits (Nienhuis *et al.*, 1987; Paterson *et al.*, 1988).

---

1 (602) 621-8746, FAX: (602) 621-7186, E-mail: HELNJARS@ARIZRVAX

2 RFLPs = Restriction Fragment Length Polymorphisms, PCR = Polymerase Chain Reaction, RAPDs = Randomly Amplified Polymorphic DNAs

This latter application is particularly interesting for a number of reasons. Many, if not most, of our important agronomic and horticultural traits, such as yield, maturity, seed or fruit composition, stress resistance, and height, demonstrate complex inheritance. These are also often the most difficult traits to manipulate by classical breeding approaches due to their genetic complexity. Additionally, these traits have not been studied very intensively from either a molecular of genetic perspective due to our inability to apply Mendelian principles (Falconer, 1961). Consequently, when one discusses the application of molecular genetic approaches or even genetic engineering to agriculture, it is usually assumed that this is only practical with "single-gene" traits. This assumption is only natural in some respects as conventional wisdom states that quantitative traits are due to "hundreds" of genes, each with insignificant individual effect. There is actually little direct evidence to support the latter supposition, its premise based predominantly on a hypothesis that was convenient for statistical reasons in promulgating quantitative genetics theory.

With the application of molecular markers to "dissect" quantitative traits into major loci (Stuber *et al.*, 1982; Stuber *et al.* submitted; Nienhuis *et al.*, ibid.; Paterson *et al.*, ibid.), it has been conclusively shown that there are regions of the genome that are able to account for large portions of the phenotypic variation observed for many quantitative traits. Whether this fraction of the variation is due to the actions of single genes or large clusters of genes was not specifically addressed in these studies; however, it is intriguing that some of these obtained results that suggested that single genes may be responsible for these major effects. For instance it was noted in one study (Edwards *et al.*, in press) that many major loci for plant height were located near genes known to carry alleles with drastic effects on plant form. Specifically a major factor for plant height in this study (R^2 = 0.27) was located quite close to a GA[3]-biosynthetic dwarf gene, *d3*, which maps near the centromere of chromosome 9.

This latter result is reminiscent of a theory proposed by D. Robertson (1985), in which he stated that quantitative variation and qualitative variation might be due to different alleles at the same loci. Alleles with extreme effects would result in very disturbed phenotypes (i.e. dwarf plants, defective embryos), whereas "wild-type" alleles with different levels of gene activity would result in normal phenotypes that vary "quantitatively" (i.e. plant height or yield). Accordingly it has been noted that several alleles of a dwarf gene, *d1*, when inserted into the same genetic background, yield plants of various heights, some quite extreme and some which are within the range of normal "wild-type" phenotypes (R. Winkler, unpubl. results). Robertson's idea is also supported by studies in other organisms, where significant portions of variation were putatively ascribed to the actions of known genes (Cowan *et al.*, 1990, Winkelmann et al., 1990). If true, this idea is appealing because it allows one to identify important QTLs[4] by relating them to genes with alleles with very extreme phenotypic alterations. While it can be difficult to reliably detect the effect of a gene accounting for 10% of the variation in some field situations, finding that this same gene could also cause

---

[3] GA = Gibberellic acid
[4] QTLs = quantitative trait loci

a major disruption, such as dwarfism or embryo lethality, would greatly increase our ability to detect its function in genetic analyses or even cloning strategies.

This theory offers a realistic approach for identifying, cloning, and even engineering genes functioning as important quantitative factors. We have also recognized an important addendum to Robertson's theory, in that RFLP analysis may allow one to sort through the large number of loci for those that are actually located near major QTLs. This allows one to prioritize those most promising candidate loci for further analysis. Consequently we have dedicated a significant research effort towards verifying several of this theory's implications and testing whether it provides information that would be useful in identifying and cloning genes acting as major QTLs. We have chosen to work with maize as our model system and to utilize height as our quantitative trait. There are several good arguments for this particular case. Maize has both a highly-detailed molecular marker map and a good set of well-characterized extreme mutants, many of which have been mapped to their genomic locations. Height is also an appropriate choice as it can easily be measured on single plants (as it has a high heritability) and much is known about it from both a physiological and developmental perspective. We were particularly intrigued by the relationship between height, heterosis for yield, and GA-levels observed by Rood et al. (1988) and Pinthus and Gale (1990).

One of the first questions concerning this hypothesis is whether the variation ascribed to a single genomic region is in fact due to the action of a single gene. This requires the ability to accurately discern the activity at a single location. For instance one piece of corroborating evidence might be that the major QTL and an extreme mutant map precisely to the same genomic location. Determination of the precise position of QTLs is quite difficult, as the influence of the rest of the genome, which is usually segregating in the populations used to examine quantitative variation, severely compromises the precision of the positional information derived in RFLP analyses. One can note from published studies that the locations of these QTLs is usually defined to within 15-20 cM, a value hardly useful in proving coidentity of two loci.

To approach this particular problem, we have adopted the strategy of using near-isogenic lines (NILs, Young *et al.*, 1988). By introgressing two variant "alleles" of a single genomic region in the same genetic background, we believe that we can more reliably discern the action of a single locus, evaluate its genetic complexity, and determine its map location. In a sense, we are attempting to convert a situation where a region which is contributing to quantitative variation to a simpler situation where its effect can be monitored as qualitative variation. We are currently in the process of developing NILs for the cross, Mo17 X B73, in which a number of these QTLs have been analyzed. Unfortunately, this process can take several generations. We have meanwhile taken advantage of NILs that were developed by others that we believe will contain important QTLs. For instance many *waxy* conversion lines have been developed by breeders because waxy corn is a small but important niche market for a number of industrial applications. The *waxy* gene is located only 3 cM away from a gene that we believe functions as an important QTL for both height and yield, *d3*, so we believed that a significant number of these conversion lines would also contain the original *d3*-allele from the donor line due to linkage drag, but now contained in a number of new genetic backgrounds. In a sense, these conversion lines are also NILs for this region and we have collected over 40 independently-derived *waxy*-conversion lines. We are currently mapping the extent of the donor segments in these conversion lines and testing these lines for phenotypic differences that can be attributed to this particular region.

Besides data which is essentially positional in nature, it would be desirable to obtain other corroborating evidence that would confirm coidentity of QTLs and Mendelian loci. There has been considerable physiological study of the GA-biosynthetic loci, which makes these loci good models for the study the quantitative control plant height. Therefore, we are investigating a strategy whereby addition of GA to plants would abrogate the effects of different alleles of GA-biosynthetic genes within inbred lines. In this case we are again using our *waxy*-conversion lines, which we presume to only differ by a single effective factor in this region, the *d3* GA-biosynthetic locus. If the phenotypic differences between recipient and conversion lines are due to allelic differences in the *d3* gene's ability to modulate GA levels, then we may be able to confirm this relationship by demonstrating that exogenously added GA removes these differences by supplying all of the GA that these plants can respond to. In this case, one might expect that recipient lines and conversion lines would then exhibit no significant phenotypic differences, following an application of hormone. In preliminary studies, we have shown that we can accurately measure the phenotypic response of inbred lines to exogenously added GA in the greenhouse. Variation between treatments was significantly greater than what was observed within treatments. We are now in the process of determining if we can accurately apportion the treatment variance to differences in the *d3* region.

Other efforts are ongoing in support of this strategy. While we were originally concerned with designing cloning strategies for the candidate loci, we now recognize that if these QTLs do indeed represent alternative alleles of many Mendelian loci, particularly those involved with hormone biosynthesis and response, there are already significant ongoing efforts to clone these genes in many labs and with many species. Some excellent candidates have already been isolated, such as *ga-1* from *Arabidopsis*, several phytochrome genes, and an auxin-binding protein from various species, etc. We are now in the process of obtaining these clones from the original investigators and fine-mapping them and other uncloned Mendelian loci to determine their relevance as good candidates. In fact, the challenging aspect of this strategy may not be to clone genes once they are determined to be good candidates for major QTLs, but to confirm previously-isolated clones as actually being responsible for controlling quantitative variation. We are not focusing narrowly on verifying the control of quantitative variation for plant height by GA, but are attempting to develop a general strategy for obtaining clones of QTLs for many different types of traits. It is our hope that once clones are obtained for QTLs, they can be used to clarify how changes in the expression of single genes can so profoundly affect the entire organism's morphology in some cases. Are these changes predominantly due to alterations in transcriptional activity or to differences in gene product activity? We know that environment has a tremendous effect upon quantitative variation, how is this interacting with single gene expression? It may even be possible in the very near term to create new types of quantitative variation through genetic engineering of single or very few loci. The example in which a growth hormone gene was inserted into mice resulting in an unnaturally large animal is one example of altering an obviously quantitative trait, body size, through genetic engineering. Use of molecular markers to pinpoint which candidate loci are most receptive to this type of engineering may facilitate the application of genetic engineering to those traits of most agronomic importance.

Bernatsky R, Tanksley SD (1986) Toward a saturated linkage map in tomato based on isozymes and random cDNA sequences. Genetics 112:887-898.

Botstein D, White R, Skolnick M, Davis RW (1980) Construction of a genetic linkage map in man using restriction fragment length polymorphisms. Am. J. Hum. Genet. 32:314-331.

Cowan CM, Dentine MR, Ax RL, Schuler LA (1990) Structural variation around prolactin gene linked to quantitative traits in an elite Holstein sire family. Theor. Appl. Genet. 79:577-582.

Edwards MD, Helentjaris T, Wright S, Stuber CW (in press) Molecular-marker-facilitated investigations of quantitative trait loci in maize: IV. Analysis based on genome saturation with isozyme and restriction fragment length polymorphism markers. Crop Sci.

Falconer DS (1961) Introduction to Quantitative Genetics. Ronald Press, New York.

Helentjaris T, Slocum M, Wright S, Schaefer A, Nienhuis J (1986) Construction of genetic linkage maps in maize and tomato using restriction fragment length polymorphisms. Theor. Appl. Genet. 72:761-769.

Keim P, Diers BW, Palmer RG, Shoemaker RC (1989) Mapping the soybean genome with RFLP markers. In: Pascale AJ (ed) Proc. World Soybean Res. Conf. IV, Buenos Aires, Argentina, pp. 1246-1251.

Landry BS, Kesseli RV, Farrara B, Michelmore RW (1987) A genetic map of lettuce (*Lactuca sativa* L.) with restriction fragment length polymorphisms, isozyme, disease resistance, and morphological markers. Genetics 116:331-337.

McCouch SR, Kochert G, Yu Z, Wang Z, Khush GS, Coffman WR, Tanksley SD (1988) Molecular Mapping of rice chromosomes. Theor. Appl. Genet. 76:815-829.

McMullen MD, Louie R (1989) The linkage of molecular markers to a gene controlling the symptom response in maize to Maize Dwarf Mosaic Virus. Mol. Plant-Micro. Interact. 2:309-314.

Nienhuis J, Helentjaris T, Slocum M, Ruggero B, Schaefer A (1987) Restriction fragment length polymorphism analysis of loci associated with insect resistance in tomato. Crop Sci. 27:797-803.

Paterson AH, Lander ES, Hewitt JD, Peterson S, Lincoln SE, Tanksley SD (1988) Resolution of quantitative traits into Mendelian factors by using a complete linkage map of restriction fragment length polymorphisms. Nature 335:721-726.

Pinthus MJ, Gale MD (1990) The effects of 'gibberellin-insensitive' dwarfing alleles in wheat on grain weight and protein content. Theor. Appl. Genet. 79:108-112.

Robertson D (1985) A possible technique for isolating genic DNA for quantitative traits in plants. J. Theor. Biol. 117:1-10.

Rood SB, Buzzell RI, Mander LN, Pearce D, Pharis RP (1988) Gibberellins: a phytohormonal basis for heterosis in maize. Science 241:1216-1218.

Smith OS, Smith JSC, Bowen SL, Tenborg RA, Wall SJ (1990) Similarities among a group of elite maize inbreds as measured by pedigree, F_1 grain yield, grain yield, heterosis, and RFLPs. Theor. Appl. Genet. 80:833-840.

Stuber CW, Goodman MM, Moll RH (1982) Improvement of yield and ear number resulting from selection at allozyme loci in a maize population. Crop Sci. 22:737-740.

Stuber CW, Lincoln SE, Wolff DW, Helentjaris T, Lander ES (submitted) Identification of genetic factors contributing to heterosis in a hybrid from two elite maize inbred lines using molecular markers.

Submitted to Genetics.

Walton M, Helentjaris T (1987) Application of restriction fragment length polymorphism (RFLP) technology to maize breeding. In: Wilkinson D (ed) Proc. of the 42nd Ann. Corn and Sorghum Indus. Res. Conf., pp. 48-75.

Williams JGK, Kubelik AR, Livak KJ, Rafalski JA, Tingey SV (1990) DNA Polymorphisms amplified by arbitrary primers are useful as genetic markers. Nucl. Acids Res. 18:6531-6535.

Winkelmann DC, Querengesser LD, Hodgetts RB (1990) Growth hormone restriction fragment length polymorphisms that segregate with 42-day live weight of mice. Genome 33:235-239.

Young ND, Zamir D, M. Ganal MW, Tanksley SD (1988) Use of isogenic lines and simultaneous probing to identify DNA markers tightly linked to the *Tm-2a* gene in tomato. Genetics 120:579-585.

From Tradition to Modernity: Which Right for Which Varieties?

Nicole BUSTIN
General Secretary of Plant Breeders's Rights Board
11 rue Jean Nicot
75007 Paris
FRANCE

Besides the general laws regulating our social life (property rights, penal law etc..), several specific laws can step in practice of the work of a plant breeder. Chronologically, these laws are patent law, trade marks law, protection of know-how and industrial secrecy and finally protection of plant breeders' rights.

Wanting to explain all those legal systems would be too much ambitious, all the better than some of them are not concerned for the time being by specific disputes. Only patent and plant breeders' rights laws were actually disturbed in their traditional working during the last years, so that we will today pay some attention only to these two laws.

Those rights are both based on the same way of thinking : to ensure the initiator of technical progress (inventor or breeder), within the general interest, an equitable remuneration of his reseach efforts through an exclusive exploitation of his original process or product during a limited period.

Traditionally, Paris Convention of 1883 which built international harmonization of patent laws didn't exclude, on the contrary, theoric possibility to patent new varieties of plant. But, on one hand, national laws passed in conformity with this Convention could decide such an exclusion, and so did some laws, depriving plant breeders of any possibility to have their rights protected. On an other hand, efficient practice of such a right, when available, was particularly difficult for plant breeders. In fact, criteria for patentability are non-disclosure of the invention before the date of application, sufficient description of the inventive process in view to allow a specialist to repeat the invention on this sole basis, sufficient inventive step and industrial applicability of the invention. As far as agriculture was declared to be an industry, this last criteria didn't prevent breeders from patenting their varieties. Nevertheless, it was and it is still very difficult to maintain absolute secrecy on a new variety before filing an application for protection ; it was and it is still often impossible, even for a super-specialist, to obtain identically a variety only by repeating the described breeding process, in despite of a very precise description ; finally, estimation of inventive step in the field of varieties wouldn't have been favorable to breeders, particularly in case of mutations. These different non adequate criteria of patent to plant varieties prevented many breeders from a protection to which they might normally accede. The most tenacious finally mastered all the obstacles and first patents had been granted for rose varieties and "seeds" of lucern (France, Germany,...).

But, after granting, the possibilitily to enforce patent rights in the field of plant varieties was still to be demonstrated.

Unfortunatly, the owners of such patents who have sued

presumed infringers for damages in case of unfair commercial exploitation of their patented varieties didn't win their cases : the Courts have said that in consideration of the matter on which the patent applied there was no way to prove that the presumed infringing varieties were the same that the patented ones.

The plant breeders' rights system has been built with taking in account the different obstacles encountered in the patent field and the specificities of the agricultural industry. At the same time, patent law also evolved together with scientific knowledge and practice of new technologies. Traditional refusal for patenting "natural nature", and consequently refusal of patentability for essentially biologic process and for products obtained from such process, had yet been matter of modernist interpretation when granting the first patent for a variety of plant bred by hybridization. It has been decided that a work in which human intervention existed for the choice of parents and the control of cross-fertilization might not be considered as a pure natural work of nature. Another stage has been reach with the Budapest Treaty on 1977 ; this Treaty deals with proceedings making available the possibility to deposit a sample of a microorganism in an international recognized conservatory instead of describing it within the description of the inventive process. Patent practice so recognized that an invention can, among others, apply to a biological process ; this new step implied the necessity to find a substitute to the complete and precise description which became impossible to realize or would have been too heavy for a good communication of the invention.
In view to make clear the essential difference beetween matters which must remain free for public exploitation

(work of natural nature) and patentable matters in the field of biology, the European patent convention and the national laws which, as a principle, excluded patentability of essentially biological inventions, contain a specific disposition by which micro-biologic process and products obtained but such process are patentable. This modification was only the first step of a long dispute relating to the necessary development of patent law for ensuring its applicability to living and complex organisms.

The discussions which have taken place during the last years on the international level (World Intellectual Property Organization -WIPO- International Union for Protection of Plant Varieties -UPOV-) or on a regional level (EEC, European Patent Office -EPO-) are based on the social request. Great strides of scientifics in the different area of genetic and molecular biology allowed better mastering of nature ; also natural phenomena previously ununderstood have been scientifically explained and new technologies have been improved which permit to program precise modifications within plant or animal genotypes. To mark the evolution of traditional biologic science to its new approach, a neologism was born : the biotechnologies. Such an expedient vocabulary is not efficient and it is remarkable that the notion of "new biotechnologies" has appeared in the course of dispute. The only good question for avoiding this semantic would have been to ask if living matter might or not be patentable and we can regret that economic debate has taken place before the ethic one.

First of all, it has appeared from discussions that no more ground exists for the exclusion of living matter from patentabilily. The only and evident ethic exclusion is relating to the human being.

Neither any part of the plant kingdom nor any process applied to the plant kingdom, breeding process or production process or use process, is excluded a priori from patentability.

As a consequence of this position, responsible authorities must solve several difficulties :

- which are the necessary adaptations of patentability criteria ?
- must the scope of the rights be specially adapted ?
- is there some parts of the animal kingdom or of the plant kingdom which must be excluded ?
- may a variety of plant be dependant on different industrial property rights applying to similar or different elements ?
- is it necessary to restrict specifically free exercice of patent rights ?
- is it necessary to provide inter-dependency dispositions for the relations beetween owner of patent and owner of breeders' rights ?

Some Countries let to the breeders the free choice of the protection system (United States for exemple) and, for the time being, leave to practice, administration, doctrine and courts the definition of necessary evolution of the law. A formal new Bill will finaly be usefull if too much contorsions of the law are necessary.

Other important industrial Countries, like Japan, are investigating in view to studdy the proposed solutions examined by their governmental partners.

EEC member Countries are on the way of a direct modification of their Laws for the purpose of a uniform evolution on all the territories of the Unic Market which will be in force on 1993. The EEC Commission so presented a proposal of Directive for the protection of biotechnologic inventions. The EEC Commission tries to answer the questions we put above trough the proposed Directive.

According to this text, a process wont't be qualified as essentially biologic as far as, considered as a whole, it doesn't exist in the nature and is not only a process for obtaining natural product. Moreover, the fact that a product could have been found in the nature is not, as such, an obstacle to its novelty or to sufficient inventive step. Finally, if the invention is relating to a biologic material which is not of public disposal and which description is impossible, this material shall be deposit in an agreed conservatory.
This proposal intends to adapt patent criteria of non-disclosure, inventive step and sufficient description. One more evolution indeed, with the substitution of a deposit to part of the description : now a date, only a microorganism used for the purpose of the inventive process can be deposit ; from the proposed text, it seems that the matter itself of the invention, if it is a new undescriptible product, can be deposit. The main contribution of the proposed text is relating to novelty of a genetical product yet existing in the nature ; in the meantime, it must be clear that the inventor shall have put the product out of its context and shall have made it active out of its natural conditions. So, it has been necessary to modernize traditional interpretation of patent criteria.

Moreover, patent law has not been supposed to apply to self-reproductive matters (like cells for example). It is the reason why once a patentee had received a royalty for one unit of a the patented product, he ough to stop to exercise his rights on this unit. This theory is perfectly working as far as one needs to follow the inventive process for the reproduction of the patented product in view to produce new units. But living matters often have a tendency to self-reproduction. A particular gene introduced in a variety of plant not only will be reproduced within other varieties under breeder's knowledge (even this gene is not essential in the breeding program) but also will be naturally transmitted to other varieties by ways out of human control (pollen...). The proposal provides consequently that the traditional rule of exhaustion of the patentee rights will not apply ; on one hand, the rights will extend to any material obtained by sexual or vegetative propagation of the patented product and having essentially the same characteristics, on the other hand, in case the matter of the patent is a genetic information, the rights also extend to any product including or expressing this genetic information. The author of the text wishes to establish that the patent right can be exercised on each propagating cycle. The scope of the rights has so been modified for being closer to a wellknown mecanism in the field of plant breeder's rights. A last modification would be very important in the law organization of the established ways by which a patentee can inforce his rights : when a patent will relate to a production-process, any product similar to the one implying the use of the patented process will be presumed to have been obtained by using the process ; that is to say that the defendant (presumed infringer) will be bound to prove that he has not used

the patended process. The normal way is that the patentee has to prove the infrigement.
Except the derogations laid down but the proposed Directive, biotechnologic inventions shall fulfil all other ordinary patent rules.
Most of the Countries in which improvement on plant breeding is an important and performing subject for research are UPOV members and protect varieties of plant trough a specific law. This situation is the most common within Europa where, on the contrary, a specific system for animal doesn't exist. The Commission's proposal provides the exclusion of animals together with varieties of plants ; we will only deal with exclusion of varieties of plant which shall not be, as such, patentable matters. But, any other part of the vegetable kingdom will be patentable. A variety which comprises a biotechnologic invention would be simultaneously bound by a plant breeder's right and, for one or more of its components, by one or several patent rights.
Dependency of a second invention on a previous one is classical in patent law ; dependency of an essentially derived variety on an original one has just been introduced in plant breeder's rights. The proposed Directive for biotechnologies tries to establish a system by which a breeder would be in position to claim a licence against the patentee's decision in case the variety for the production of which he is interested to use the patent constitutes a major technical progress in the public interest (compulsory licence).
Otherwise, no more specicific exception to patent rights is provided. Such a patent will be, as usual, free to others for research or experimental purposes and also for private non-commercial use.
Another EEC Commission's proposal is currently examined for the purpose of plant breeder's rights.

Due to common volonty of the Commission and the Member States, this text will finally be conform to the revised UPOV Convention previously exposed. It includes two very specific dispositions by comparison with patent law : the free commercialization of new varieties bred by using protected varieties (excepted in case of repeated use of inbred lines for producing commercial hybrids and when new notion of essentially derived variety applies) and the free disposal to the farmer on his own premises of the seeds he obtains from his own havested material. The first exception is justified by the nature itself of plant breeding activities, but the second one, more political, is based on public-order requirements and it is one of the reasons why it has been considered very important to apply a unic system for the protection of all kind of varieties.

Traditionally, one breeder used the whole genetic variability he could dispose of, applied genetic knowledge and bred a new variety which he immediatly delivered to users (after complying some administrative proceedings when it occured). If another breeder wanted to plangiarize the new bred variety, the very simple law applying to the matter made the things very easy for such a plagiarist. The adopted modification in plant breeder's rights area are supposed to preserve plant breeding trades together with the introduction of more economical equity. Nethertheless, all the troubles caused by legal innovations are frightening breeders, who think they will loose a part of autonomy, but also other industrials in biotechnologies who are anxious to receive lower remuneration for their very expensive research. Not only those two categories are anxious. Environment responsibles also are frightened with the possible consequences of dissemination of genetically modified organisms (such as varieties bred by genetic engineering) and, may be, the consumer will also have to be informed about "novel food".

Let us shortly imagine the life of a breeder using in his programs protected varieties from others ; in view to reduce the time of its traditional breeding process, he applies patented process to such varieties ; after this, he finally introduce in the bred variety a ....patented gene. The genetically modified organism he will have obtained is of course to be used in agro-food industry. Beyong the authorizations he has to obtain from the owners of plant variety rights and patents on which he is dependant, this unfortunate breeder would have to make the demonstration that his variety is not dangerous for the environment and to obtain an administrative authorization to grow it in the open fields. This is not the end of the story, because the agro-food marketing this "novel food" will be bound by specific obligations. Please, just think a moment about economical market criteria : whether the final marketed product would be too much expensive due to administrative fees added to royalties, or the sale price would be competitive, with subsequent consequences on the level of remuneration of each participant in the product. If the theorical possibilities are applied without enough judgment, such products won't never been sold because of lake of economic interest for the consumer or for the industries. There is a double purpose in this very bad story : the first is to draw the attention of breeders on the necessity to improve their administrative structures, although they think such an investissement is unusefull and too expensive, the second is to try to explain that a reasonable practice will naturally exists in this economic area because the increase of value of the concerned products cannot be compared to the normal increases in other parts of industry ; the modifications in the agricultural world and the new compulsions bending on the market will also have incidences.

From Tradition to Modernity: Which Right for Which Varieties?

Bernard LE BUANEC
Head of Research Programmes
Limagrain Group
BP 1
63720 CHAPPES
FRANCE

As far as plant varieties are concerned, the "tradition" of the protection of intellectual property is fairly recent since, apart from a pontifical law going back to September the 3rd 1833, which was never applied, the oldest laws giving any real protection to breeders are all less than a hundred years old. The first, passed in the United States in 1930, only concerned plants with vegetative propagation, and with a host of exceptions.

In Holland in 1942, and in Germany in 1953, specific laws were passed, but only concerning a limited number of varieties. In fact it is only since 1961, with the signing of the UPOV Convention, that there has been a truly efficient framework for the protection of plant varieties. In several countries which have signed the Convention, the laws have been applied for less than 25 years. So any tradition really is recent ; indeed can we use the word tradition in this context ?

Why did we have to wait so long to obtain the protection of plant varieties when there have been laws on patents for more than two hundred years, and the protection of inventions started in Venice more than 500 years ago with the law known as Parte Veneziana, which was passed on March the 19th 1474 ?

To my mind there are two reasons for this. First there was no real seed industry until the first quarter of this century ; and then the creation of varieties was something very empirical, which explains why there was no seed industry. From an economic point of view there were no real vested interests to defend. Technologically speaking there wasn't enough description or stability of the object to be protected.

As from the beginning of this century progress in the science of genetics was applied to improve plants in a systematic way, after the "rediscovery" of the laws of Mendel by H. DE VRIES, K.E. CORRENS and E. VON TSCHERNA. This led to systematic plant breeding, particularly by state-sponsored research to start with. Private companies gradually took over since "it has been gradually realized that the efforts of the state were not enough" as the representative for the French Ministry of Agriculture pointed out in Paris during the celebrations held in 1986 to mark the 25th anniversary of the UPOV. This industry developed and got organized, and in Amsterdam, on November the 17th 1938, the International Breeders' Association for the Protection of Plant Varieties known as ASSINSEL, was founded. This association, whose first congress was held in Paris on July the 8th and 9th 1939, soon became very active, and in 1956 suggested that an international conference be held, which, in 1961, led to the creation of the UPOV and to the adoption of the first "International Convention for the Protection of New Varieties of Plants" which had 20 participating countries in 1990. The Convention of 1961 was revised in 1972 and in 1978, but without any change in the basic principles.

These principles were as follows :

* The breeder benefits from the protection provided by the Convention if the variety :
  + is clearly distinguishable from any other well-known variety through one or serveral characteristics, and whatever its origin.
  + is new.
  + is sufficiently homogeneous.
  + is stable.
  + has received a denomination in accordance with the rules of the Convention.

* The protected rights and scope of the protection are defined thus :

  + The breeder's authorization is required for :
    . any production for commercial purposes
    . all sales
    . all marketing

    of any material for plant reproduction or multiplication as such, of the variety

  + The breeder's authorization is not necessary for use of the variety as an initial source of variation in order to create other varieties, nor for the marketing of these new varieties. Nevertheless their approval is required if the variety is needed repetitively in order to produce another variety commercially.

  + Each member state of the Union can, for certain botanical genii or species, grant wider rights than those defined above, and these rights can extend to marketed products.

* Generally speaking the Convention applies to all botanical genii and species, but a State can limit its protection to a minimum of twenty four genii or species altogether.

* If a national law recognizes a breeder's rights for a given species in the framework of the UPOV Convention, it must exclude patent protection for the same species unless the state concerned notifies the General Secretary of the Union when the Convention is signed.

This breeder's "traditional" right is a considerable step forward compared with the previous legal void, and it gives general cause for satisfaction to most breeders. However, in its present state it is far from perfect, and is no longer suited to the technical and economic conditions prevailing at the end of the twentieth century, since it was based on those of the fifties.

First of all there are some failings :

* A state can restrict protection to a total number of 24 genii or species. This figure is too low, and protection should be extended to all genii or species.

* The special case of hybrids is not taken into account enough, and for certain species at least, the hybrid should be defined in relation to its parental constituents and the formula associating them. A solution should also be found for the protection of parental lines which are used as "production tools" and not as marketed varieties.

* Finally, the application of minimum distances, which are the keystone to applying the Convention itself, is made all the more difficult because of conditions defined by the Convention : "two varieties are distinct if they show marked differences in one or several important characteristics". Some breeders have felt with increasing frequency in recent years that this definition, combined with free access to genetic variability, has encouraged the plagiarism of existing varieties, even though rules for measuring distance have become more and more complex and sophisticated, as is the case for maize in France for example.

Besides these imperfections in the Convention, or the difficulties met in applying certain points, there have been a certain number of important technical and economic changes over the last fifteen years.

From a technical point of view we have seen a host of new tools become available to breeders as a result of the considerable progress made in cellular and molecular biology : the culture of immature embryos, androgenesis, gynogenesis, cell culture and protoplast fusion, genetic engineering, molecular markers such as RFLP, RAPD,... Some were already being developed at the beginning of the decade, others were at the initial research stage, while others had not yet been imagined. All these tools help us to move in the same direction : to forge ahead more quickly with greater precision and leaving less and less to chance. Two of them have already been of particular importance to the evolution of variety protection :

* Genetic engineering which means a gene can be added to a variety, and which has sparked off the following reactions :
  + Biotechnicians feared that their inventions, particularly their new genes, would be "plundered" after their introduction into a variety, because of the rule of free access to genetic variability.
  + Traditional breeders who, because of this same free access rule, feared that the inventors of genes would get hold of their protected varieties if these patented genes were inserted into these varieties.

* The development of the RFLP technique or substitution techniques allowing for more and more precise knowledge about the genome itself, and not only its expression through morphological or physiological characters. This has facilitated more efficient breeding work at precise stages and descriptions for protection which are harder to contest.

This technical evolution alone would have justified a revision of the Convention. At the same time there have been far-reaching socio-economic changes in the interlinked world of seed-producers and agriculture. Four facts seem to deserve particular attention :

* There has been greater and greater concentration of companies from the seeds sector which had otherwise been extremely dispersed. This process began at the end of 70's and has continued up to today. Concentration became necessary above all because of the considerable increase in the cost of variety creation.

* The main companies have become multinational, so that cultures in the area of intellectual property have been brought together, particularly those of Europe and North America.

* The powerful arrival on the scene of new groups, using different methods :
  + Complete integration by the main chemical or pharmaceutical groups : of the five biggest seed producers in the world three are major industrial groups.
  + Agrochemists have bought up a minority, but significant share in the capital of major seed producers.
  + Powerful agrochemists have developed research into plant biology and genetic engineering, thus becoming potential partners of plant breeders.
  + Finally the development of biotechnology companies which, like the companies mentioned above, are working upstream of variety creation and will become key partners.

* In this evolving socio-economic panorama the fourth important fact in my opinion is the fundamental change taking place in agriculture in developed countries, where ten years ago there was a balance, or even a shortage, in terms of production and consumption, to a surplus situation. One of the solutions to come through this crisis will be the use of agricultural products for targeted agro-industrial purposes, whether it be in food, chemicals, or the production of energy. Agriculture should be thinking in terms of outlets and markets, and this approach will be all the more successful either in the frame of "filieres" or by contractual agreements between producers and users.

Of course breeders must also be integrated into this approach, and they are beginning to create varieties for the production of specific products which will enter the market either directly, or after different transformation stages. It is thus important for them to have some say in what happens to these specific products.

All these changes, which I have described in some length in order to understand today's changes better, and the shortcomings of the UPOV Convention which I developed earlier, were behind the movement which, after five years of often heated debate, and concertation too, led to the revision of the UPOV Convention, which was adopted on March the 19th 1991. Just like any compromise solution involving several interested parties, this new revision does not satisfy anybody one hundred per cent. It is nevertheless a significant step forward compared with the previous version.

So what are the main changes ?

* All species and all plant genii, with no exceptions, can be protected in a state within 10 years after the state becoming a member of the Convention.

* The breeder's rights have been extended, and his authorization is now required for :
  + production or reproduction (multiplication),
  + conditioning for the purpose of propagation,
  + offering for sale,
  + selling or other marketing,
  + exporting,
  + importing
  + stocking for any of the above-mentioned purposes,

but also, for these different acts, in respect of harvested material, whole plants or parts of plants, and of products made directly from a harvested material of the protected variety. On this last point each member state is free to grant this right or not.

* The notion of a variety which is essentially derived from a protected variety has also been introduced. The definition of such a variety is that it is a distinct variety but which keeps the expressions of the essential characteristics resulting from the genotype of the initial variety. An essentially derived variety can for example be obtained through selecting a natural or induced mutant, or from a somaclonal variant, either through back-crosses or through transformations by genetic engineering. An essentially derived variety is considered to be dependent on the initial variety and cannot be exploited commercially without the authorization of its owner.

  Apart from this dependence situation with defined conditions, the breeder's rights do not extend to action taken to create new varieties or to trading in new varieties. Free access to genetic variability is thus maintained in the framework of the UPOV Convention.

* "Double protection" is no longer prohibited ; each member state is free to choose the solution best suited to its conditions.

As can be seen this new revision of the Convention takes into account the main points raised by the evolution of a technical nature and in the socio-economic environment.

Personally I am fully in favour of the changes which I have just described, with the exception of the last one, namely the fact that "double protection" is no longer prohibited, which I am against for two reasons :

* The possibility of protecting plant varieties by patent may well weaken the UPOV Convention which, as we know, has been of considerable value to breeders.
* Moreover the danger of limiting access to genetic variability is great.

It would certainly be harmful particularly to our profession and to agriculture in general.

Within the framework of the new revision of the Convention, three other points are worth mentioning :

* An exception to the breeder's rights has been introduced into the new text : this exception concerns farm seeds, often known as "farmers' privilege". This introduction is certainly unfavourable to breeders' interests. However, to the extent that the text indicates that this exception should be kept within reasonable limits, and safeguard the legitimate interests of the breeder, this change may in fact be positive in the frequent cases where this "privilege" was common practice, yet often illegal.

* As in the case with the previous revision, the new revision does not take into account the specific cases of hybrids and their parental lines to a sufficient extent.

* Finally, even if the spirit of the Convention would seem to be clear as far as distinction and essential derivation are concerned, the text which was adopted, following a Japanese amendment, is somewhat confusing.

It states that :

+ By a "variety" we mean a plant grouping within a single botanical taxon of the lowest known rank which (...) can be
  - defined by the expression of the characteristics resulting from a given genotype or combination of genotypes,
  - distinguished from any other plant grouping by the expression of at least one of the said characteristics and
  - considered as a unit with regard to its suitability for being propagated unchanged.

+ A variety is said to be distinct if it is clearly distinguishable from any other variety (whose existence...).

+ A variety is said to be essentially derived from another variety ("initial variety") if
  - it is predominantly derived from the initial variety while retaining the expression of the essential characteristics that result from the genotype or combination of genotypes of the initial variety,
  - it is clearly distinguishable from the initial variety and
  - except for the differences which result from the act of derivation, it conforms to the initial variety in the expression of the essential characteristics that result from the genotype or combination of genotypes of the initial variety.

The preparatory text of the diplomatic conference (DC/93/3 of November 9th 1990) read, as far as the last derivation condition was concerned, as follows : "if it conforms to the genotype of the initial variety, except for any differences resulting from the derivation method used". A UPOV working document (UPOV/3196V of February 23rd 1990) to prepare an informal meeting of experts on dependence, indicated that : "The expression "essentially derived" has seemingly been used thus far to designate a variety which is distinct from an earlier protected variety from which it is derived and which retains the whole of the genome(s) of the first variety with the exception of specific and deliberate modifications, whether by way of addition, substitution or substraction".

We must therefore be careful in our application of the new revision of the Convention, justifying distinction on the expression of characteristics resulting from the genotype, and justifying dependence on variations of the genotype itself.

Otherwise we shall end up with confusion between the two concepts and dependence will become second level distinction, which would remove any meaning from the notion of essentially derived varieties. It will also be necessary to define, species by species, dependence thresholds according to genotype similarity. This is a major task for all private and public actors of our sector of activity.

I hope that in this presentation I have come up to the expectations of the organizers who asked me to develop the theme "From tradition to modernity : which right for which varieties ?".

Just as at the beginning of my presentation, I said that I was rather reluctant to use the term "tradition" in the area of plant varieties, it seems we should be careful with the word "modern". Besides certain technical aspects of the new Convention which I have developed at some length in order to stress their importance, my wish was to show that rights in the domain of plant varieties, (and it is the same thing for all other rights) are something totally relative.

It should be in harmony with the state of technology and the socio-economic environment. As these factors evolve, rights should adapt. It is clear that in our sector of activity there are quick and far-reaching evolutions. Rights in the domain of plant varieties have certainly not stopped evolving, and what seems to be modern today will certainly be considered to be traditional in a few years' time.

B I B L I O G R A P H Y

* UPOV DOCUMENTS :
  . International Convention for the Protection of New Varieties of Plants", December 2nd 1961, revised in Geneva on November 10th 1972 and October 23rd 1978.

  . Subgroup meeting to discuss technical aspects of minimum distances and dependency. Annoted Agenda (UPOV/3196V.1990.02.03.)

  . Proposition de base pour un nouvel acte de la Convention Internationale pour la protection des obtentions végétales (DC/91/3 du 9 novembre 1990).

  . International Convention for the Protection of New Varieties of Plants", December 2nd 1961, revised in Geneva on November 10th 1972 and October 23rd 1978 and March 19th 1991.

  . Les vingt-cinq premières années de la Convention Internationale pour la protection des obtentions végétales, Geneva 1987.

* Plant breeding perspectives, Centennial publication of Koninklijk Kweekbedsijf en Zaadhandel, D.J. van der Have, 1979, Centre for agricultural publishing and documentation, Wageningen, pp 321-369.

* Hermitte M.-A. : Le droit du génie génétique végétal. Librairie Techniques, 1987.

* Le Buanec B. : La protection des obtentions végétales in Biofutur, October 1987, pp 49-53.

* Le Buanec B. : Variety creation and intellectual property, Plant Variety Protection N°60, January 1991, pp 19-30.

* Le Buanec B. : Nouvelle révision de la Convention UPOV : un grand pas dans la bonne direction. In Biofutur, May 1991, page 5.

Subject Index

Taxon Index

MIX
Papier aus verantwortungsvollen Quellen
Paper from responsible sources
FSC® C105338

If you have any concerns about our products,
you can contact us on
ProductSafety@springernature.com

In case Publisher is established outside the EU,
the EU authorized representative is:
Springer Nature Customer Service Center GmbH
Europaplatz 3, 69115 Heidelberg, Germany

Printed by Libri Plureos GmbH
in Hamburg, Germany